5th International Conference on Theoretical and Applied Physics 2015

AA005878

Journal of Physics: Conference Series Volume 846

Kendari, Indonesia
23 – 25 November 2015

ISBN: 978-1-5108-4190-1
ISSN: 1742-6588

Printed from e-media with permission by:

Curran Associates, Inc.
57 Morehouse Lane
Red Hook, NY 12571

proceedings.com

Some format issues inherent in the e-media version may also appear in this print version.

Copyright© (2015) by the Institute of Physics
All rights reserved. The material featured in this book is subject to
IOP copyright protection, unless otherwise indicated.

Printed by Curran Associates, Inc. (2017)

For permission requests, please contact the Institute of Physics
at the address below.

Institute of Physics
Dirac House, Temple Back
Bristol BS1 6BE UK

Phone: 44 1 17 929 7481
Fax: 44 1 17 920 0979

techtracking@iop.org

Additional copies of this publication are available from:

Curran Associates, Inc.
57 Morehouse Lane
Red Hook, NY 12571 USA
Phone: 845-758-0400
Fax: 845-758-2633
Email: curran@proceedings.com
Web: www.proceedings.com

Table of contents

Volume 846

The 5th International Conference on Theoretical and Applied Physics 2015
23–25 November 2015, University of Halu Oleo, Indonesia

Accepted papers received: 11 May 2017
Published online: 2 June 2017

Preface

011001
OPEN ACCESS
The 5th International Conference on Theoretical and Applied Physics 2015

011002
OPEN ACCESS
Peer review statement

Papers

012001
OPEN ACCESS
Utilization of GPS Tropospheric Delays for Climate Research

Wayan Suparta.....1

012002
OPEN ACCESS
Location and Pressures Change Prediction of Bromo Volcano Magma Chamber Using Inversion Scheme

Ratih Kumalasari, Wahyu Srigutomo and Mitra Djamal.....9

012003
OPEN ACCESS
Utilization of Shrimp Skin Waste (Sea Lobster) As Raw Material for the Membrane Filtration

Ni Nyoman Rupiasih, Made Sumadiyasa, Hery Suyanto and Putri Windari.....15

012004
OPEN ACCESS
Comparison of Magnetohydrodynamics Simulation And Observasional Result to Solar Coronal Helmet Streamer

Sujito, Hari Wisodo and Bambang Setiahadi.....20

012005
OPEN ACCESS
Determination of Relaxation Time in Warming (Cooling) Processes of Atmosphere-Surface Thermal Interaction Based on A Serial RC Circuit Model

Arsali, Octavianus Cakra Satya and Supardi.....26

012006
OPEN ACCESS
Twisting Motion Frequency Dependent I-V Characteristics of 102 Base Pairs Poly(dG)-Poly(dC) DNA Molecule

Efta Yudiarsah.....31

012007
OPEN ACCESS
The Role of Fe_2O_3 and Light Induced on Dielectric Properties of Borosilicate Glass

Markus Diantoro, Norma Dian Prastiwi, Ahmad Taufiq, Nurul Hidayat, Nandang Mufti and Arif Hidayat.....36

012008
OPEN ACCESS
On Heat Transfer - Stress Analysis of Modified Brick (Reed Filler) Upon Its Production Stage

Kurniati Ornam, Masykur Kimsan, Cadas Teguh Prakasa, La Ode Ngkoimani and Santi.....41

012009
OPEN ACCESS
Physical Properties of Nyamplung Oil (*Calophyllum inophyllum L*) for Biodiesel Production

Syamsir Dewang, Suriani, Siti Hadriani, Bannu and B. Abdullah.....48

012010
OPEN ACCESS
Application of Local Adsorbant From Southeast Sulawesi Clay Immobilized *Saccharomyces Cerevisiae* Bread's Yeast Biomass for Adsorption Of Mn(Ii) Metal Ion

Halimahtussaddiyah R, Mashuni and Budiarni.....52

012011
OPEN ACCESS
Microwave-Assisted Hydrolysis of Chitosan from Shrimp Shell Waste for Glucosammine Hydrochlorid Production

Ahmad Zaeni, Endang Safitri, Badrotul Fuadah and I Nyoman Sudiana.....59

012012
OPEN ACCESS
Characterization of Briquette from the Corncob Charcoal and Sago Stem Alloys

Lina Lestari, Viska Inda Variani, I Nyoman Sudiana, Dewi Purnama Sari, Wa Ode Sitti Ilmawati and Erzam Sahaluddin Hasan 67

012013
OPEN ACCESS
Fast Microwave-assisted Pretreatment for Bioconversion of Sawdust Lignocellulose to Glucose

I Nyoman Sudiana, Seitaro Mitsudo, Prima Endang Susilowati, Desak Ketut Sutiari, Made Widana Arsana, Muhammad Zamrun Firihu, La Ode Ngkoimani, La Aba, Erzam Sahaluddin Hasan, Edi Cahyono, Svilen Sabchevski, Haji Aripin and Komang Gde Suastika.....73

012014
OPEN ACCESS
Air Pollutants Minimalization of Pollutant Absorber with Condensation System

Yayat Ruhiat, Firmanul Catur Wibowo and Yuvita Oktarisa.....79

012015
OPEN ACCESS
Estimation of Coal Distribution in Tawanga Village, East Kolaka Regency, Southeast Sulawesi by Using DC Resistivity Method

Laode Ngkoimani, Laode Saleh Isa, Jahidin, Syamsurizal, Suryawan Asfar and Andi Makkawaru.....86

012016
OPEN ACCESS
Near-Surface Fault Structures of the Seulimuem Segment Based on Electrical Resistivity Model

Nazli Ismail, Muhammad Yanis, Syafrizal Idris, Faisal Abdullah and Bukhari Hanafiah.....92

012017
OPEN ACCESS
Deformation Pattern of Nickel Slag Bonding on the Development of Concrete Construction

E. H. Sujiono, H. Husain, M. Mulyadi, S. Samnur and M. Arsyad.....97

012018
OPEN ACCESS
Typology of Slum Management in Coastal Settlement as a Reference of Neighborhood Planning in Konawe

Santi, Ratna Bachrun and Kurniati Ornam.....101

012019
OPEN ACCESS
Tank Model Application for Runoff and Infiltration Analysis on Sub-Watersheds in Lalindu River in South East Sulawesi Indonesia

Sitti Wirdhana Ahmad.....110

012020
OPEN ACCESS
Study on Emission Spectral Lines of Iron, Fe in Laser-Induced Breakdown Spectroscopy (LIBS) on Soil Samples

Nasrullah Idris, Kurnia Lahna, Fadhli and Muliadi Ramli.....118

012021
OPEN ACCESS
Study of Hydrogen and Oxygen and Its Reaction With Host Elements in Sandstone by Laser-Induced Breakdown Spectroscopy (LIBS)

Hery Suyanto.....125

012022
OPEN ACCESS
Proximate and The Calorific Value Analysis of Brown Coal for High-Calorie Hybrid Briquette Application

Erzam Sahaluddin Hasan, Muhammad Jahiding, Mashuni, WOS Ilmawati, Wa Wati and I Nyoman Sudiana.....130

012023
OPEN ACCESS
Fast Drying of Agriculture Commodities by Using Microwave

La Ode Ngkoimani, Megawati, Gde Purwana Saputra, Edi Cahyono, Haji Aripin, Komang Gde Suastika and I Nyoman Sudiana.....137

012024
OPEN ACCESS
Characterization of Central Kalimantan's Amethysts by Using X-Ray Diffraction

Komang Gde Suastika, Lila Yuwana, Luqman Hakim, Darmaji and Didik Khusnul.....145

012025
OPEN ACCESS
Characterization of Coconut Shell Liquid Volatile Matter (CS-LVM) by Using Gas Chomatroghaphy

Muhammad Jahiding, Mashuni, WOS Ilmawati, Ermawati, Rahmat, Jumiati Arsyad and S S Riskayanti.....150

012026
OPEN ACCESS
Characterization of Liquid Volatile Matter (LVM) of Cashew Nut Shell using Pyrolysis and Gas Chomatroghaphy

Mashuni, Muhammad Jahiding, Waode Sitti Ilmawati, Ita Kurniasih, Wa Wati, Muzirah and Muniati Burhan.....157

012027
OPEN ACCESS
Geometric and Road Environmental Effects against Total Number of Traffic Accidents in Kendari

M. Akbar Kurdin, La Welendo and Nur Annisa.....163

012028
OPEN ACCESS
Development of Integrated Natural Science Teaching Materials Webbed Type with Applying Discourse Analysis on Students Grade VIII in Physics Class

Luh Sukariasih.....169

012029
OPEN ACCESS
Analysis of Learning Tools in the study of Developmental of Interactive Multimedia Based Physic Learning Charged in Problem Solving

Sondang Manurung and Deo Demonta Pangabean.....177

012030
OPEN ACCESS
Effect of Tutorial Giving on The Topic of Special Theory of Relativity in Modern Physics Course Towards Students' Problem-Solving Ability

Hartatiek, Yudyanto and Dwi Haryoto.....184

IOP Conf. Series: Journal of Physics: Conf. Series **846** (2017) 011001 doi:10.1088/1742-6596/846/1/011001

Preface: Proceedings of the 5th International Conference on Theoretical and Applied Physics (ICTAP) 2015

The 5th International Conference on Theoretical and Applied Physics (ICTAP) 2015 and the 28th National Physics Symposium (SFN XXVIII) was held in Universitas Halu Oleo, Kendari, Southeast Sulawesi, Indonesia, November 23-25, 2015. The Conference was organized by Physics Department, Universitas Halu Oleo (UHO) and supported by Indonesian Physical Society (HFI). These meeting are aimed to provide the much needed forum of scientific communication and interaction among distinguished scientists working in the field of physics, physics applications. The conference covered the area of theoretical physics and computation, advanced material and nanotechnology, biophysics and medical physics, nuclear physics, geophysics, astrophysics, instrumental physics, laser and optoelectronics, energy and environment physics, interdisciplinary physics, and physic education.

On the behalf of Organizing Committee, we are very appreciated by all participation in these meeting as invited speakers, contributors and participants. We have 12 invited speakers, about 42 papers for ICTAP 2015 and 44 papers for SFN XXVIII, which will be presented during two days of meeting. All submitted papers were reviewed by these members about 25 members. The selected papers and articles examine wide ranging and cutting edge developments in various areas of physics and its applications. We feel the variety of topics will be of interest to both undergraduate and graduate students as well as researchers.

We believe these events will never come to happen without your participation and contribution. We would like to thank very much to all of your uncountable support in organizing these important event. We hope that the new ideas that are presented in this conference could catalyze future advances in the field of physics.

Kendari, 4 April 2017
On behalf of the ICTAP2015 Editorial Board

Ass. Prof. Dr. I Nyoman SUDIANA
(Head of Department of Physics, UHO)

Content from this work may be used under the terms of the Creative Commons Attribution 3.0 licence. Any further distribution of this work must maintain attribution to the author(s) and the title of the work, journal citation and DOI.
Published under licence by IOP Publishing Ltd

Peer review statement

All papers published in this volume of *Journal of Physics: Conference Series* have been peer reviewed through processes administered by the proceedings Editors. Reviews were conducted by expert referees to the professional and scientific standards expected of a proceedings journal published by IOP Publishing.

Content from this work may be used under the terms of the Creative Commons Attribution 3.0 licence. Any further distribution of this work must maintain attribution to the author(s) and the title of the work, journal citation and DOI.
Published under licence by IOP Publishing Ltd

IOP Conf. Series: Journal of Physics: Conf. Series **846** (2017) 012001 doi:10.1088/1742-6596/846/1/012001

Utilization of GPS Tropospheric Delays for Climate Research

Wayan Suparta
Space Science Centre (ANGKASA), Institute of Climate Change
Universiti Kebangsaan Malaysia, 43600 Bangi, Selangor Darul Ehsan, Malaysia

E-mail: wayan@ukm.edu.my

Abstract. The tropospheric delay is one of the main error sources in Global Positioning Systems (GPS) and its impact plays a crucial role in near real-time weather forecasting. Accessibility and accurate estimation of this parameter are essential for weather and climate research. Advances in GPS application has allowed the measurements of zenith tropospheric delay (ZTD) in all weather conditions and on a global scale with fine temporal and spatial resolution. In addition to the rapid advancement of GPS technology and informatics and the development of research in the field of Earth and Planetary Sciences, the GPS data has been available free of charge. Now only required sophisticated processing techniques but user friendly. On the other hand, the ZTD parameter obtained from the models or measurements needs to be converted into precipitable water vapor (PWV) to make it more useful as a component of weather forecasting and analysis atmospheric hazards such as tropical storms, flash floods, landslide, pollution, and earthquake as well as for climate change studies. This paper addresses the determination of ZTD as a signal error or delay source during the propagation from the satellite to a receiver on the ground and is a key driving force behind the atmospheric events. Some results in terms of ZTD and PWV will be highlighted in this paper.

1. Introduction

One of the most atmospheric parameters that play a key role in the Earth's atmosphere and a highly variable atmospheric constituent in determining the accuracy of weather forecasts as well as the positioning accuracy in altitude determination is the total tropospheric delays. Tropospheric delay value represents the total amount of signal delay that would result from dry gases (~90%) presents at the troposphere (N_2, O_2, Ar, etc.) and the water vapor condensed in form of clouds (~10%) at a particular time and over a given location from the surface of Earth to the top of the atmosphere. On the other hand, the tropospheric delay amount depends on the signal path through the neutral atmosphere. The current state analysis revealed that the tropospheric delay is retrieved from the *Global Navigation Satellite System* (*GNSS*) receivers, which known as the total troposphere zenith path delay (ZPD), or often referred to as zenith tropospheric delay (ZTD) [1]. In this method, GNSS (e.g., Global Positioning System (GPS) for simplicity) satellite sends electromagnetic signals through the atmosphere to a receiver on the ground at a fixed location. Total delay in the GPS signals is measured, and the ZTD is obtained from a summation of dry and wet components [2].

Since the ZTD data is strongly significant in meteorological studies such for improving numerical weather prediction (NWP) models, to benefit, their value needs to be converted into the amount of water vapor in the atmosphere, which so-called the precipitable water vapor (PWV) [3]. PWV is one indicator of the critical components of our atmosphere, where their distribution and content are critical variables that can explain the evolution of various physical processes in the atmosphere. With the availability of PWV data at fine temporal and spatial resolution, unpredictable phenomena such as extreme weather and

Content from this work may be used under the terms of the Creative Commons Attribution 3.0 licence. Any further distribution of this work must maintain attribution to the author(s) and the title of the work, journal citation and DOI.
Published under licence by IOP Publishing Ltd

IOP Conf. Series: Journal of Physics: Conf. Series **846** (2017) 012001 doi:10.1088/1742-6596/846/1/012001

climate change at a specific location can be mitigated its impacts before they happen, and this, in turn, will maintain the resilience of the system either in local, regional, and global scales. Hence, this paper aimed at determination of ZTD for climate studies. Conversion of ZTD into PWV will be discussed. Quantification of both parameters and interpretation of their physical characters are crucial in understanding on how the global climate system such as global warming changing over time. Based on that basis, some results from Antarctica research will be presented.

2. Experimental procedure

2.1. Determination of Tropospheric Delays

When the radio signals traveling through the atmosphere to a receiver, it is retarded and affected significantly by the presence of free electrons in the ionosphere and the refractivity of gases, hydrometeors, and other particulates in the neutral atmosphere. During the propagation, the time at which the signal received is given by clock in a receiver, and the difference between transmits time and receiver time is so-called pseudorange. Figure 1 shows the GPS signals propagates through the atmosphere. The difference between the actual path of the carrier (S) and the straight-line (rectilinear) path in a free-space (G) will obtaine dt,

$$dt = \int_S \frac{1}{v} dS - \int_G \frac{1}{c} dG \text{, with } S = \int_{P_1}^{P_2} n(s)\, dS = G + D^{Trop} \text{, and } G = \int_{P_1}^{P_2} dG \qquad (1)$$

where dS and dG are an infinitesimal part of the path length, v and c are speed of the radio signals in medium and in free-space respectively. From the Fermat's principle, the transmission will follow the path between two points (P_1 and P_2 are the initial and end point of the S, respectively). Let D be the delay of the radio wave, with θ is the elevation angle of the direction from which the wave arrives, S be the true path along which the radio wave propagates, and G the shortest geometric path along which the signal would transverse, assuming as $n = 1$. Then

$$D = c\, dt = \int_S n(s)\, dS - G = \int_S \left[n(s) - 1\right] dS + \left[S - G\right] \qquad (2)$$

In the first term Eq. (2), the integral is performed along the line increment dS of straight ray path (excess path delay) from a receiver to a GPS satellite. The second term indicates the geometry delay due to ray path bending.

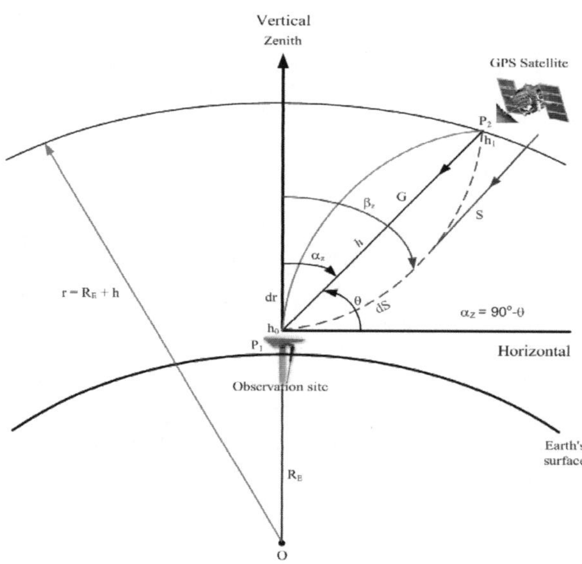

FIGURE 1. GPS signals travel through Earth's atmosphere and tropospheric path delay geometry [4]

As the electromagnetic wave propagates in the atmosphere, it is continuously refracted due to the varying index of refraction of the air between the top of the atmosphere and the ground. Basically, it has two effects on a ray path: bending and retarding, which both produce an excess path length with respect to propagation in a free-space. Excess path length due to signal retarding in the troposphere is expressed as [5]

$$\Delta D = \int_S [n(s)-1]dS = 10^{-6}\int_{P_1}^{P_2} N(s)dS \tag{3}$$

where ΔD is the difference delay between S and G. However in the neutral atmosphere several simplifying assumptions can be made to simplify Eq. (3). First, the atmosphere is assumed to be spherically symmetric, where the Earth is a sphere and the properties of the atmospheric vary only with geometric radius. In this way the atmosphere can be considered layered with a refractive index characterizing each layer. Second, the atmosphere is usually assumed to be azimuthally symmetric; that is with no variation in azimuth in each layer. In this way the electromagnetic ray is confined to a plane defined by the start and end points of the ray and the geocentric. These assumptions allow us to represent the refractive index profile as a function of geocentric radial distance only, $n(h)$. With the application of the refractivity expression (3), Eq. (2) for S can then written as

$$G + D^{Trop} = \int_{h_0}^{h_1} n(h)\sec\beta_z(h)\,dh \tag{4}$$

where the refractive index is integrated along the path between points h_0 and h_1 which are the geocentric distance of the user's antenna and the geocentric distance of the 'top' of neutral atmosphere, respectively. Angle β_z is the actual (refracted) zenith angle of the ray path at distance h (see Fig. 1). The path delay is caused by variation of n from unity, hence:

$$G + D^{Trop} = \int_{h_0}^{h_1}[n(h)-1]\sec\beta_z(h)\,dh + \int_{h_0}^{h_1}\sec\beta_z(h)\,dh \tag{5}$$

This gives in the first term the excess delay equivalent path length and in the second term the geometric length of the curved path. To obtain the total tropospheric delay, we can subtract the delay in free-space distance to get the following integral equation [6]:

$$D_s^{Trop} = \underbrace{\int_{h_0}^{h_1}[n(h)-1]\sec\beta_z(h)\,dh}_{symmetric} + \left[\underbrace{\int_{h_0}^{h_1}\sec\beta_z(h)\,dh - \int_{h_0}^{h_1}\sec\alpha_z dh}_{\delta-asymmmtric}\right] \tag{6}$$

To summarize, the first integral accounts for the difference between the electromagnetic and geometric distances along the ray path and the bracketed integrals account for the curvature of the ray path, i.e. the difference between the refracted and rectilinear geometric distances. Angle α_z is the true (unrefracted) satellite zenith angle and hence constant along the unrefracted path. The total tropospheric delay, can be simplified as

$$D_s^{Trop} = 10^{-6}\int_{actual} N(h)dh + \delta = \left(10^{-6}\int N_{hyd}\,dh + 10^{-6}\int N_{wet}\,dh\right) + \delta \tag{7}$$

or symbolically,

$$D_s^{Trop} = \left(D_{hyd} + D_{wet}\right) + \delta \tag{8}$$

where δ is the tropospheric correction (recently, known as gradient tropospheric delay, if $\delta = 0$, the asymmetric components are neglected). Propagation delays at arbitrary elevation angles are determined from the zenith delays and so-called "mapping functions". Tropospheric delays increase with decreasing satellite elevation angle. This is accounted for multiplying the zenith delays by a correction factor, *m*. In general, the total tropospheric delay from (8) following Davis et al. [5] can be rewritten as

$$ZTD = m_h(\theta)\, ZHD + m_w(\theta)\, ZWD \tag{9}$$

where ZTD is the zenith tropospheric delay or equal to the total tropospheric delay (D_s^{Trop}), *ZHD* is the hydrostatic zenith delay, *ZWD* is the wet zenith delay, θ is the satellite elevation angle that can be extracted from the GPS signals [7] and $m_h(\theta)$ and $m_w(\theta)$ are hydrostatic and wet mapping functions, respectively.

2.2. Determination of Precipitable Water Vapor

Figure 2 shows the flow diagram for the determination of PWV from GPS signals and surface meteorological measurements [7]. The PWV is extracted from ZWD. ZTD is a main key in determining the PWV, which is composed of ZHD and ZWD.

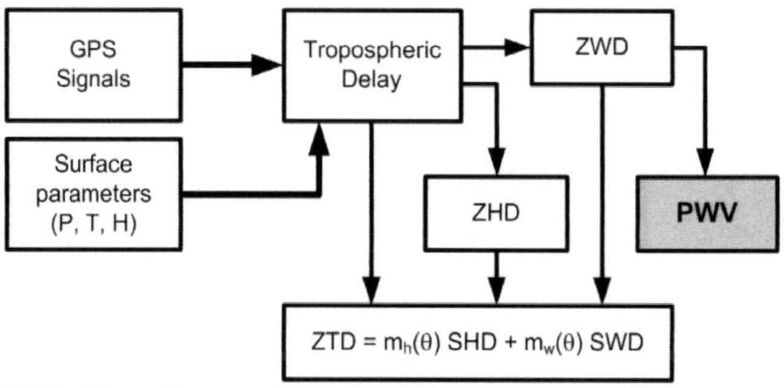

FIGURE 2. Flow diagram showing the determination of ZTD and PWV

In this work, the ZTD is computed using the Modified Hopfield model that can be expressed as [2],

$$ZTD_j^{Trop}(\varepsilon) = 10^{-12}\, N_{j,0}^{Trop}\left(\sum_{k=1}^{9} \frac{\beta_{k,j}}{k} r_j^k\right) \tag{10}$$

where for the hydrostatic component subscript *j* is replaced by *h* and for the wet component subscript *j* is replaced by *w* respectively. $N_{j,0}^{Trop}$ is the total refractivity at the surface of the Earth and is given as:

$$N_{j,0}^{Trop} = k_1 \frac{P}{T_K} + k_2' \frac{P_w}{T_K} Z_w^{-1} k_3 + \frac{P_w}{T_K^2} Z_w^{-1} \tag{11}$$

where *P* is the measured total atmospheric pressure in mbar. The refraction constants k_1, k_2' and k_3 are computed empirically by Bevis et al. [8]. In (11), a constant k_2' was introduced by Davis et al. [5] and is given as $k_2' = k_2 - (M_w / M_d) k_1$, with M_w and M_d are the molar masses of water vapor and dry air, respectively. In this work, we corrected the refractivity in (10) by taking the wet inverse compressibility factor (Z_w^{-1}) into account for the determination of $N_{j,0}^{Trop}$ assuming that a non-ideal gas represents the neutral atmosphere layer. The formula for Z_w^{-1} have been determined empirically by Owens [9] as follows:

$$Z_w^{-1} = 1 + 1650 \frac{P_w}{T_K^3}\left(1 - 0.01317T + 1.75x10^{-4} T^2 + 1.44x10^{-6} T^3\right) \tag{12}$$

where *T* and T_K are the surface air temperatures and both are in Celsius and Kelvin respectively. The P_w (in mbar) is partial pressures of water vapor and is obtained from relative humidity (*H*) with using the formula as recommended by WMO Technical Note No. 8.

$$P_w = \frac{H}{100}\exp\left(-37.2465 + 0.213166T_K - 2.56908x10^{-4}T_K^2\right) \quad (13)$$

The ZHD is calculated using Saastamoinen (SAAS) model [10]. This model is most precise that uses the surface pressure measurements and a correction factor to correct the local gravitational acceleration at the center mass of the atmospheric column. The ZHD from SAAS can be expressed as follows and is close to unity,

$$ZHD_{SAAS}\,(P,\lambda,h) = \frac{(2.2768 \pm 0.0024)P_{surface}}{(1 - 0.00266\cos(2\lambda) - 0.00028h)} \quad (14)$$

where λ is the station latitude (in degree) and h is the height of the site above the ellipsoid (in km) measured by GPS. In this work, the hydrostatic Niell mapping function, $m_h(\theta)$ is employed to reduce the dependence of the zenith delay (ZTD) to the satellite elevation angle. This mapping function has been normalized to yield a value of unity at the zenith [11]:

$$m_h(\theta) = m_{hyd}(\theta) + \Delta m(\theta) = \frac{1 + (a/(1 + b/(1 + c)))}{\sin\theta + (a/\sin\theta + (b/\sin\theta + c))} + \left[\frac{1}{\sin\theta} - f(\theta, a_{ht}, b_{ht}, c_{ht})\right]h \quad (15)$$

where h is the height of the site above geoid in meters. Here, $f(\theta, a_{ht}, b_{ht}, c_{ht})$ represents the three-term continued fraction in the Marini mapping function,

$$f(\theta, a_{ht}, b_{ht}, c_{ht}) = \frac{1 + (a_{ht}/(1 + b_{ht}/(1 + c_{ht})))}{\sin\theta + (\sin\theta + a_{ht}/(\sin\theta + b_{ht}/(\sin\theta + b_{ht})))} \quad (16)$$

In the eq. (16), the coefficients $a_{ht} = 2.53 \times 10^{-5}$, $b_{ht} = 5.49 \times 10^{-3}$ and $c_{ht} = 1.14 \times 10^{-3}$ were determined by a least-squares to fit the height corrections at nine elevation angles.

In modern mapping function, Boehm et al. [12] has updated the "b" and "c" coefficients of the Marini continued fraction form, as shown in (16). However, the hydrostatic "a" coefficients are still valid for zero heights. This mapping function is later called the hydrostatic Vienna Mapping Function (VMF1). To increase the precision, it is using the European Centre for Medium-Range Weather Forecasts (ECMWF) instead of the Numerical Weather Models (NWMs). More than that, the coefficient c shown in (15) is now modeled to remove the systematic errors as

$$c = c_0 + \left[\left(\cos\left(\frac{DoY - 28}{365}2\pi + \psi\right) + 1\right)\frac{c_{11}}{2} + c_{10}\right](1 - \cos\varphi) \quad (17)$$

where DoY is the day of the year and 28 January has been adopted as the reference epoch [11], φ is latitude (deg), and ψ specifies the northern (0) and southern hemisphere (π). On the other hand, the ψ value was selected $\pi/2$, $c_0 = 0.062$, $c_{10} = 0.0015$, and $c_{11} = 0.006$. Detail parameters for c_0, c_{10}, c_{11} and ψ needed for computing the coefficient c of the hydrostatic mapping function can be found in [12]).

Since the ZWD cannot be modeled by using meteorological parameters, it is calculated from subtracting ZHD from ZTD. PWV now can be calculated as given by Bevis et al. [8]

$$PWV = \pi(T_m)ZWD_{GPS} \quad (18)$$

where the dimensionless $\pi(T_m)$ parameter is a conversion factor that varies with the summation on the local climate (e.g. location, elevation, season and weather) and depend on a weighted mean temperature (T_m) as given by

$$\pi(T_m) = \left[\rho_{lw}R_v\left(k_2' + k_3/T_m\right)\right]^{-1}10^6 \text{ and } T_m = 70.2 + 0.72T_K \quad (19)$$

where ρ_{lw} is the liquid water density in 1000 kg m^{-3}, R_v is the universal dry gas constant (287.054 J Kg^{-1}K^{-1}), k_2' (22.1 K mbar^{-1}) and k_3 (3.701×10^5 K^2 mbar^{-1})) are the refraction constants. The mean refraction constants and temperature T_m are estimated by Bevis et al. [8]. T_m for this work is calculated from the surface temperature (T) measured at the particular site.

3. Experimental Results and Discussion

3.1. ZTD Results

Figure 3 shows the example of ZTD results for Davis station (DAV1: 68.58°S, 77.95°E, and ellipsoid of 44.4 m) in Antarctica at one hour resolution. The results are compared to the Crustal Dynamics Data Information System (CDDIS) NASA. The ZTD in this work was computed using a Tropospheric Water Vapor Program (*TroWav*) developed using Matlab source codes [4, 7, 13]. The program is capable of computing ZHD, ZWD, ZTD, and PWV.

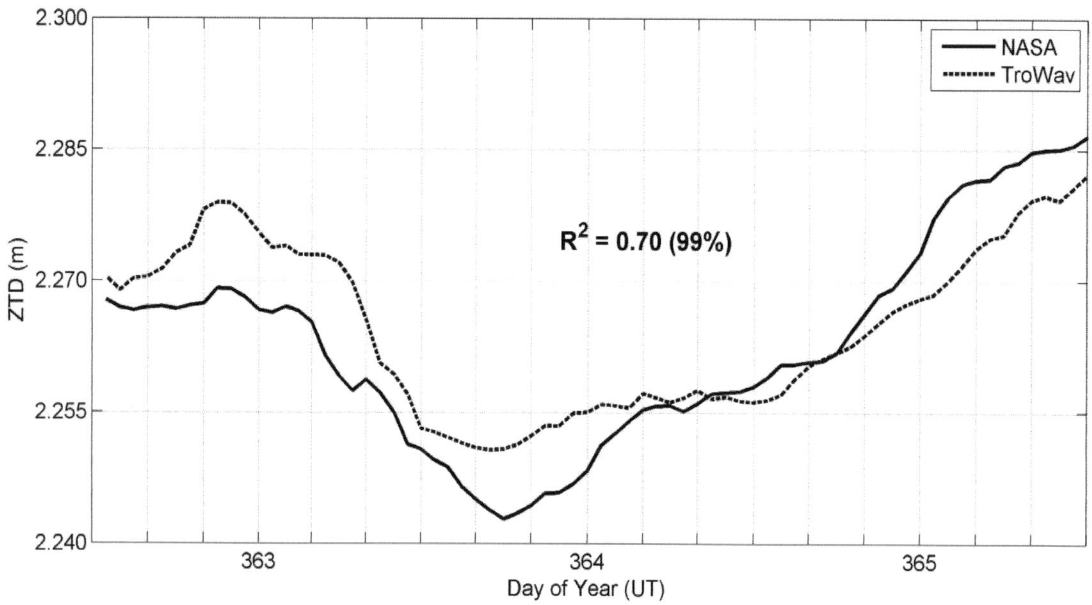

FIGURE 3. The ZTD from *TroWav* compared to the CDDIS NASA ZTD product in one-hour resolution for the period of 29-31 December 2015 (DoY: 363-365).

As shown in the figure, a small drift between *TroWav* ZTD and NASA ZTD exhibited for two peaks with one valley, where *TroWav* ZTD reached its highest value on the DoY = 363. This drift for three days is about 1.02 mm higher from their mean difference. The GPS ZTD at DAV1 exhibit almost similar fluctuation compared to the ZTD from NASA and shown a strong correlation (R^2 = 0.70 at the 99% confidence level). The mean difference values between *TroWav* ZTD and NASA ZTD is 2.04 mm. On the other hand, the ZTD average from *TroWav* at DAV1 for the period of 29-31 December 2015 is 2.246 ± 0.009 m while ZTD obtained from NASA is 2.262 ± 0.011 m. Another work to compare the ZTD results between *TroWav* and GeoForschungsZentrum (GFZ) and Center for Orbit Determination in Europe (CODE) products has reported by Suparta and Kemal [1]. One can be found from this observation is that the variation of both ZTDs in this region is strongly following the pattern of surface pressure.

3.2. PWV Results

After the ZTD computed by *TroWav* program and the accuracy obtained are reasonable, then the PWV value can be retrieved. Figure 4 shows the PWV variability in monthly average for four selected stations in Antarctica. In this paper the data is collected and processed for Davis (DAV1: 68.58°S, 77.95°E and ellipsoid of 44.4 m), McMurdo (MCM4: 77.85°S, 166.67°E and a height of 98.01 m), Syowa (SYOG: 69.00°S, 39.58°E and ellipsoid of 45 m), and Amundsen-Scott South Pole (SPO: 90°S, 0°W). The blank plot shows no data at a particular station.

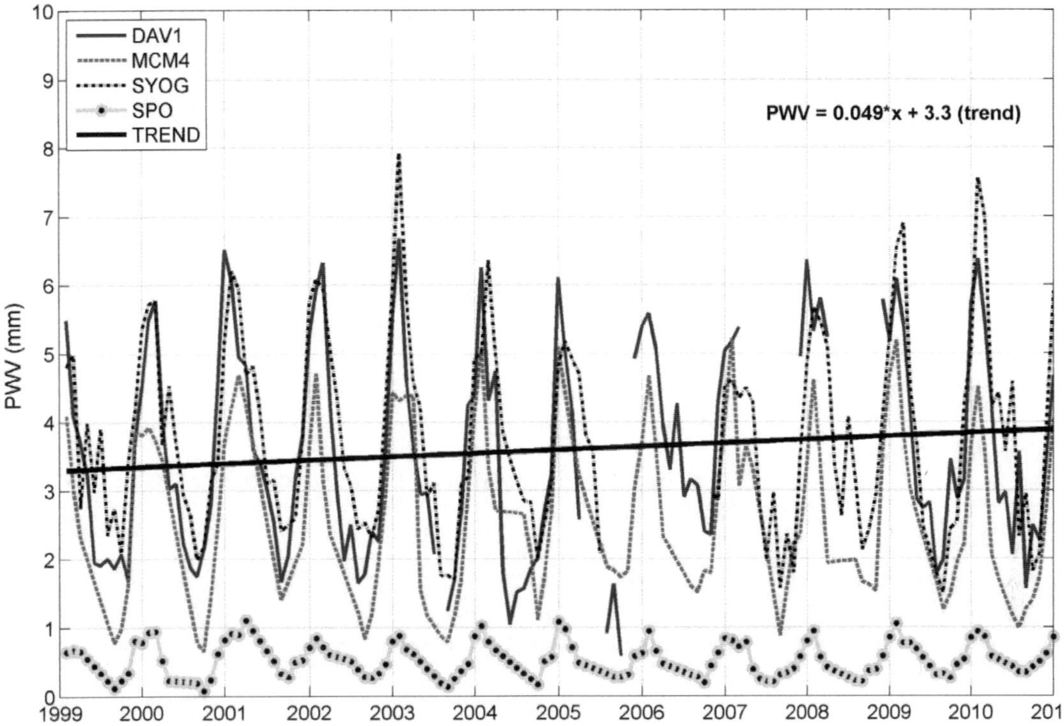

FIGURE 4. PWV variability at four selected stations in Antarctica for the period of 2000 – 2011 (12 years)

As shown in the figure, the lowest PWV was obtained at Amundsen-Scott South Pole (SPO) with the average about 0.5 mm. This indicates that although SPO is located at the pole, their area still to have recorded well of water vapor although it is very small. On the other hand, the characteristics of PWV in Antarctica showed a seasonal variation: largest in summer (December, January, and February, DJF) and lowest in winter (June, July, and August, JJA) which formed V-shape that correlated with the sun activities. The PWV variability was observed closely following the surface temperature pattern [14]. During the 12-year observation period, the PWV trends shown increases of about 0.05 mm, which indicates that global warming effects can be detected by the GPS measurements.

4. Conclusion and Future Work

The tropospheric delays (ZTD) have been successfully retrieved using ground-based GPS meteorology in Antarctica. The quality of ZTD was compared with GPS ZTD from IGS and found that accuracy is bounded by a root mean square (RMS) of 0.95 cm. Furthermore, the quantity of PWV to indicate the climate parameter is also succeeded computed using *TroWav* program and found that the accuracy was within 2-3 mm. The PWV variability obtained shows that the PWV can be used as a comprehensive tool for monitoring of climate change. However, to comprehensively understand how the physical processes of the Antarctic region can drive the climate and the ecology of our planet and comparing its result with another parameter at different regions is suggested to be investigated for future work.

Acknowledgments

This research is supported by the Ministry of Education Malaysia (MoE) under grant FRGS/2/2013/SG02/UKM/02/3 and Ministry of Science, Innovation and Technology Malaysia (MOSTI) under Flagship Program: ZF-2014-016. The author acknowledges the Crustal Dynamics Data Information System (CDDIS) NASA for archiving the ZTD. My appreciation also goes to SOPAC for GPS data archived, the Australian Antarctic Data Center (AADC) and the British Antarctic Survey (BAS) for meteorological data archive.

IOP Conf. Series: Journal of Physics: Conf. Series **846** (2017) 012001 doi:10.1088/1742-6596/846/1/012001

References

1. W. Suparta And K. M. Alhasa, *Modeling of Tropospheric Delays Using ANFIS* (Springer International, Netherlands, 2016), pp. 19-52.
2. B. Hofmann-Wellenhof, H. Lichtenegger and J. Collins, *Atmospheric effect on the Global Positioning System, Theory and Practice* (Springer, Berlin, 2001).
3. M. Bevis, S. Businger S, T.A. Herring, C. Rocken, R. A. Anthes and R. H. Ware , J. Geophys. Res-Atmos. **97**, 15787-15801 (1992).
4. W. Suparta, "Characterization and analysis of atmospheric water vapour using ground-based GPS receivers at Antarctica", Ph.D. thesis, Universiti Kebangsaan Malaysia, 2008.
5. J. L. Davis, T. A. Herring, I. I. Shapiro, A. E. E. Rogers and G. Elgered, Radio Sci **20**, 1593–1607 (1985).
6. R. B. Langley, A. Kleusberg and P. J. G. Teunissen, Lecture Notes in Earth Sciences (Berlin Heidelberg, Springer-Verlag, 1996), pp. 103-140.
7. W. Suparta, Z. A. Abdul Rashid, M. A. Mohd Ali, B. Yatim and G. J. Fraser, J. Atmos. Sol-Terr. Phys. **70**,1419-1447 (2008).
8. M. Bevis M, S. Businger S, T.A. Herring TA, C. Rocken, R. A. Anthes, R. H. Ware and S. Chiswell S, J. Appl. Meteorol. **33**, 379-386 (1994).
9. J. C. Owens, Appl. Opt. **6**, 51–58 (1967).
10. J. Saastamoinen, *Bul. Geodes*. **106**, 383-397 (1972).
11. A. E. Niell, J. Geophys. Res. **101**(B2): 3197-3246 (1996).
12. J. Boehm, B. Werl and H. Schuh, J. Geophys. Res. **111**, B02406 (2006).
13. W. Suparta, J. Phys. Conf. Ser. **495** (2014).
14. W. Suparta, Z. A. Abdul Rashid, M. A. Mohd Ali, B. Yatim and G. J. Fraser, Phys. Chem. Earth, Parts A/B/C **34**, 72-87 (2009).

IOP Conf. Series: Journal of Physics: Conf. Series **846** (2017) 012002 doi:10.1088/1742-6596/846/1/012002

Location and Pressures Change Prediction of Bromo Volcano Magma Chamber Using Inversion Scheme

Ratih Kumalasari[1], Wahyu Srigutomo[2], Mitra Djamal[3,*]

Earth Physics and Complex System, Faculty of Mathematics and Natural Sciences, Bandung Institute of Technology, Jl. Ganesa 10 Bandung, Indonesia 40132

E-mail :mitra@fi.itb.ac.id

Abstract. Bromo volcano is one of active volcanoes in Indonesia. It has erupted at least 50 times since 1775 and has been monitored by Global Positioning System (GPS) since 1989. We applied the Levenberg-Marquardt inversion scheme to estimate the physical parameters contributing to the surface deformation. Physical parameters obtained by the inversion scheme such as magma chamber location and volume change are useful in monitoring and predicting the activity of Bromo volcano. From our calculation it is revealed that the depth of the magma chamber d = 6307.6 m, radius of magma chamber α = 1098.6 m and pressure change $\Delta P \approx 1.0$ MPa.

1. Introduction

Bromo volcano is one of active volcanoes type A in Indonesia, which has erupted at least 50 times since 1775[1], for 2 decades Bromo volcano has erupted 5 times, in 1995, 2000, 2004, 2010, and 2011. Bromo volcano has been monitored by GPS and EDM since 1989. Abidin, et al.[1] has published report on surface deformation of Bromo volcano as detected by GPS and henceforth there is no more analysis and publication about Bromo volcano.

On catalog book published by PVMBG (Center for Volcanology and Geological Hazard Mitigation) we found Map of GPS distribution on 2008, which shows 8 point positions of GPS on Bromo volcano. We tried to generate deformation data based on the map using point pressure source model or Mogi model[2], then we applied the Levenberg-Marquardt (LM) scheme to predict physical parameters contributed to deformation of Bromo such as magma chamber location and the volume change[7].

2. Basic Theory

2.1. Bromo Volcano

Based on historical records, volcanic eruption of Bromo volcano has been recorded since 1804. Types of its eruption are effusive and explosive from the central crater. Over two decades, Bromo volcano has erupted five times in 1995, 2000, 2004, 2010, and 2011.Bromo volcano has been monitored using GPS and EDM since 1989.

Content from this work may be used under the terms of the Creative Commons Attribution 3.0 licence. Any further distribution of this work must maintain attribution to the author(s) and the title of the work, journal citation and DOI.
Published under licence by IOP Publishing Ltd

Figure 1. Bromo volcano in 2010 (picture by Fajidi)[10]

Bromo activity mainly is characterized by formation of thin white smoke blowing with height of about 50 m to 100m from the crater and generally spread to the west and northwest direction. The height and intensity of the blown volcanic gas usually increase under heavy rain fall which is related to characteristics of phreatic eruption as a result of heating of surface and hydrothermal water by heat rocks or magma (Fig. 1).

2.2. Point Pressure source

Surface deformation of a volcano is controlled by the shape and size of the source, the increment of pressure, and the elastic properties of the medium. The deformation is proportional to the ratio of the cavity pressure change to the half-space elastic modulus $\Delta P/G$, and Poisson's ratio[2]. Mogi model assumes that the crust is half-elastic medium and surface deformation caused by pressure source in the form of a spherical of magma which is locate data certain depth. If changes in hydrostatic pressure take place on chamber, the deformation will occur symmetrically. The model input required in the model Mogi point pressure source are the depth (*d*), the radial distance to the source of pressure monitoring points (*r*), and changes in volume (*ΔV*).

$$\begin{pmatrix} u \\ v \\ w \end{pmatrix} = \alpha^3 \Delta P \frac{(1-\upsilon)}{G} \begin{pmatrix} \frac{x}{R^3} \\ \frac{y}{R^3} \\ \frac{d}{R^3} \end{pmatrix} \tag{1}$$

Where υ = poisons ratio, G = shear modulus, and ΔP = Pressure change. u, v, w are the changes of positional the point $x, y, 0$ respectively and $R = \sqrt{x^2 + y^2 + d^2}$ is the radial distance from a point to the surface. It is illustrated graphically in Fig. 2. An example of surface deformation based on Mogi model is shown in Fig. 3 and Fig. 4 in term of horizontal and vertical displacements. The input are a = 1000 m, d = 1000 m , *x-min*= -1000, *x-max* = -1000, $\Delta P /G$ = 0.001 , and υ = 0.25 respectively. The displacement vector which has radial direction to the source has a magnitude

$$U_r = \sqrt{u^2 + v^2 + w^2} = \alpha^3 \Delta P \frac{(1-\upsilon)}{G} \frac{1}{R^2} \tag{2}$$

Its magnitude varies inversely with squared distance from the center of cavity. The location of the source depth can be estimated by calculating $\left({}^{w}\!/_{U_r} \right) r$ where $U_r = \sqrt{u^2 + v^2}$. Pressure changes can be estimated using the relationship between volume of surface deformation and speed of lava discharge. Relationship between of surface deformation (ΔV_{uplift}) with the change of volume on the center of pressure $(\Delta V_{injection})$is expressed as the equation $V_{uplift} = 2(1-v)\Delta V_{injection}$ with v is poisons' ratio. Mogi model has been used for examples by Murray et al.[4], Fialko[5], Larson[6], Materlark[8], and Srigutomo[9].

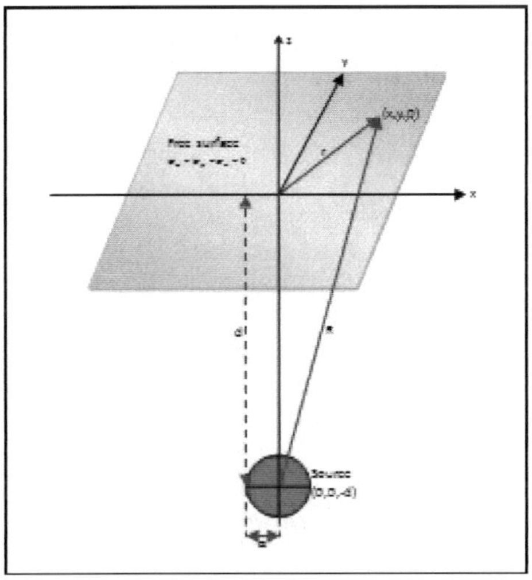

Figure2. Geometry of source in elastic half space (redrawn based on Lisowski[2])

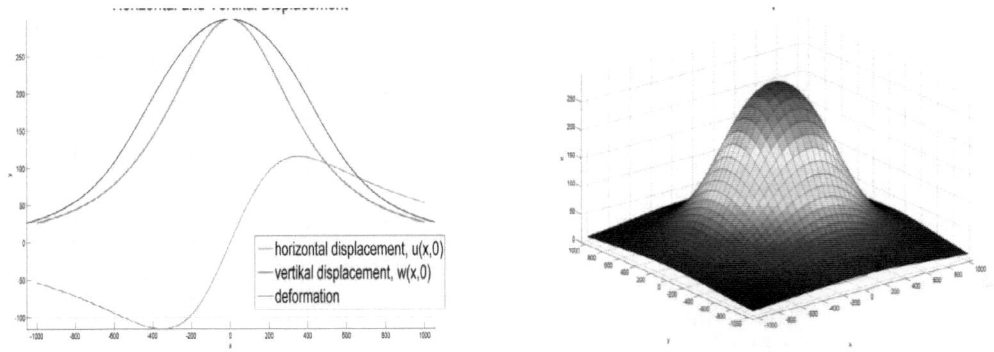

(a) (b)

Figure3. (a) Horizontal and vertical displacements using Mogi model. (b) 3D surface displacement

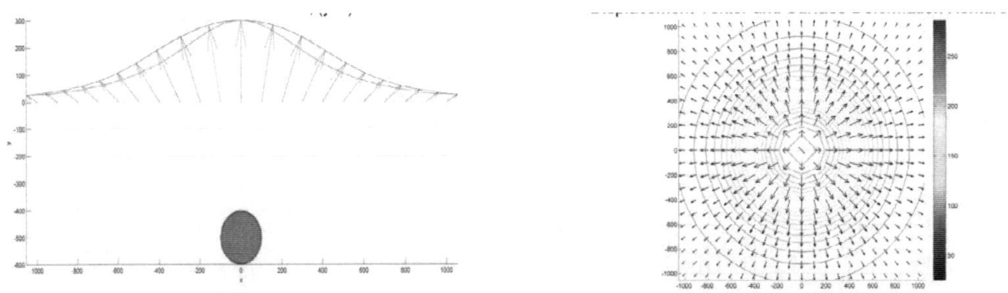

(a) (b)

Figure 4. (a) Surface deformation. (b) Displacement vectors using Mogi model.

Inversion Scheme

Based on the observation of a system in term of data, inversion is a mathematical technique to estimate the physical parameters that characterize the system[7]. For a non-linear inversion scheme, we can use relation between the forward operator ***G***, the model parameters **m**, and the data **d** as

$$\mathbf{G}(\mathbf{m}) = \mathbf{d} \tag{3}$$

In the Mogi model, the parameters that we seek are (α, ΔP, υ) which generate calculated data that are best fitted to the observed data. By assuming the measurement errors are normally distributed, we minimize the sum of the squared error that normalized by the maximum likelihood principle, respective to their standard deviations σ,

$$F(\mathbf{m}) = \sum_{i=1}^{m} \left(\frac{G(m)_i - d_i}{\sigma_i} \right)^2 . \tag{4}$$

If we let

$$\mathbf{f}(\mathbf{m}) = \begin{bmatrix} f_1(\mathbf{m}) \\ . \\ . \\ . \\ f_m(\mathbf{m}) \end{bmatrix} \tag{5}$$

and $\mathbf{J}(\mathbf{m})$ is the Jacobin matrix,

$$\mathbf{J}(\mathbf{m}) = \begin{bmatrix} \frac{\partial f_i(\mathbf{m})}{\partial m_i} & \cdots & \frac{\partial f_i(\mathbf{m})}{\partial m_n} \\ \vdots & \ddots & \vdots \\ \frac{\partial f_m(\mathbf{m})}{\partial m_1} & \cdots & \frac{\partial f_m(\mathbf{m})}{\partial m_n} \end{bmatrix} \tag{6}$$

In the Levenberg-Marquardt (LM) scheme, the update of model parameters at n iteration is found by solving the equation

$$\left(\mathbf{J}(\mathbf{m}^k)^T \mathbf{J}(\mathbf{m}^k) + \lambda \mathbf{I}\right)\Delta \mathbf{m} = -\mathbf{J}(\mathbf{m}^k)^T \mathbf{F}(\mathbf{m}^k) \tag{7}$$

where λ is a dumping parameter which should be adjusted during the inversion until convergence is achieved.

3. Results and Discussion

PVMBG has placed several GPS stations at Bromo volcano since 2007. The locations of these stations are shown in Fig. 5a. We use data of deformation recorded at these stations in 2008 published by PVMBG[3].

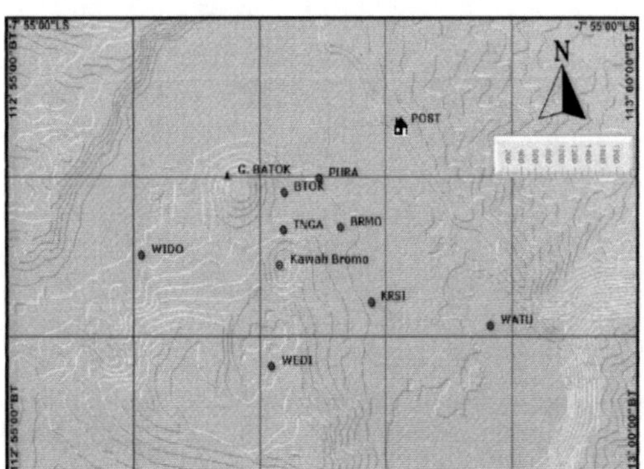

(a)

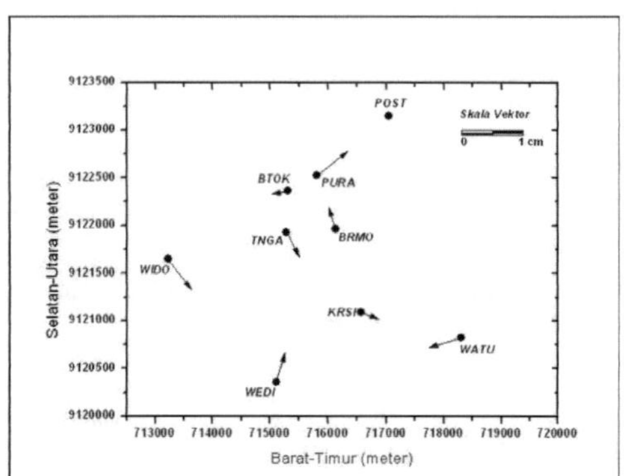

(b)

Figure 5. (a) Coordinate of GPS stations placed at Bromo volcano in 2007. (b) Deformations detected by GPS stations at Bromo volcano in 2008[3]

The LM inversion scheme was applied to the 2008 data of deformations recorded at 7 GPS stations. First we tried to reveal models parameter that would be resulted if we used only horizontal displacement components. It is revealed that the depth of the magma chamber d = 6330.6 m, radius of magma chamber α = 1097.4 m and pressure change $\Delta P \approx$ 1.0 MPa. The comparison between the observed and calculated data is shown in Fig. 6.

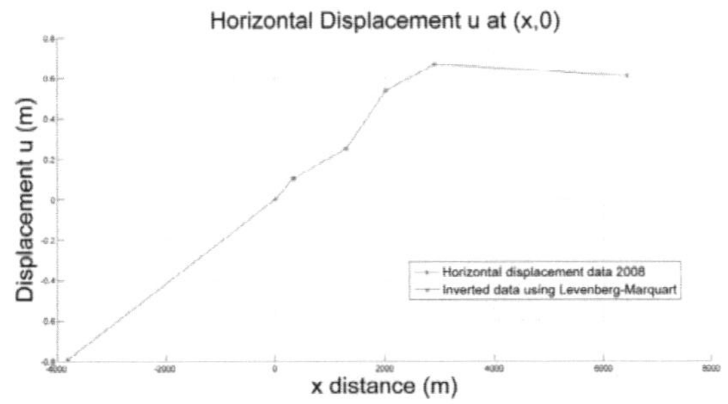

Figure 6. Horizontal Displacement u at (x,0)

Fig. 7 depicts the comparison between the observed and inverted data upon using only horizontal displacement data in y-direction. Depth of the magma chamber d = 6308.3 m, radius of magma chamber α = 1098.2 m and pressure change $\Delta P \approx$ 1.0 MPa.

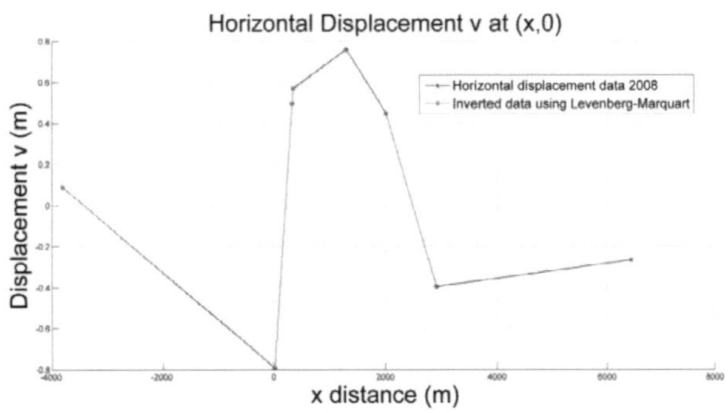

Figure 7. Horisontal Displacement v at (x,0)

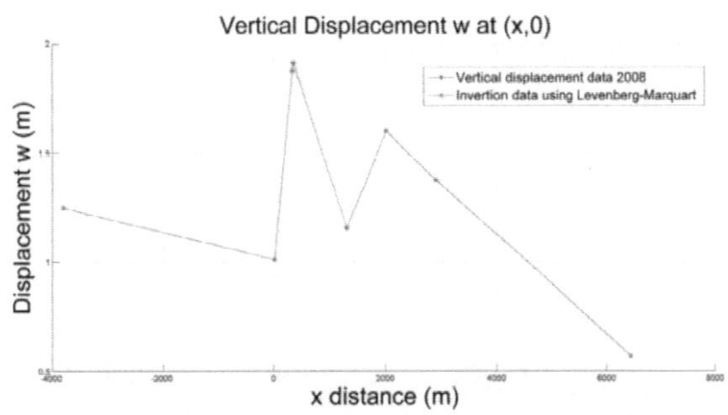

Figure 8. Vertical Displacement w at (x,0)

Implementation of the LM scheme using only vertical components was also conducted and it is revealed that the depth of magma chamber d = 6307.6 m, radius of magma chamber α = 1098.5 m and the involved pressure change $\Delta P \approx$ 1.0 MPa.

4. Conclusion and Future Work

We have used Mogi model to observe deformation of Bromo volcano in 2008. The implementation of Levenberg-Marquardt inversion into the separated horizontal and vertical components of the deformation data suggested that the depth of magma chamber and pressure change in the chamber can be summarized in Table 1.

Table 1. Model Parameters of Displacement

Models	Horisontal displacement u	Horisontal displacement v	Vertikal displacement w
d(m)	6330.6	6308.3	6307.6
a(m)	1097.4	1098.1	1098.5
ΔP(MPa)	1.0	1.0	1.0

It is observed that within the scope of this study the utilization of each component of displacement separately produced almost similar values of model parameters.

References

1. Hasanuddin Z. Abidin, H. Andreas, M. Gamal, M. Hendrasto, Ony K. Suganda, M.A. Purbawinata, Irwan Meilano and F. Kimata. 2005. *The Deformation of Bromo Volcano (Indonesia) as Detected by GPS Surveys Method*. Journal of Global Positioning System vol.03, No.1-2:16-24
2. Lisowski, M., *Analytical volcano deformation source models,* in Dzurisin, D. (ed.): Volcano DeformationGeodetic Moitoring Techniques, 2007.Springer-Praxis Publishing, Chichester, UK.
3. Pusat Vulkanologi dan Mitigasi Bencana (PVMBG). 2010. *Data Dasar Gunung Api di Indonesia.* Pusat Vulkanologi dan Mitigasi Bencana (PVMBG)
4. Srigutomo, W. and Agusta, S.U., 2012. *Application of Levenberg-Marquardt inversion to microgravity data for investigation of shallow volcanic magma chamber deformation*, AIP Conf. Proc. 1454, pp. 126-129
5. Materlark, Timothy. 2007. *Magma Instrusion and Deformation Predictions: sensitives to the Mogi Assumptions.* Journal of Geophysical Research, Vol. 112, B06419, doi:10.1029/2006JB004860
6. Murray, J.B., Rymer, H., and Locke, C.A., 2000. *GroundDeformation,Gravity, andMagnetics*, in Sigurdsson, H. et al. (eds.): Encyclopedia of volcanoes, Academic Press, Toronto
7. Menke, William. 1984. *Geophysical Data Analysis: Discrete Inverse Theory*, Academic Press, New York.
8. Fialko, Yuri, Khazan, Yakov, and Mark, Simons. 2001. Deformation to a Pressurized Horizontal Circular Crack in an Elastic Half-Space, with Application to Volcano Geodesy. Geophys. J. Int. (2001) 146, 181-190
9. Larson, Kristine M., Cervelli, Peter, Liswoski, Michael, Miklius, Asta, Segall, Paul, Owen, Susan. 2001. *Volcano Monitoring Using the Global Positioning System : Filtering Strategies*. Journal of Geophysical Research, Vol.106, No. 0, Page 1-12.
10. Zaenudin, Akhmad. 2011. *Perbandingan antara erupsi Gunung Bromo Tahun 2010 – 2011 dan erupsi Kompleks Gunung Tengger.*Jurnal Lingkungan dan Bencana Geologi, Vol. 2 No. 1 April 2011: 21 - 37

Utilization of Shrimp Skin Waste (Sea Lobster) As Raw Material for the Membrane Filtration

Ni Nyoman Rupiasih[1,*], Made Sumadiyasa[2], Hery Suyanto[2], Putri Windari[1]

[1]Biophysics Laboratory, [2]Applied Physics Laboratory, Department of Physics, Faculty of Mathematics and Natural Sciences, Udayana University, Denpasar, 80361, Indonesia

* E-mail: rupiasih@gmail.com

Abstract. In view of the increasing littering of the sea banks by shells of crustaceans, a study was carried out to investigate the extraction and characterization of chitosan from skin waste of sea lobster i.e. 'Bamboo Lobster' (*Panulirus versicolor*). Chitosan was extracted using conventional methods such as pretreatment, demineralization, deprotienization, and deacetylation. The result showed that the degree of deacetylation of chitosan obtained is 70.02%. The FTIR spectra of the chitosan gave a characteristic of $-NH_2$ band at 3447 cm^{-1} and carbonyl group band at 1655 cm^{-1}. This chitosan has been used to prepare membrane. The chitosan membrane 2% has been prepared using phase inversion method with precipitation by solvent evaporation. The membranes were characterized by FTIR spectrophotometer, *Nova 1200e* using BJH method, and filtration method. The results show that thickness of the membrane is about 134 μm. The FTIR spectra show that functional groups present in the membrane are -NH, -CH, C=O, and -OH. Using BJH method obtained that the pore diameter is 3.382 nm with pore density is 8.95×10^5 pores/m^3. By filtration method obtained that pure water flux (*PWF*) of the membrane are 386.662 and 489.627 $l/m^2.h$ at pressure 80-85 kPa and 90-100 kPa, respectively. These results show that skin waste of sea lobster was discovered as a raw material to prepare chitosan membrane. The membrane obtained is belonged to mesoporous group which may use in microfiltration process.

1. Introduction

The processing of frozen lobster will produce solid waste, namely skin and head. During the time, solid waste is not fully utilized, which are usually thrown into the sea or just left on the edge of the beach. This create problem that must be faced by the community, especially coastal communities. In terms of the usefulness of such waste the parts that can be further harnessed into useful products, such as chitin and chitosan.

Chitin is the second most abundant polysaccharide (next to cellulose) synthesized by a large number of living organisms [1]. In nature, chitin is found as structural components in the exoskeleton of arthropods or in the cell walls of fungi and yeast. It is also produced by a number of other living organisms in the lower plant and animal kingdoms. Meanwhile, chitosan is produced only by some fungi from the family *Mucoraceae* [2].

Chitin is a high molecular weight linear homopolymer of β-(1,4) linked N-acetylglucosamine (Nacetyl-2-amino-2-deoxy-D-glucopyranose) units. Chitin samples contain a high content of N-acetylglucosamine units; hence, they are insoluble in water and common organic solvents. On the other hand, they dissolve only in solvents such as *N,N*dimethylacetamide, hexafluoroacetone or hexafluoro-2-

Content from this work may be used under the terms of the Creative Commons Attribution 3.0 licence. Any further distribution of this work must maintain attribution to the author(s) and the title of the work, journal citation and DOI.
Published under licence by IOP Publishing Ltd

propanol. Chitosan, a copolymer of glucosamine and N-acetyl glucosamine units linked by 1-4 glucosidic bonds, is a cationic polysaccharide [3,4].

Industrially, chitosan is usually produced by de-*N*-acetylation of chitin. When the degree of *N*-acetylation is less than 50%, chitin becomes soluble in aqueous acidic solutions (pH < 6.0) and is called chitosan [3,4,5]. This means that chitosan represents a group of fully and partially deacetylated chitins, but a rigid nomenclature with respect to the degree of *N*deacetylation between chitin and chitosan has not been established [6].

Chitin and chitosan have great economic value because of their versatile biological activities such as biocompatibility, biodegradability, non-toxicity and adsorptive abilities, as well as chemical applications, mainly in the medical and pharmaceutical fields [7,8,9]. The biological properties of these compounds depend closely on their physicochemical parameters, especially their solubility in water and other commonly used solvents.

The purpose of this study was to isolate chitosan from skin waste of sea lobster that is 'Bamboo Lobster' (*Panulirus versicolor*). The product was used as a matrix material to synthesize filtration membrane.

2. Basic Theory

2.1. Extraction and Characterization of Chitosan

The Bamboo Lobster (*Panulirus versicolor*) skin waste was obtained from Serangan, Bali. The chitosan extraction was done in four steps such as pretreatment, demineralization, deprotienization and the deacetylation. The skin waste was washed thoroughly with demineralized water (aqua-dm) and dried in an oven at 35°C, then ground and sieved. One hundred gram of the sample was taken for the extraction process. The demineralization step is the sample were soaked in 1M HCl (1:15 w/v) at 70-80°C for 4 h and stirred it at ± 50 rpm. After which it was washed in acid until no bubbles were seen and no colour changing was observed. The sample washed with aqua-dm until a relatively neutral pH obtained and dried in the oven at 70°C for 24 h. The deprotienization step is the demineralized dried sample were weighed and soaked in NaOH 3.5% solution (1:10 w/v) at 65-70°C for 4 h, stirred it at ± 50 rpm. Then the samples washed thoroughly with water followed by aqua-dm until a neutral pH obtained. The chitin was then dried and weighed. The deacetylation step is the chitin was deacetylated with NaOH 60% (1:20 w/v) for 4 h at 120°C to get chitosan. The chitosan obtained washed thoroughly with water followed by aqua-dm then dried at 70°C for 24 h. The dried chitosan is ready to characterize or use. The characterization was done by Fourier transformed infrared (FTIR) spectrophotometer (IR Prestige 21) in the range of 400 to 4000 cm^{-1}. The spectra of chitin and chitosan are shown in Figures 1.

2.2. Preparation of Chitosan Membrane

The chitosan solution of 2% (w/v) was prepared by dissolving 5 g of the chitosan powder that obtained in 250 ml of 1% acetic acid solution. The mixture was stirred for ± 4 h at room temperature to obtain dope solution. The dope solution was casted in a glass plate and dried at room temperature for 6 days. The chitosan membrane 2% obtained was dipped in NaOH 1% solution, washed with aqua-dm 3 times to eliminate the excess of NaOH, and then dried. The thickness of the membrane obtained was around 134 μm.

2.3. Characterization of Chitosan Membrane

In this study, the chitosan membrane has been characterized by FTIR spectrophotometer (IR Prestige 21), *Quantachrome instruments Nova 1200e* (*gas sorption analyzer versions 10.05*) by BJH method and filtration method.

Filtration Studies of Chitosan Membrane

The measurement of pure water flux (*PWF*) was performed using *dead-end* filtration method with variation in pressure 90-100 kPa and 80-85 kPa. Membranes in the form of circles, each of diameters approximately 4 cm, were cut from the dried membranes. The effective surface area of the membrane is around 1.023×10^{-3} m^2. They were kept in distilled water (DW) for around 5 minutes before being used as a filter. The *PWF* was calculated using formula 1 [10].

$$Flux(J) = \frac{V}{At} \quad (1)$$

Where V is volume of permeate, in l; A is area of the membrane in m^2; and t is time in h.

IOP Conf. Series: Journal of Physics: Conf. Series **846** (2017) 012003 doi:10.1088/1742-6596/846/1/012003

3. Results and Discussion

3.1. Characteristics of Chitin and Chitosan

Figures 1 shows FTIR spectra of the chitin and chitosan. The FTIR spectra gave characteristics bands of $-NH_2$ at around 3447 cm^{-1} and carbonyl group at around 1655 cm^{-1}. The degree of deacetylation (*DD*) was calculated using equation (2) by Domszy dan Robers (2008) [11]. The wavenumber used are 1655 and 3450 cm^{-1}.

$$DD=\left[1-\left[\frac{A_{1655}}{A_{3450}}\times\frac{1}{1,33}\right]\right]\times 100\% \qquad (2)$$

Where A_{1655} and A_{3450} are the absorbance at wavenumber of 1655 cm^{-1} and 3450 cm^{-1}, respectively. The degree of deacetylation (*DD*) of chitosan obtained is 70.02%.

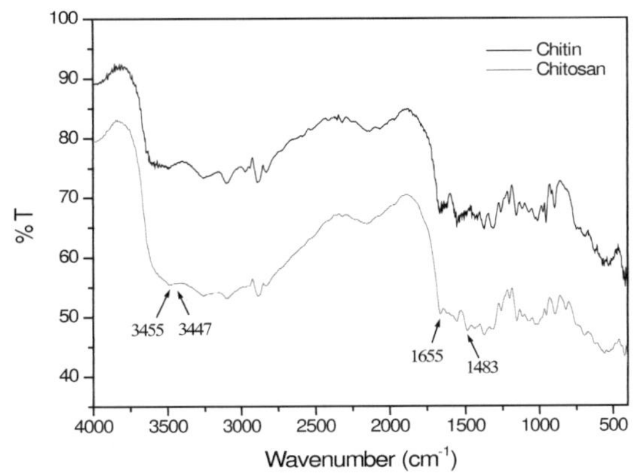

FIGURE 1. FTIR spectra of Chitin and Chitosan.

3.2. Characteristics of Chitosan Membrane

3.2.1 FTIR Characteristics of Chitosan Membrane

The FTIR spectrum of chitosan membrane is presented in Fig. 2. The characteristic absorption of the chitosan is a band at around 3310 cm^{-1} which is assigned to the NH bending and OH stretching vibration of amino group. A band at around 2884 cm^{-1} is assigned to vibration of C-H stretching. A band at around 1666 cm^{-1} and 1564 cm^{-1} are due to C=O amida and N-H, respectively [3,12,13].

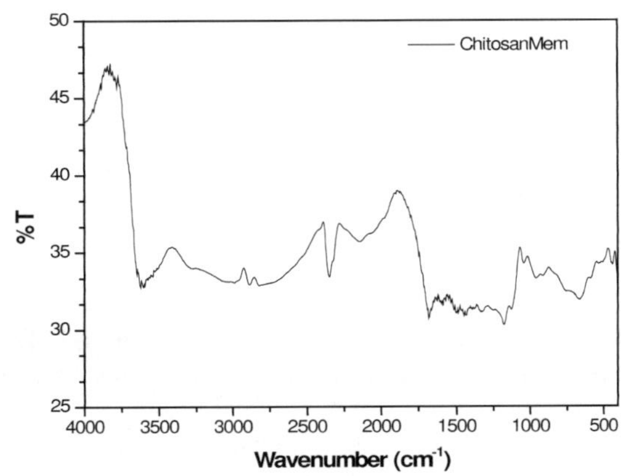

FIGURE 2. The FTIR spectrum of Chitosan biomembrane.

3.2.2 Pore Structure of Chitosan Membrane by BJH Method Using Nova 1200e

The physical characteristics of pore structure of the membrane obtained using *Quantachrome instruments Nova 1200e* (*any gas sorption analyzer versions 10.05*) by BJH method. The results obtained are the surface area, pore volume, pore diameter, pore number, and pore density of chitosan membrane 2% which is presented in Table 1.

TABLE 1. The surface area, pore volume, pore radius, pore number, and pore density of chitosan membrane 2% obtained by *Nova 1200e* using BJH method.

Membrane	***Surface Area*** **(m^2/g)**	***Pore Volume*** **(cc/g)**	***Pore Diameter*** **(nm)**	**Pore Number (pores/g)**	**Pore Density (pores/m^3)**
2%	1.816	0.003	3.382	1.79×10^9	8.95×10^5

Based on IUPAC (*convention for nomenclature of porosity,* 1972), the membrane obtained is belonged to mesoporous group which may use as filtration membrane such as microfiltration [10,14].

3.2.3 Filtration Studies of Chitosan Membrane

The average value of pure water flux (*PWF*) obtained is calculated using equation (1) and the results are presented in Table 2. The result shows that *PWF* increased as increasing the pressure.

TABLE 2. The average value of pure water flux (*PWF*) for each operational pressure: 90-100 kPa and 80-85 kPa.

Membrane	***Pure Water Flux*** **(l/m^2.h)**	
	90-100 kPa	**80-85 kPa**
2%	489.627 ± 33.389	386.662 ± 14.779

4. Conclusion

From the experimental data, the study has demonstrated that skin waste of sea lobster i.e. Bamboo Lobster (*Panulirus versicolor*) was discovered as a raw material to prepare chitosan membrane. The membrane obtained is belonged to mesoporous group which may use in microfiltration process.

Acknowledgments

The work was financially supported by the Ministry of National Education, Republic of Indonesia, Fundamental Research Grant 2013 (Hibah Fundamental), Udayana University which is gratefully acknowledged.

References

1. Muzzarelli, R.A.A; Barontini, G. and Rocchetti, R. “Immobilization of Enzym on Chitosan Colums”. J. Acid Phosphates and α-chmotripsip, Biotechnology Bioengineering, 18, 1986: 1445.
2. Roberts, G.A.F. “Chitin Chemistry”, 2nd ed. MacMillan, London, 1998.
3. Pillai, C.K.S.; Paul, W. & Sharma, C.P. “Chitin and chitosan polymers: Chemistry, solubility and fiber formation”. Prog. Polym. Scie., 34, 2009: 641–678.
4. Jolanta Kumirska, Mirko X. Weinhold, Małgorzata Czerwicka, Zbigniew Kaczyński, Anna Bychowska, Krzysztof Brzozowski,Jorg Thöming, and Piotr Stepnowski. “Influence of the Chemical Structure and Physicochemical Properties of Chitin- and Chitosan-Based Materials on Their Biomedical Activity”. Biomedical Engineering, Trends in Materials Science, Mr Anthony Laskovski (Ed.), 2011. ISBN: 978-953-307-513-6, InTech, Available from:http://www.intechopen.com/books/biomedical-engineering-trends-in-aterialsscience/influence-of-the-chemical-structure-and-physicochemical-properties-of-chitin-and-chitosan-basedmate.
5. R. Jayakumar, M. Prabaharan, P. T. Sudheesh Kumar, S. V. Nair, T. Furuike and H. Tamura. “Novel Chitin and Chitosan Materials in Wound Dressing”. Biomedical Engineering, Trends in Materials Science, Mr Anthony Laskovski (Ed.), 2011. ISBN: 978-953-307-513-6, InTech, Available from:http://www.intechopen.com/books /biomedical-engineering-trends-in-materials-science/novel-chitin-andchitosan-materials-in-wound-dressing.
6. Ravi Kumar, M.N.V. “A review of chitin and chitosan applications”. React. Funct. Polym., 46, 2000: 1–27.

7. Murugan, R. & Ramakrishna, S. "Bioresorbable composite bone paste using polysaccharide based nanohydroxyapa-tite". Biomaterials, 25, 17, 2004: 3829-3835.
8. Rinaudo, M. "Chitin and chitosan: Properties and application". Prog. Polym. Sci., 31, 2006: 603–632.
9. Aranaz, I.; Mengíbar, M.; Harris, R.; Paños, I.; Miralles, B.; Acosta, N.; Galed, G. & Heras Á. "Characterization of Chitin and Chitosan". Curr. Chem. Biol., 3, 2009: 203–230.
10. Marcel Mulder. "Basic Prinsiples of Membrane Tecnology". Netherlands: Kluwer Academic Publisher. 1996.
11. Domsay T M, Robert. "Evaluation of Infra Red Spectroscopic Techniques for analyzing Chitosan". Macromol Chem,186, 1985: 1671.
12. Daniel. "Pembuatan dan Karakterisasi Membrane Chitosan Yang Berasal dari Kulit Udang Sungai Mahakam". Mulawarman Scientifie, 8, 1, 2009: 39-49.
13. Silverstein, Robert. M., and Francis Webster. "Spectrometric Identification of Organic Compounds". Sixth edition. John Wiley and Sons, Inc. 2004.
14. IUPAC. "Convention for Nomenclature of Porosity (IUPAC Recommendations 2005)". Cambridge (UK). 1972.

IOP Conf. Series: Journal of Physics: Conf. Series **846** (2017) 012004 doi:10.1088/1742-6596/846/1/012004

Comparison of Magnetohydrodynamics Simulation And Observasional Result to Solar Coronal Helmet Streamer

Sujito[1*], Hari Wisodo[1], Bambang Setiahadi[2]

[1]Department of Physics, Faculty of Mathematics and Natural Sciences, State University of Malang
[2] Indonesian National Institute of Aeronautics and Space (LAPAN) Watukosek Mojokerto East Java

*E-mail: Sujito.fmipa@um.ac.id

Abstract. This research is compare of magnetohydrodynamics (MHD) simulation and observational product to solar coronal helmet streamer (SCHS). Magnetohydrodynamics simulation is performed to construct evolution of a solar coronal helmet streamer attaining magnetohydrodynamics stability in solar corona. The dynamical evolution is satisfied by implementing numerical approach to reach complete self-consistent solution through modified general equation of Lagrangian-Navier-Stokes time-dependent. The density profile resulting from numerical solution is compared to the real data derived from large-angle solar coronagraph data taking by a coronagraph satellite. The research product show range of error is 10% with using measure of differential emission. We conclude that the plasma electron density distribution derived from simulation is in accordance with observational data.

1. Introduction

Helmet streamers are cusp structures seen in the corona by a coronagraph or during the total solar eclipse. They are usually situated above active regions or above remnant of old active regions [1]. These helmet streamers can be seen in white light because the electron density are about ten times higher than in the quiet corona (see fig. 1) [2]. The lower parts (arcade base) are about the scale of active regions or even larger as observed in the green line emitted by FeXIV (λ = 5303A) [3].

The plasma is a quasi-neutral gas consisting of positive ion and negative charged particles (electrons) [4]. Despite plasma having similar properties as a gas, it may form structures such as filaments, beams and double layers when in the presence of a magnetic field [1]. The presence of a charge non-negligible carriers makes the plasma electrically conductive so that it responds strongly to electromagnetic fields. Due to its attributes, plasma is sometimes considered the fourth state of matter.

The coronal plasma density distribution is not isotropic as a result of the penetration by the helmet streamer magnetic structure [1,3,4]. Figure 1 is making image of the process white light data with several helmet streamers structure (A and B) [2,3].

Below the cusp structure of the helmet streamers are frequently become inflated and all the structure seem to loss their equilibrium [1]. Study of the helmet streamers formation in the solar corona is very important to have some plausible physical condition prior a launch of energetic solar CMEs since the CMEs mostly emerge from the helmet streamers [3,5].

Content from this work may be used under the terms of the Creative Commons Attribution 3.0 licence. Any further distribution of this work must maintain attribution to the author(s) and the title of the work, journal citation and DOI.
Published under licence by IOP Publishing Ltd

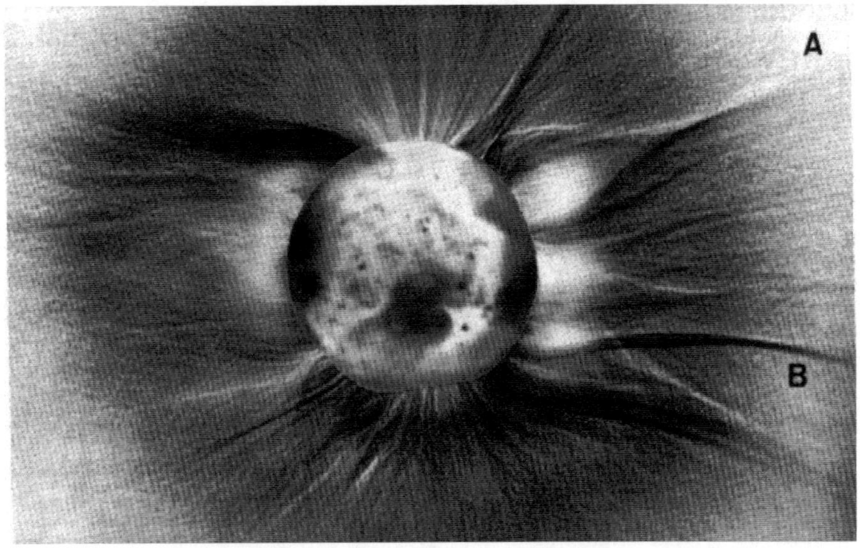

Figure 1: Helmet Streamer Structure

The idea of research is magnetic fields can induce currents in a moving conductive fluid, which create forces on the fluid. The set of equations which describe this condition are combination of the Navier-Stokes equations of fluid dynamics and Maxwell's equations of electromagnetism [3,6,7]. Magnetohydro-dynamics equation was used in reference to the transfer of momentum from the Sun to the planets, but it's eventually expanded upon to include the current field. These differential equations can be solved simultaneously, either analytically or numerically.

In this paper, simulate the formation of a helmet streamer under gravitationally stratified and isothermal atmosphere. The helmet streamer formation closely to a realistic helmet streamer formation in the real solar corona. The atmosphere is initially in hydrostatic balance and has an isothermal temperature distribution at 2×10^6 K [3].

The simulation to this project was run up to 1000 time-steps to obtain a numerically initial hydrostatic balance and isothermal condition. The arcade magnetic field structure is then permeates the atmosphere most instantly as a result of penetration of the arcade magnetic fields from a remnant, of an old active region [2,4,8]. The comparison is made in a situation when "dynamical equilibrium" is satisfied by the simulated helmet streamer.

2. Basic MHD Equations

Magnetohydrodynamics transport equation used to study process the helmet streamer formation [4,6,7]. The equations is

Mass transport:

$$\frac{\partial \rho}{\partial t} + \nabla . \rho \vec{V} = 0 \tag{1}$$

Momentum transport:

$$\frac{\partial \rho \vec{V}}{\partial t} + \nabla . \rho \vec{V} \vec{V} = -\nabla P + \left(\nabla \times \vec{B} \right) \times \vec{B} + \rho \vec{G} \tag{2}$$

Magnetic transport:

$$\frac{\partial \vec{B}}{\partial t} + \nabla . \vec{B} \vec{V} = -\vec{B} \left(\nabla . \vec{V} \right) \tag{3}$$

Pressure transport:

$$\frac{\partial P}{\partial t} + \nabla . P \vec{V} = -(\gamma - 1) P \left(\nabla . \vec{V} \right) \tag{4}$$

And the gas ideal relation

$$P = \rho T \tag{5}$$

These equations are solved numerically in a rectangular two-dimensional space by the *Shasta-FCT* algorithm [6,8]. Equations (1) to (4) can be solved in GLNS as invented and implemented by [6]. We solve in a way such that the solution has second order accuracy in space and in time. Adiabatic environment (γ = 5/3) is assumed during the helmet streamer formation [7,9]. The plasma pressure *P* and plasma density ρ are directly computed through equations (1) and (4), while the plasma temperature *T* is derived from equation (5). Since we assumed the plasma is non-degenerate material such that the Maxwell-Boltzman statistics is not violated then the relation in equation (5) is presumably held.

Initial conditions

Computational domain is a rectangular which represents a region in the corona. The base of the corona lies along the horizontal *x*-axis which runs parallel to the solar surface. The vertical *y*-axis is perpendicular to the solar surface and points upwards such that *y*=0 represents the solar surface. The solar gravitation vector runs parallel with the y-axis and points downward. We assume that gravitation is due to the whole body of the sun so that we do not taken into account the self-gravity of the plasma.

In top, left, and right of the computational domain we apply free boundaries where the plasma can pass the boundaries freely. The bottom boundary is hard boundary which makes the coronal magnetic fields in line-tied condition. The plasma can penetrate the bottom boundary from below as a result of ever escaping plasma from the photosphere and the chromospheres [6,7,9].

The solar corona is initially in hydrostatic equilibrium with the temperature is uniform every where (or isothermal) and takes a value of 2 x10^6 K [8].. In the initial configuration the pressure is derived from the equation of hydrostatic balance as follows [6],

$$\nabla_y P(y) = \rho(y)\, g_y(y) \qquad (6)$$

where we assign that

$$\rho(y) = \rho_0 \exp\left[-\left(\frac{g_0 R_0^2}{\Re T}\right)\frac{y}{(R_0 + y)^2}\right] \qquad (7)$$

And

$$g_y(y) = g_0 \frac{R_0^2}{(R_0 + y)^2} \qquad (8)$$

All symbols have their usual meanings. This initial analytical condition is tested to 1000 cycles to check whether this solution can reach (numerical) equilibrium before penetrated by a magnetic field. Equations (6) and (7) express that our initial model is held as our planet's atmosphere with the exception that the material is purely electron plasma. While equation (8) indicates the gravitational scale exceeds great height from the solar surface into interplanetary space extension.

A remnant but still evolving old active region then penetrates this initial atmosphere almost instantly by a magnetic field of the following type [1],

$$B_x(x, y) = B_0 \cos\left(\frac{\pi x}{2x_0}\right)\exp\left(\frac{-\pi y}{2x_0}\right) \qquad (9)$$

and

$$B_y(x, y) = -B_0 \sin\left(\frac{\pi x}{2x_0}\right)\exp\left(\frac{-\pi y}{2x_0}\right) \qquad (10)$$

where B_0 is arbitrary coronal base magnetic strength. In this case we put a value of 10 Gauss. The $2x_0$ is the length of the helmet streamer's arcade base. The negative value of $B_y x, y)$ will result in a total inward force.

This force is necessary to balance the pressure gradient force. The magnetic fields structure in equations (9) and (10) exhibit that initially a primitive helmet streamer structure penetrates lower corona and attains equilibrium with exponential decreasing magnetic strength [3,6].

IOP Conf. Series: Journal of Physics: Conf. Series **846** (2017) 012004 doi:10.1088/1742-6596/846/1/012004

3. Results and Discussion

After the magnetic field structure penetrated the hydrostatic and isothermal atmosphere, we run simulation up to 2,000 cycle or about 11.13 minutes of real time. Before cycle 1,000 (or about 6.35 minutes of real time) was attained a noticeable variation of the physical parameters could be seen easily at every 100 cycles (see Fig. 2). After cycle 1,000 up to the last of cycle 2,000 there was no much change on the pressure, density, and magnetic pressure profiles as well. It can be inspected in the Fig. 2 that the profiles in the axis of symmetry superimposed each other. Evolution of the pressure and the density of plasma along axis of symmetry. The superimpose profiles are obvious after cycle 1000.

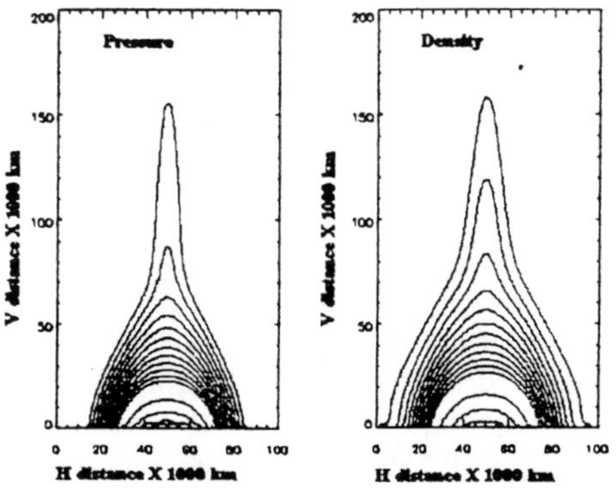

Figure 2. This is The Signature of New Equilibrium Onset In The Helmet Streamer Structure In The Solar Corona.

In the last cycle 2,000 it can be seen on Fig. 3 that the pressure contour and the density contour resemble nicely the cusp type helmet streamers as observed in white light with a coronagraph. We concluded that the system satisfied new dynamical equilibrium. The term "dynamical" means that even the helmet streamer attained its new equilibrium state, there were still plasma flow upward in the neighboring regions [1,5].

From Fig. 2 we can see that, on the axis of symmetry, the superimposing profiles in general satisfied approximately the profiles of the magnetohydrostatic balance as derived by Setiahadi *et al.* (1998). Even though, it should be noted that we did not include the solar gravitational acceleration formally in the previous work. Nevertheless, within a scale height smaller than that of the corona (<50,000 km), these profiles (with and without gravitation) should approach each other.

During the first 1000 cycles, along the axis of symmetry, a relatively high plasma speed was blown upward. After cycle 1,000 the plasma speed at the symmetrical axis ceased and at the last cycle 2,000 the speed distribution was much less than that of the quiet corona (< 140 km s^{-1}). Though, there were relatively high speeds of plasma still persisted around the axis of symmetry.

This high speed of plasma came from below and flew along the boundaries of the helmet streamer structure (see Fig. 3). At the last of cycle 2,000 the cusp type of the pressure distribution and the density distribution are clearly reproduced. Note the density output matches with the density calculated from the real data using differential emission measure equation. If we continue the simulation this situation will not change much and the flow will exist for a long time. From the observational works, it is suggested that the continuous radio burst of type III may be originated from the boundaries of the helmet structure [2].

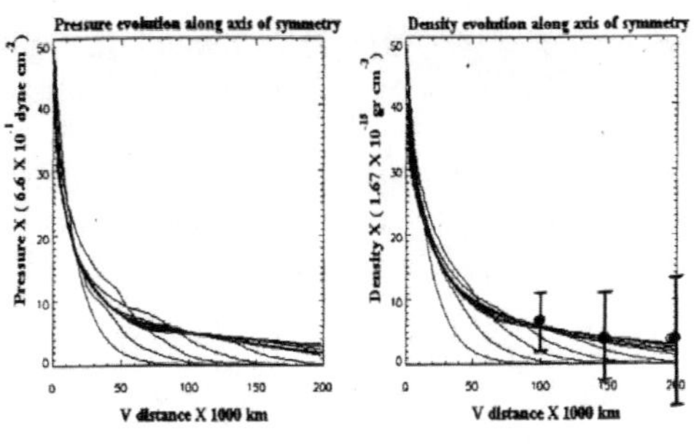

Figure 3. The Density Inside of This Structure Is, on The Average, Ten Times Higher Then In The Outside.

Though we started with a comparable plasma $\boldsymbol{\beta}$ =1 as an initial state, the plasma $\boldsymbol{\beta}$ became higher at all points in the helmet streamer at the last of cycle 2,000. We suggest that the evolution to higher plasma $\boldsymbol{\beta}$ is due to plasma flow-trapping by the magnetic field of the helmet streamer during its way to reach a new equilibrium [2,5]. Outside the helmet streamer the plasma is freely to flow along the boundaries and forms a high speed solar wind. Thus, finally we have a helmet streamer which is dominated by plasma pressure. Evolution to other state is then controlled by the plasma pressure evolution in the helmet streamer [9].

The only way for validation with real physical world is accomplished through density output from computation with the density provided from observational data, since the helmet streamer from real corona is nothing than the electron density scattered by Thomson electron scattering mechanism. The formula to convert from intensity data to density data is written as [7]

$$\rho = k\ I \tag{11}$$

Where $\boldsymbol{\rho}$ is the plasma density derived from observation, $\boldsymbol{k}$ is the conversion that accommodates the surface electron intensity to plasma density, and $\boldsymbol{I}$ is the electron intensity due to the Thomson electron scattering mechanism in solar corona. Through the above formulation we may eventually compare the density from MHD output with electron density in a real helmet streamer.

From Fig. 3 it is noted the density output matches with the density calculated from the real data using differential emission measure equation (1). The electron scattering data is very important since it provides us a gate to compare indirectly the MHD computation output with real physical world.

4. Conclusion

Helmet streamers are seen unchanged during several days and even weeks before they loss their equilibrium to launch the solar CMEs. It means that the magnetic field in the helmet streamers structure may have a unique physical condition which axe able to resist their stability against disturbances from below [1]. The most probably stabilizing mechanism is the magnetic line-tied condition. Any change of the line-tied condition by some process may change the stability on the magnetic field.

The line-tied condition may change due to shearing motion on the base of the helmet streamer [5]. The change should result as increasing magnetic energy in the helmet streamers and should attain a critical state so that the helmet streamers will loss their equilibrium and will release the magnetic energy almost instantly. Research on the problem is still uncertain now. Some authors concluded that after the magnetic field of the helmet streamers attain their critical state they will loss their equilibrium [4]. In the other hand, some authors found that after attaining a critical state, the magnetic field of the helmet streamers will not loss their equilibrium but attain other equilibrium sequence as discussed by [6] and [12]. Other possibility of losing equilibrium may be caused by plasma pressure evolution in the base of the helmet streamers as a result of ever escaping plasma flow from the photosphere and the chromosphere. Results of ideal MHD numerical experiments [9,10,11] pointed out that the loss of equilibrium may be caused by pressure evolution on the coronal base.

References

[1] Zuccarello, F. Balmaceda, L. Cessateur, G. Cremades, H. Guglielmino, S. Lilensten, J. Kretzschmar, M. Fernando M. Lopez, Mierla M, Parenti S. Pomoell, S. Romano, P., (2013), Solar activity and its evolution across the corona: recent advances, *J. Space Weather Space Clim* A18, 10. 1051/sws c/201303 9

[2] Setiahadi, B., Anwar, B., Sakurai, T. (1995)*, 2D MHD Simulation of CME Generated Interplanetary Shock Waves,* The Ninth International Symposium on Equatorial Aeronomy (ISEA), Bali, Indonesia, March 20-24, p. D-10, 1995

[3] F. Goryaev1, V. Slemzin1, L. Vainshtein1, and David R. Williams, (2014), Study of Extreme-Ultraviolet Emission And Properties Of A Coronal Streamer From Proba2/Swap, Hinode/Eis And Mauna Loa Mk4 Observations, *The Astrophysical Journal, Volume 781, Number 2.* The American Astronomical Society

[4] Jokers, K. (2008*), Solar Physics*, **56,** p. 37.

[5] Setiahadi, B., Anwar, B., Akioka, M., Sakurai, T. (1998), *Non-linear Evolution of Erupting Coronal Magnetic Fields,* Proceedings of Yohkoh 5th Anniversary Symposium, p. 345.

[6] Fan, Y. and Gibson, S., E. (2004), Numerical Simulations of Three-dimensional Coronal Magnetic Fields Resulting from the Emergence of Twisted Magnetic Flux Tubes, *The Astrophysical Journal, Volume 609, Number 2*, The American Astronomical Society.
[7] Madhavan, S., Gandhinagar, M., V., Majalee, A., (2005), Two-Dimensional MHD Simulation of Quasi-Spherical and Cylindrical Metallic Liner Acceleration. *Pulsed Power Conference, 2005 IEEE* Inst. for Plasma Res.,
[8] Priest, E.R. (1982), *Solar Magnetohydrodynamics,* D. Reidel, Dordrecht, Holland, p. 149.
[9] Kliem, B. Van Ballegooijen, and Deluca, E. (2013), Magnetohydrodynamic Modeling of The Solar Eruption On 2010 April 8, *Astro-Phy.SR*

IOP Conf. Series: Journal of Physics: Conf. Series **846** (2017) 012005 doi:10.1088/1742-6596/846/1/012005

Determination of Relaxation Time in Warming (Cooling) Processes of Atmosphere-Surface Thermal Interaction Based on A Serial RC Circuit Model

Arsali[1*], Octavianus Cakra Satya[1], Supardi[1]
Physics Department, University of Sriwijaya, Indralaya-Ogan Ilir, South of Sumatra

[*] Email: arsali_99@yahoo.com

Abstract. An inverse calculation method of a simple serial RC circuit model (RSRC model) has been implemented to construct a more appropriate model in order to explain the process of warming (cooling) of the atmosphere and the surface in a period of day and night, as well as to determine the relaxation time related to it. Testing of the model involving only radiation mechanisms lead to the conclusion that the better values of relaxation time can only be obtained when the interaction between the atmosphere and the surface are taken into account. Meanwhile, with a few flaws that still encountered, the results of the study recommends the need for the development of a model with a mechanism involving convection and phase transformation, while also increase a physical dimension of the depth of the surface layer.

1. Introduction

There are great many similarities between the variations of air temperature (and surface temperature) within a period of day and night with the variation of charges inside the capacitor in an RC circuit under the influence of the periodic voltage battery. In simple terms the things above happens because the air and the soil (as well as the water) have the ability to store the thermal energy that they have been received, which partly expressed as an increase in temperature, then it can be released back when the energy supply from the outside (especially the solar radiation) stopped, which partially of them are represented in the form of a decrease in it's temperature.

For the majority of the university students of science and engineering, they have also known that the charge inside capacitor, as a function of time, *Q(t)* on a simple serial RC circuits can be expressed as [1], [2]

$$\frac{dQ}{dt} = -\frac{1}{RC}(Q - C\epsilon) \qquad (1)$$

with R and C denote the resistance and capacitance of the system, while ε stated the voltage source. In this case $\tau = RC$ is the time constant which, for a discussion here, because its value can change with time, we call it as the relaxation time.

Equation (1) is a mathematical formula of the Serial RC Circuit Model (RSRC Model). They are in the form of issue related to *standards problem* in order to the determination of *Q(t)* as the solution of differential equations [1], [3], but in many applications it often appears as an *inverse problem* [4].

Heating and cooling processes have been studied empirically for a long time, which generally started by Newton (1701), Dulong and Petit (1817), and later by Stefan (1879). Afterwards, the theoretical study on this subject conducted by Boltzmann who then made publicly in 1884. However, in its development, various studies on the laws of heating and cooling experiencing a critical appraisal and revision, given that the issue actually difficult to separate the mechanisms of heat transfer [5], [6].

The simplest concept of heating/cooling is the transfer or exchange of thermal energy between the two objects (systems) as well as between the system with the environment, either through the mechanism

Content from this work may be used under the terms of the Creative Commons Attribution 3.0 licence. Any further distribution of this work must maintain attribution to the author(s) and the title of the work, journal citation and DOI.
Published under licence by IOP Publishing Ltd

IOP Conf. Series: Journal of Physics: Conf. Series **846** (2017) 012005 doi:10.1088/1742-6596/846/1/012005

of radiation [7] [8] [9], convection [8], conduction [9], and the phase transformation. In the latter case the process also involves an exchange (transfer) of mass.

By limiting the process that only involves radiation mechanism, the formula to express changes in air temperature and surface temperature, which is identical to the RSRC model, can be approximated based on the following case (model):

Case#1: The Self Heating/Cooling Model (The PDM Model)

In this case the warming or cooling process is considered takes place separately, no interaction between the atmosphere and the surface so the statement that is identical to equation (1) can be written as

$$\frac{dT_x}{dt} = -\frac{e_x\sigma T_x^3}{\rho_x c_x z_x}\left(T_x - \frac{\gamma_x}{e_x\sigma T_x^3} I_G\right) \tag{2}$$

the index $_x$ expresses the atmosphere ($_a$) or the surface ($_s$); σ is the Stefan-Boltzmann constant (5.6704x10^{-8} W/m^{2}K^4); e is emissivity; ρ is density; c is specific heat; and z is the layer thickness, the zero point on the surface. T and I_G are the temperature and global radiation intensity respectively; and γ is the absorption coefficient of the I_G.

When T_x and I_G in equation (2) is identical with Q and ε in equation (1) respectively, the relaxation time, equivalent resistance and equivalent capacitance can be expressed as

$$\tau'_x = \frac{\rho_x c_x z_x}{e_x\sigma T_x^3} \tag{3a}$$

$$R'_{x_{eq}} = \frac{\rho_x c_x z_x}{\gamma_x} \tag{3b}$$

$$C'_{x_{eq}} = \frac{\gamma_x}{e_x\sigma T_x^3} \tag{3c}$$

Case#2: The Mutual Warming/Cooling Model (PDB Model)

If, in the reality, there is interaction (thermal energy exchange) between the atmosphere and the surface, equation (2) should be formulated as

$$\frac{dT_x}{dt} = -\frac{\sigma}{\rho_x c_x z_x} e_{xy}\left(T_x - T_y - \frac{\gamma_x}{\sigma e_{xy}} I_G\right) \tag{4}$$

the indices $_x$ and $_y$ express the atmosphere ($_a$) and the surface ($_s$) or vice versa; while e_{xy} expressed by

$$e_{as} = \frac{e_a T_a^4 - a_{sa} e_s T_s^4}{T_a - T_s} \tag{5a}$$

for the atmosphere, and

$$e_{sa} = \frac{e_s T_s^4 - a_{sa} e_a T_a^4}{T_s - T_a} \tag{5b}$$

for the surface. Meanwhile a_{xy} is the absorption coefficient of $_x$ in relation to the radiation of $_y$. In this case, the relaxation time, equivalent resistance and equivalent capacitance can be expressed as

$$\tau''_x = \frac{\rho_x c_x z_x}{\sigma e_{xy}} \tag{6a}$$

$$R''_{x_{eq}} = \frac{\rho_x c_x z_x I_G}{\sigma T_y e_{xy} - \gamma_x I_G} \tag{6b}$$

$$C''_{x_{eq}} = \frac{\sigma T_y e_{xy} - \gamma_x I_G}{\sigma e_{xy} I_G} \tag{6c}$$

The fact that the problem faced is the inverse problem [10], the scope and purpose of this paper is how to get τ, through the time series data of T and I_G, for the atmosphere and the surface respectively.

2. Material and Method

2.1. Data

The main data of this study are the time series data of air temperature, surface temperature, and global radiation at intervals of every 3 minutes recording on the observation of Mauna Loa Observatory (MLO), Hawaii [11]. Because of this study is still at an early stage, the data used is chosen arbitrarily, limited only from observations during the month of February 2014 alone.

IOP Conf. Series: Journal of Physics: Conf. Series **846** (2017) 012005 doi:10.1088/1742-6596/846/1/012005

Although they have been averaging for as much as 28 days of observation, in order to be processed further the preparation to every kind of the data above was done by the "5-stage running mean", each for the five data before and after. Furthermore, in order that the data can be treated as like as a truly daily cycle events, the running mean calculation is first done by connecting the beginning and the end of the data in question.

2.2. Method

As already stated above, this study is limited only to the two cases (models) that involve only the mechanism of heat exchange by radiation. The first step, the determination of C_{eq} is done by using the information about the extreme condition of T and implementation or the Inverse Calculation Method (MPT) that have been studied and tested on RSRC Model [10]. The difference here is that the function of the C-Test has been adapted to the shape function C_{eq} as stated in the equation (3c) and (6c), and by setting the values of physical parameters of the calculation resulted by the other methods [12] [13] and/or by other rational approach.

The next step, the relaxation time τ can be directly determined by the equation (2) and/or (4) without calculating first the R_{eq}, by equation (3b) and/or (6b). This can be done because if C_{eq} has been obtained then all the information needed to get the τ has been fulfilled by equation (2) and (4).

3. Results and Discussion

The graph in Figure 1 shows the pattern of global radiation, air temperature, and surface temperature resulted by preparation of the related data from observation in the MLO during February 2014. Note that the pattern of T-I_G relationships are very similar to the pattern relations of Q-ε in the RSRC Model [10]. Figure 1 reveal solar radiation began to emerge ($I_G \geq 0.1$ Wm^{-2}) at ± 5:30 LST (Local Standard Time), peaked at ± 12:00 LST to the intensity of 788.74 Wm^{-2} and sinks ($I_G \leq 0.1$ Wm^{-2}) at ± 19:45 LST. The existence of points representing the extreme conditions of T_a (6:04 LST, 276.49 K & 12:21 LST, 281.77 K), T_s (5:54 LST, 270.87 K & 12:31 LST, 302.41 K), and the points of intersection of T_a and T_s (8:11 LST, 278.62 K & 18:25 LST, 279.13 K) was decisive in the calculation of C_{eq} and τ [11].

Case#1: The PDM Model Results

Figure 2 shows the pattern of τ for the Case # 1 (PDM model) by setting $e_a = 0.7$ and $e_s = 0.9$. Rational values of τ obtained mainly in the region $I_G = 0$, generally with $\tau_a > \tau_s$. This can be understood by remembering that the surface (soil) have Diurnal Temperature Range (DTR) greater than atmosphere's DTR that obviously have to experience rate (changes in) temperature faster.

However, Figure 2 also shows the negative values and/or too large values of τ in the region $I_G \neq 0$, which indicates that the model in this case failed for region in question. Some critical point in the case where the values of τ deviate from rational estimate is precisely at points of extreme and points of intersection of T_a and T_s.

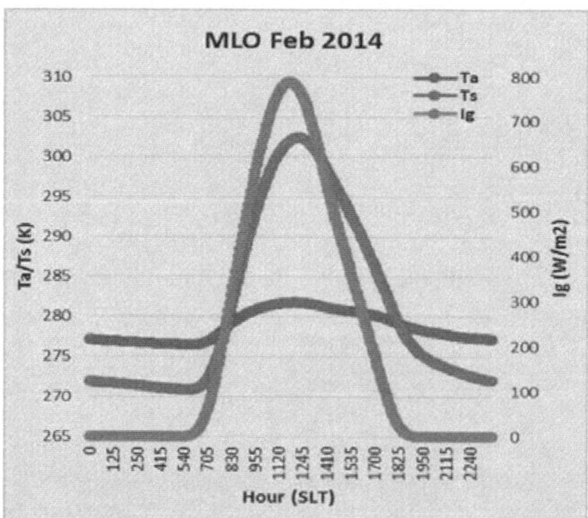

Figure 1. Typical form of atmosphere and surface temperature in relation to global solar radiation represented after data preparation (Data taken from MLO, 2014).

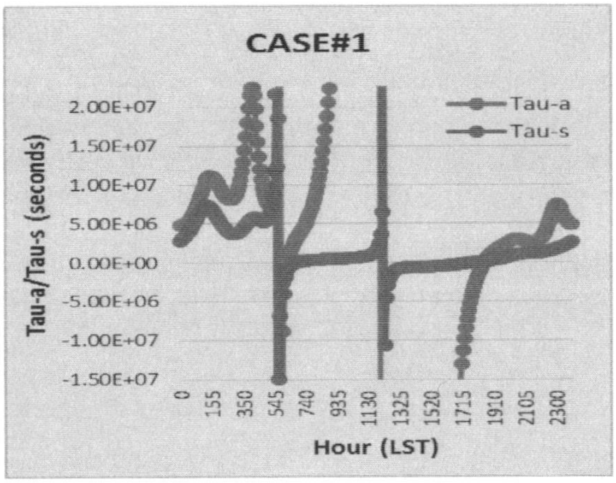

Figure 2. Relaxation time for Case-1 by setting $e_a = 0.7$ and $e_s = 0.9$. Asymptotic values are not shown. The negative and very height values of τ at $I_G \neq 0$ indicate the model is failed for that region.

Case#2: The PDB Model Results

The τ chart for Case # 2 (PDB Model), with $e_a = 0.7$ and $e_s = 0.9$, the same as in Case # 1, is shown in Figure 3. Note that the pattern of τ in the the region with the $I_G = 0$, are similar in both cases, but with values of τ in the Case #1 are 50 times greater than the values of the corresponding τ in the Case # 2.

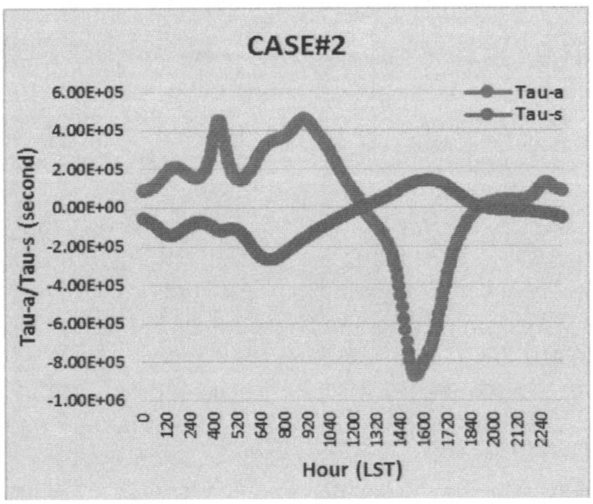

Figure 3. Relaxation time for Case-2 with $e_a = 0.7$ and $e_s = 0.9$. τ_a is reasonable, except in the second half of region with $I_G \neq 0$, whereas the τ_s is not.

Figure 3 also shows the pattern and value of τ_a that fairly rational in the half of the region with $I_G \neq 0$, but still led to a weakness when the values of τ_a are negative in the other half of region in question. Theoretically these problems arise due to the sameness signs regarding the value of the component parts of the left and right inside the brackets of equation (4). The solution of this problem is to add the mechanism of conduction, convection, and/or phase transformation on the part inside the brackets.

Another weakness of Case # 2 (PDB Model) is shown by the τ_s graph which in most of the region is worth negative. This weakness has been identified when T_s is still down on the condition that $I_G = 0$ because when $T_a > T_s$, in accordance with the PDB Model, T_s should be increase. Physically T_s can still go down if we add the 'depth component', with a lower temperature than the surface temperature, to the surface sub-system.

4. Conclusion

Although based on a model that is very simple but, as generally the inverse problem, it is very difficult to determine exactly just one parameter associated with a process or characteristic of the system. On the one hand this is because the parameter is changed in value at any time, but in the other hand due to the process or the mechanisms associated with many components that are intertwined and with only

involves the mechanism of heat transfer by radiation, the relaxation time in the warming/cooling of the atmosphere and surface still can't be determined well except for the condition without external radiation (specially the solar radiation), in which case the PDM Model and PDB Model produce similar pattern of the relaxation time but with a degree of rationality better value on the PDB Model.

From better calculation result of the PDB Model we also concluded that the process of warming (cooling) must involve the interaction, as well as mutual influences between the atmosphere and the surface. Meanwhile, the failure of the PDB Model in determining the relaxation time related to the atmosphere in the second half of the region with $I_G \neq 0$ and the failure in determining the relaxation time associated to the surface on most of the region suggests the need for the development of this model through inclusion the role of the other heat transfer mechanism, and put a layer of depth surface as an integral part of the system.

Acknowledgments
This study and paper can be realized on the support of Prime Competitive Research granted (Penelitian Unggulan Kompetitif) by University of Sriwijaya Fiscal Year 2015, by the Decree of Rector of the University of Sriwijaya, No. 0066/UN9/KP/2015.

References

1. Giancoli, D.C., *PHYSICS, Principles with Applications*, 6th Edition, Pearson Prentice Hall, Pearson Education Inc. Upper Saddle River NJ, 07458, 2005.
2. https://en.wikipedia.org/wiki/**RC_circuit**
3. Boas, M.L., *Mathematical Methods in the Physical Sciences*, 3rd Edition, John Wiley & Sons, New Jersey, USA, 2006.
4. Tarantola, A., *Inverse Problem Theory and Methods for Model Parameter Estimation*, SIAM, 2005.
5. O'Sullivan, C.T., "Newton's Law of Cooling-A Critical Assessment", *Am. J. Phys. 58 (10): 956-960,* 1990.
6. Vollmer, M., "Newton's Law of Cooling Revisited", *Eur. J. Phys. 30: 1063-1084*, doi: 10.1088/0143-0807/30/5/013, 2009.
7. Twomey P, C. O'Sullivan & J. O'Riordan, "An Experimental Investigation of The Role of Radiation in Laboratory Bench-Top Experiments in Thermal Physics", *Eur. J. Phys.* **30** 559–66, 2009.
8. Besson, U., "Cooling and Warming Laws: An Exact Analytical Solution", *Eur. J. Phys. 31: 1107-1121*, doi: 10.1088/0143-0807/31/5/013, 2010.
9. Battaglia O.R., C. Fazio, N. Pizzolato, & R.M.Sperandeo-Mineo, "An Investigation of Environmental Temperature Effects on Energy Exchange by Thermal Radiation", *Am. J. Phys.* 81 (12), 923-928, 2013.
10. Arsali, O.C. Satya, & Supardi, "Determination of System Dynamic Characteristics Based on A Serial RC Circuit Model", paper on the *International Conference of Mathematics, Science, Education, and Technology*, Padang State University, Padang, 2015.
11. http://www1.ncdc.noaa.gov/pub/data/uscrn/products/subhourly01
12. Arsali, O.C. Satya, & S. Hamdi, "Uji Formula Brutsaert-Crawford pada Perhitungan Radiasi Gelombang Panjang Atmosfer", Proceeding of *Simposium Fisika Nasional (SFN) XXVII*, Universitas Udayana, Denpasar-Bali, 2014
13. Arsali, O.C. Satya, & S. Hamdi, "Perlakuan Mirip Koreksi Cosinus pada Model Parametrik Permukaan untuk Perbaikan Estimasi Radiasi Balik Gelombang Panjang Atmosfer", Paper and Poster on *Seminar Nasional Sains Atmosfer (SNSA) 2015*, LAPAN, Bandung, 2015.

IOP Conf. Series: Journal of Physics: Conf. Series **846** (2017) 012006 doi:10.1088/1742-6596/846/1/012006

Twisting Motion Frequency Dependent I-V Characteristics of 102 Base Pairs Poly(dG)-Poly(dC) DNA Molecule

Efta Yudiarsah[1*]
[1]Departemen Fisika, FMIPA, Universitas Indonesia, Depok 16424

* E-mail: e.yudiarsah@sci.ui.ac.id

Abstract. The I-V characteristic of 102 base pair Poly(dG)-Poly(dC) DNA molecule have been calculated for several base pairs twisting motion frequencies. The calculation is carried out on doubled-stranded DNA model sandwiched in between two metallic electrodes. The effect is studied by taking into account twisting angle dependent on-site energy and hopping constant in the tight binding Hamiltonian of double-strand DNA model. We use semi-empirical Slater-Koster theory in the twisting angle dependent intra- and inter-strand hopping constant. We consider the temperature dependent sugar-phosphate backbone on-site energy by employing random energy disorder using uniform distribution function. The standard deviation of twisting angle is obtained by assuming that the average kinetic energy of twisting motions is proportional to system temperature. The transfer and scattering matrix methods are used simultaneously in calculating the transmission probability of charge on the molecule. We choose the contacts between molecule and both electrodes such that the main features of transport properties of the molecule do not change much by the presence of metallic electrodes. By assuming the voltage drops symmetrically at the contacts, Landauer-Buttiker Formalism is used in calculating the I-V characteristic of the molecule from transmission probability. The results show that the magnitude of current increases by twisting motion frequency increment. Larger current magnitude increment is observed at higher voltage. The influence of twisting motion frequency on the I-V characteristic is stronger at higher temperature, in the range of considered temperature.

1. Introduction

In the last decades, DNA molecules have attracted much attention of physicists. This interest stems from the hope of using DNA molecules as an active element in electronic devices, such as biosensor [1] and molecular transistor [2]. Research done by Eley and Spivey in 1962 [3] shows that DNA can conduct current. It opens an opportunity of using DNA molecule in electronic devises. DNA molecule also has a potential to be used as a template in fabricating nanoscale wire [4] for its long nature. Experimental results that show electrical conduction through DNA may involve in several biological processes, such as DNA mutation [5], damage detection and self-repairing [6, 7], as well as sensor and signaling system [8, 9], also increases the interest of physicist.

Unfortunately, different groups measuring conductivity of DNA molecule using different types of DNA reports different results. The transport property of DNA can span from insulator to proximity-induced superconductor [10]. Its results in a controversy about the transport property of DNA molecules. Beside due to the different in DNA sequence [11, 12] used, other factors such as medium or substrate [13, 14], base pairing mismatch [15], and internal motion of DNA [16, 11] may lead to this controversy.

Since DNA is a long and flexible molecule, this report will be focused on the effects of base pair twisting motion on the I-V characteristic of 102 base pairs Poly(dG)-Poly(dC) DNA molecule. Semi-empirical theory of Slater-Koster [10] has been used in representing twisting motion dependent electron

Content from this work may be used under the terms of the Creative Commons Attribution 3.0 licence. Any further distribution of this work must maintain attribution to the author(s) and the title of the work, journal citation and DOI.
Published under licence by IOP Publishing Ltd

IOP Conf. Series: Journal of Physics: Conf. Series **846** (2017) 012006 doi:10.1088/1742-6596/846/1/012006

hopping constant. Landaur-Büttiker formalism [17] has been used in calculating current from transmission probability which is calculated using transfer matrix technique and scattering matrix method, simultaneously [18]. The calculation is done by assuming potential drops at contact symmetrically. The results show the twisting motion influences the current such that the magnitude of current increases with twisting motion frequency increment.

The organization of this report is as follow. In the next section, DNA molecule model and calculation method will be explained. Then, calculation results will be discussed. In the last section, a conclusion will be stated.

2. Material and Method

In the model we used, we consider that electron can hop from one base to other base and to sugar-phosphate back bone vice versa. However, the model exclude hopping from one sugar-phosphate back bode to the other. Inter-base hopping can occur from one base to other base in the same strand (intra-strand hopping), to its pair in Watson-Crick base pair (Watson-Crick pair hopping), and to other base in other strand (inter-strand hopping). Tight binding Hamiltonian of the model can be written as follow.

$$H_{dna} = \sum_{i=1}^{N}\sum_{j=1}^{2}\left\{\varepsilon_i^j c_{i,j}^+ c_{i,j} + t_{i,i}^{j,j+1}\left(c_{i,j}^+ c_{i,j+1} + c_{i,j+1}^+ c_{i,j}\right)\right\} + \sum_{i=1}^{N}\sum_{j=1}^{2}\sum_{k=u,d}\left\{B_i^k c_{i,k}^+ c_{i,k} + t_{i,i}^{j,k}\left(c_{i,j}^+ c_{i,k} + c_{i,k}^+ c_{i,j}\right)\right\}$$
$$+ \sum_{i=1}^{N-1}\sum_{j,k=1}^{2} t_{i,i+1}^{j,k}\left(c_{i,j}^+ c_{i+1,k} + c_{i+1,k}^+ c_{i,j}\right) \qquad (1)$$

In Eq. (1), ε, *B*, and *t*, consecutively, represent base on-site energy, sugar-phosphate backbone on-site energy, and electron hopping constant from one base to the other or to sugar-phosphate backbone. *N* is the length of DNA molecule, in this case $N = 102$ base pairs.

Thermal fluctuation can cause the base pair to twist [13, 19]. The twisting motion influences base on-site energy according to [20]

$$\varepsilon_i^j = \phi_i^j + b_{i-1,1}^j\left(\cos\Delta_{i-1,i}^{j,j} - \cos\varphi_{eq}\right) + b_{i,1+1}^j\left(\cos\Delta_{i,i+1}^{j,j} - \cos\varphi_{eq}\right) \qquad (2)$$

In Eq. (2), $\Delta_{i,i+1}^{j,k} = \varphi_{eq} - \left(\varphi_i^j - \varphi_{i+1}^k\right)$ and ϕ is base on-site energy at equilibrium, 7.84 eV for G and 8.85 eV for C. *b* is coupling parameter between charge and base pair twisting motion. φ is twisting angle of base relative to base position at equilibrium. The twist angle is assumed following Gaussian distribution with zero average and $\sqrt{k_b T/\left(I_n \omega^2\right)}$ deviation standard, where *T* is temperature, k_b is Boltzmann constant, I_n is base moment of inertia, and ω is twisting motion frequency. I_n is assumed to be in the order of theoretical value, i.e. 1.0×10^{-44} kgm^2 [22]. φ_{eq} is angle between two neighboring base pairs at equilibrium and assumed to be 36°.

Angle dependent intra-strand, Watson-Crick pair hopping constant, and inter-strand obtain by expanding semi-empirical theory of Slater-Koster [10]. They are written, consecutively, as

$t_{i,i+1}^{j,j} = t_{i,i+1}^{pa}\left[1 - \alpha_{i,i+1}^{pa}\sin^2\left(\Delta_{i,i+1}^{j,j}/2\right)\right]/\left[1 - \alpha_{i,i+1}^{pa}\sin^2\left(\varphi_{eq}/2\right)\right]$, $t_{i,i}^{j,k} = t_{i,i}^{pp}\left[1 + \alpha_{i,i}^{pp}\sin^2\left\{\left(\varphi_i^j - \varphi_i^k\right)/2\right\}\right]$, and

$$t_{i,i+1}^{j,k} = t_{i,i+1}^{sl}\left[1 - \alpha_{i,i+1}^{sl}\cos^2\left(\Delta_{i,i+1}^{j,k}/2\right)\right]/\left[1 - \alpha_{i,i+1}^{pa}\cos^2\left(\varphi_{eq}/2\right)\right], \qquad (3)$$

where $\alpha^{pa} = 2.00$, $\alpha^{sl} = 0.42$, and $\alpha^{pp} = 1.00$. The values of parameter *b* and $t_{i,j}$ can be seen in table 1.

In this study, the effect of medium and temperature on the back-bone onsite energy is involved as back bone energy disorder (*W*) and back bone on-site energy *B* is assumed to be distributed uniformly in the range [*D-W*, *D+W*]. *D* is chosen to be equal to 9.36 eV, back bone on-site energy without disorder. Other parameters used in this study are metal on-site energy 8.45 eV and electron hopping constant in metal 0.75 eV. Moreover, in calculating current, Fermi energy is set equal to 8.45 eV.

Table 1. Parameter for electron hopping constants ($t_{x,y}$) and coupling constant between charge and base pair twisting motion ($b_{x,y}$) for tight binding Hamiltonian. [20, 21].

Base (X,Y)	$t_{x,y}^{pa}$ (eV)	$t_{x,y}^{sl}$ (eV)	$t_{x,y}^{pp}$ (eV)	$b_{x,y}$ (eV)
G,G	0.30	0.14	-	0.0334
C,C	0.23	0.28	-	0.0341
C,G	0.14	0.27	0.57	0.0299

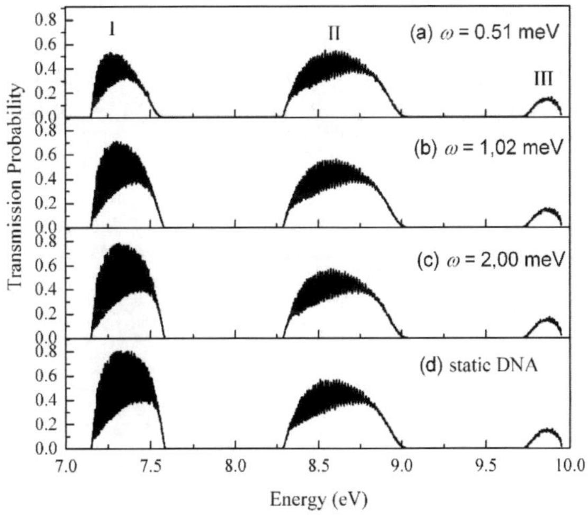

FIGURE 1. Transmission probabilities as a function of energy on 102 base pairs long poly(dG)-poly(dC) DNA molecule at 4.2 K and W = 0.3 eV for various twisting frequencies (ω), (a) 0.51 meV, (b) 1.02 meV, (c) 2.00 meV and a static DNA.

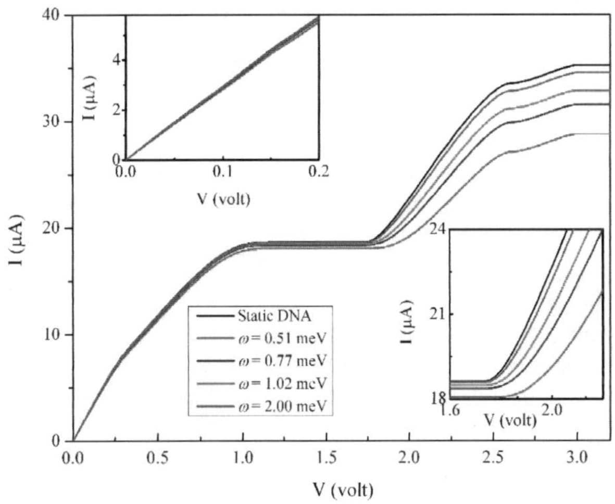

FIGURE 2. The I-V characteristic of 102 base pairs long poly(dG)-poly(dC) DNA molecule at 4.2 K and W = 0.3 eV for various twisting frequencies and a static DNA.

3. Experimental Results and Discussion

Transmission probabilities calculated at T = 4.2 K and W = 0.3 for static DNA and dynamics DNA with various frequency are shown in figure 1. This result and other calculation results is the average of 1000 twisting angle configurations. The figure shows that there are three bands of energy with transmission greater than zero. For static DNA, there is no energy with transmission maximum one, since the present of back bone disorder, destroying uniform potential profile of periodic poly(dG)-poly(dC) DNA molecule. This result is in agreement with others [23, 24]. The base pair twisting motion does not change transmission probability much, except at band I. At band I, the transmission probability increases with the twisting motion frequency.

I-V characteristic calculated at T = 4.2 K and W = 0.3 for static DNA and dynamics DNA with various frequency are shown in figure 2. Since the Fermi energy is located in band II, current increase immediately with voltages. The plateau in the figure is due to the present of two gaps, one gap between band 1 and II and the other gap between band II and III. At low voltage, current increases linearly with voltage and is only minutes affected by twisting motion frequency, as shown in the top left inset.

However, at higher voltage, especially after the plateau, current is influenced much by increases with the twisting frequency, as can be seen in bottom right inset. The inset also shows, after the plateau current increases at higher voltage for lower twisting frequency. This indicates that if Fermi energy is located in the gap, then the potential threshold will change with the twisting motion frequency.

Similar trend of change in I-V characteristic with frequency is also observed in the calculation result at T = 30 K. However, at T = 30 K the effect of twisting motion on the I-V characteristic is more prominent. Current magnitude increases as twisting frequency increases is higher than the one at T = 4.2 K. It seems to occur at all the range of voltages used.

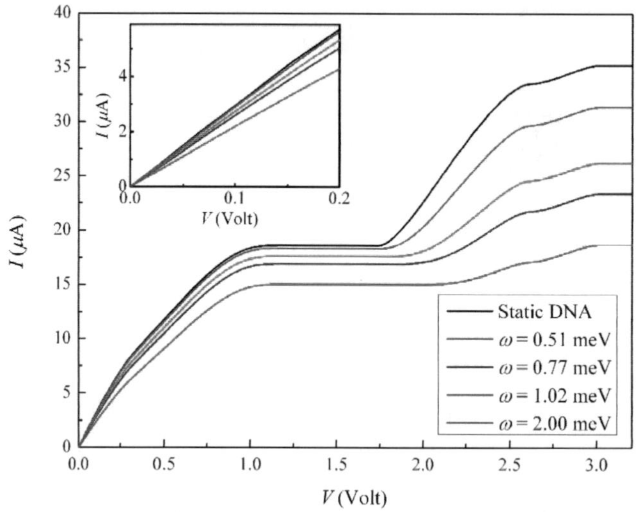

FIGURE 3. The I-V characteristic of 102 base pairs long poly(dG)-poly(dC) DNA molecule at 30 K and $W = 0.3$ eV for various twisting frequencies and a static DNA.

4. Conclusion and Future Work

In the region of low voltages, current increases linearly with voltage independent of twisting frequency and temperature. Higher twisting motion frequency results in larger current magnitude. In the range of considered temperature, twisting motion frequency influences the I-V characteristic stronger at higher temperature.

Acknowledgments

The author would like to thank UPP-IPD Universitas Indonesia for laboratory support and Mr. Daniel K. Suhendro for the parameters used in the calculation. This study is funded by Hibah Penelitian Unggulan Perguruan Tinggi (PUPT) Tahun Anggaran 2015 Nomor:0529/UN2.R12/HKP.05.00/2015.

References

1. R. L. Zaffino, M. Mir, and J. Samitier (2014), Label-free detection of DNA hybridization and single point mutations in a nano-gap biosensor, *Nanotechnology* **25**, 105501.
2. S. Nokhrin, M. Baru, and J. S. Lee (2007), A field-effect transistor from M-DNA, *Nanotechnology* **18**, 095205.
3. D. D. Eley and D. I. Spivey (1962), Semiconductivity of organic substances, *Transactional Faraday Society* **58**, 411.
4. E. Braun, Y. Eichen, U. Sivan, and G . Ben-Yoseph (1998), DNA-templated assembly and electrode attachment of a conducting silver wire, *Nature* **391**, 775.
5. J. H. Miller Jr, M. Y. S. Villagrán, S. Maric, and S. M. Briggs (2015), Normal and Impaired charge transport in biological system, *Physica B* **460**, 119.
6. P. J. Dandliker, R. E. Holmlin, and J. K. Barton (1997), Oxidative Thymine Dimer Repair in the DNA Helix, *Science* **275**, 1465.

7. E. M. Boon, A. L. Livingston, N. H. Chmiel, S. S. David, and J. K. Barton (2003), DNA-mediated charge transport for DNA repair, *Proceedings of the National Academy of Sciences of the United States of America* **100**, 12543.
8. P. A. Sontz, N. B. Muren, and J. K. Barton (2012), DNA charge transport for sensing and signaling, *Accounts of chemical research* **45**, 1792.
9. J. C. Genereux, A. K. Boal and J. K. Barton (2010), DNA-mediated charge transport in redox sensing and signaling, *Journal of the American Chemical Society* **132**, 891.
10. R. G. Endress, D. L. Cox, and R. R. P Singh (2004), Colloqium: The quest for high-conductance DNA, *Review of Modern Physics* **76**, 195.
11. Y.-S. Jo, Y. Lee, and Y. Roh (2003), Current–voltage characteristics of λ- and poly-DNA, *Materials Science and Engineering: C* **23**, 841.
12. D. Klotsa, R. A. Römer, and M. S. Turner (2005), Electronic transport in DNA, *Biophysical journal* **89**, 2187.
13. S. Roche, (2003), Sequence Dependent DNA-Mediated Conduction, *Physical Review Letters* **91**, 1081011.
14. A.-M. Guo, S.-J. Xiong, Z. Yang and H.-J. Zhu (2008), Enhancement of transport in DNA-like systems induced by backbone disorder, *Physical Review E* **78**, 0619221.
15. I. Kratochvílová, *et al.* (2010), Charge transport in DNA oligonucleotides with various base-pairing patterns, *The journal of physical chemistry. B* **114**, 5196.
16. N. B. Muren, E. D. Olmon, and J. K. Barton (2012), Solution, surface, and single molecule platforms for the study of DNA-mediated charge transport, *Physical chemistry chemical physics* **14**, 13754.
17. E. Maciá, F. Triozon, and S. Roche (2005), Contact-dependent effects and tunneling currents in DNA molecules, *Physical Review B* **71**, 1131062.
18. H. Taniyama and A. Yoshii (1996), Scattering-matrix method for the tight-binding model of heterostructure electronic states, *Physical Review B* **53**, 9993.
19. Z. Yu and X. Song (2001), Variable range hopping and electrical conductivity along the DNA double helix, *Physical Review Letters* **86**, 6018.
20. D. K. Suhendro, E. Yudiarsah, and R. Saleh (2010), Effect of phonons and backbone disorder on electronic transport in DNA, *Physica B: Condensed Matter* **405**, 4806.
21. D. K. Suhendro (2014), private communication.
22. L. V. Yakushevich, (2004), *Nonlinear physics of DNA*. 2nd ed. Weinheim: Wiley-VCH, pp. 174.
23. E. Maciá and S. Roche (2006), Backbone-induced effects in the charge transport efficiency of synthetic DNA molecules, *Nanotechnology* **17**, 3002.
24. A.-M. Guo, Z. Yang, H.-J. Zhu, and S.-J. Xiong (2010), Influence of backbone on the charge transport properties of G4-DNA molecules: a model-based calculation, *Journal of Physics: Condensed Matter* **22**,

IOP Conf. Series: Journal of Physics: Conf. Series **846** (2017) 012007 doi:10.1088/1742-6596/846/1/012007

The Role of Fe_2O_3 and Light Induced on Dielectric Properties of Borosilicate Glass

Markus Diantoro[1,*], Norma Dian Prastiwi[1], Ahmad Taufiq[1], Nurul Hidayat[1], Nandang Mufti[1], Arif Hidayat[1]

[1]Department of Physics, Faculty of Mathematics and Natural Sciences, Universitas Negeri Malang (State University of Malang), Jl. Semarang 5 Malang 65145, Indonesia

[*]email: markus.diantoro.fmipa@um.ac.id

Abstract. Functionally glass materials have been widely applied in various technological applications remarkably due to their optoelectric properties. In this present study, the glass was prepared from leaching product of local silica sands. Bi_2O_3 and Na_2CO_3 were added to reduce the melting point of silica sand to form silica glass and Fe_2O_3/B_2O_3 was incorporated to examine its effect on the crystal structure, morphology, and light-induced dielectric properties of the borosilicate-based fuctional glass. The characterizations were conducted by means of Differential Thermal Analyses (DTA), X-ray diffraction (XRD), scanning electron microscopy (SEM), Fourier transform infrared spectroscopy (FTIR), and capacitance meter under the influence of light intensity. The XRD pattern shows the absence of any sharp diffraction peak indicates the amorphous state of the borosilicate glass. While the SEM image shows that the borosilicate glass exhibited amorphous characteristic. Furthermore, the increasing of Fe_2O_3 tends to reduce the dielectric constant. On the other hand, the increase of light intensity increase the dielectric constant with a step like properties.

1. Introduction

The demand for the developement of functionally glass material for many optoelectronical technological applications increases from time to time [1]. For instance, functional glass can be applied in kitchenware, glass roof, screen, and solar smart glass. Unfortunately, many functional glass materials are fabricated from commercially high cost materials. In general, the main material of glass is silica which has melting temperature of 1715 ^{o}C [2]. In Indonesia, silica can be easily found from natural resources such as from silica sand [3]. In this study, the possibility of develop functionally glass material, borosilicate glass, from natural material be investigated.

2. Experimental Method

As alternative of the use commercial material for glass fabrication, silica sand as a natural resource from Tuban, Indonesia was chosen as the starting material. The silica sand was immersed in 2M HCl for 12 h, washed, and dried to obtain higher purity of SiO_2 powder. The powder was then characterized by X-ray flourescence (XRF) to clarify the increasing of Si content which corresponds to SiO_2. Functional glass B_2O_3-Fe_2O_3/Bi_2O_3-Na_2CO_3-SiO_2 was prepared by mixing and melting B_2O_3, Fe_2O_3, Bi_2O_3, Na_2CO_3, and SiO_2. The compositions of each samples are shown in Table 1.

Content from this work may be used under the terms of the Creative Commons Attribution 3.0 licence. Any further distribution of this work must maintain attribution to the author(s) and the title of the work, journal citation and DOI.
Published under licence by IOP Publishing Ltd

The mixtures were then annealed at 420 °C for 1 h. After processing, the samples were then characterized by using XRD, SEM, FTIR, and capacitance meter.

Table 1. Raw Material Compositions for Borosilicate Glass

Sample	Compositions (wt%)				
	Fe_2O_3	B_2O_3	SiO_2	Na_2CO_3	Bi_2O_3
Glass 1	0	18	12	10	60
Glass 2	1	17	12	10	60
Glass 3	2	16	12	10	60
Glass 4	3	15	12	10	60
Glass 5	4	14	12	10	60

3. Results and Discussion

The XRF data for the silica sand before and after leaching treatments is given in Table 2.

Table 2. Main Elemental Contents of Silica Sand

Element	Conc. (%) Before Leaching	Conc. (%) After Leaching
Si	80.7(3)	86.8(4)
K	6.22(3)	5.24(9)
Ca	4.40(3)	2.2(3)
Ti	1.40(3)	0.87(4)
Fe	4.23(07)	1.16(02)
Ni	1.96(03)	2.68(04)

From Table 2, it can be seen that leaching increases the Si content from 80.7(3)% to 86.8(4)% and reduce some elemental impurities. It means that the leaching process was effective to increase the Si content by reducing impurities. It is because Si cannot be dissolved in the HCl but impurities can.

Figure 1 shows the diffraction patterns of B_2O_3-Fe_2O_3/Bi_2O_3-Na_2CO_3-SiO_2 with 0%, 2%, and 4% of Fe_2O_3 recorded from XRD measurement. It can be clearly seen that the absence of any sharp diffraction peak indicates the amorphous state of the borosilicate glass found in this experiments.

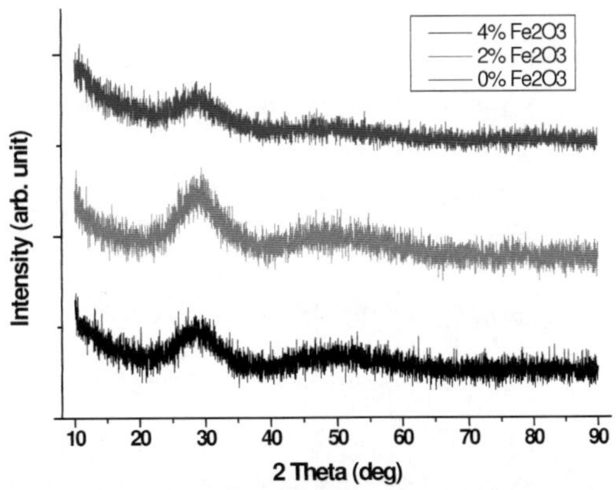

Figure 1. XRD Profiles of B_2O_3-Fe_2O_3/Bi_2O_3-Na_2CO_3-SiO_2

The SEM images of the surface of the samples show that the borosilicate glass found in this research have small grains with homogenous size and morphology. The SEM image is also show that borosilicate glass in amorphous phase as found in XRD result.

IOP Conf. Series: Journal of Physics: Conf. Series **846** (2017) 012007 doi:10.1088/1742-6596/846/1/012007

Figure 2. SEM Images for Borosilicate Glass with 1% Fe_2O_3 Incorporation.

The thermal response analyses was performed by using TGA-DSC . The result is shown in Fig. 3.There is clearly show that the linear mass change from room temperature up to 400 °C continued by relative constant mass with some small change at limited temperature. The mass change increase in higher temperatures. That can be associated to the chemical reactions. Those reaction is related to the exothermic decomposition or melting of $NaCO_3$, B_2O_3, Bi_2O_3, Fe_2O_3.

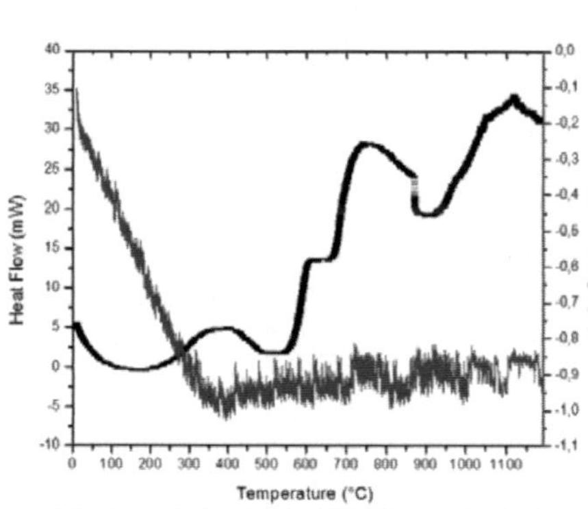

Figure 3. DTA-TG of Fe3O4 doped borosilicate glass.

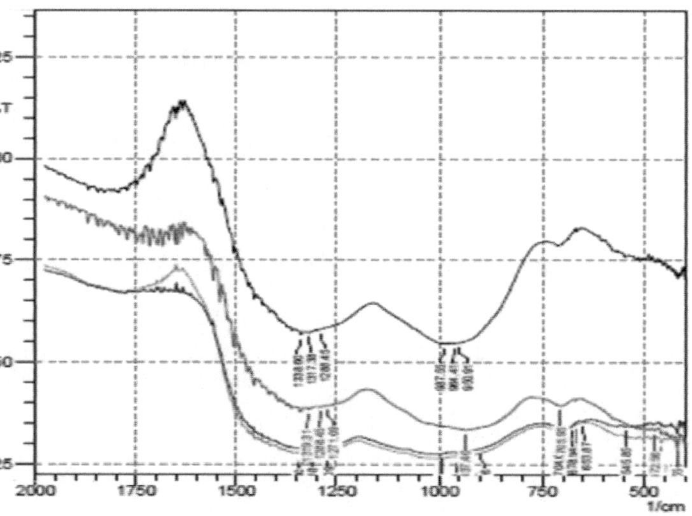

Figure 4. FTIR Spectra of Borosilicate Glass

The FTIR spectra of borosilicate glass as shwon in Fig. 4 . The figure reveals that the chemical bonding within the samples. Absorbance peak at 464-465 1/cm correspond to Fe-O formation in FeO_6 structural unit. Two absorbance peaks at 475 – 479 1/cm show that the vibration band of Si-O-Si and Bi-O. Absorbance peak at 702 - 705 1/cm represent B-O-B bonding in group BO_3. Absorbance peak at 935 - 937 1/cm is in a good agreement with streching mode of BO_4 tetrahedral unit. Absorbance peak at 1330 1/cm goes to streching of B-O bonding. This spectra is also similar to the spectra that reported by Gu et al 2014 observing at 479 1/cm, 703 1/cm, 1022 1/cm, and 1384 1/cm.

Based on the Figure 5, it can be seen that the increasing of Fe_2O_3 content tends to reduce the dielectric constant. Furthermore, the change of light intensity also reduces the dielectric constant. The light is quantized as photon which corresponds to light intensity. In this work, the change of dielectric constant due to the change of light intensity is similar to the photoelectric effect occurring on the metal surface.

Measurement of dielectric constant of samples is shown in Figure 5. It can be seen that the increasing of Fe_2O_3 content result in the decrease of their dielectric constant. Furthermore, the change of light intensity also reduces the dielectric constant. The light is quantized as photon which corresponds to light intensity. In this work, the change of dielectric constant due to the change of light intensity is similar to the photoelectric effect occurring on the metal surface. The figure also shows that the higher the Fe_2O_3 concentration. It agree with the previous result reported by Shapaan *et al.* [4]. Furthermore, Yadav *et al.* also reported that the Fe_2O_3 leads to reduce the dielectric constant [5].

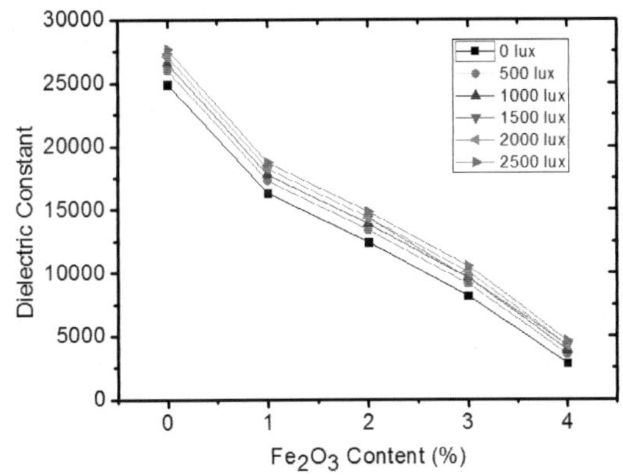

Figure 5. Dielectric Properties of Borosilicate Glass

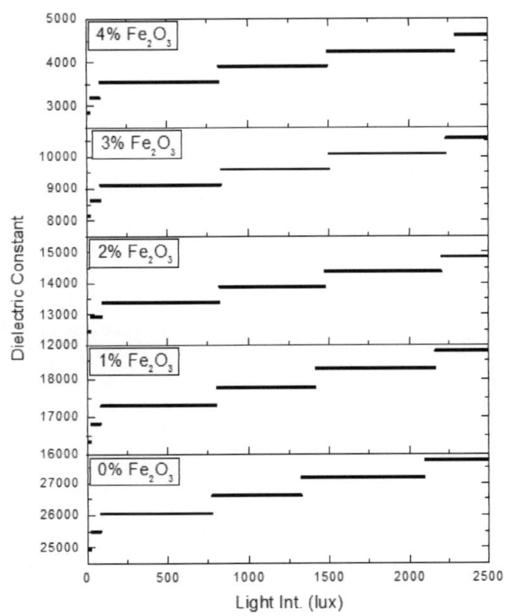

Figure 6. A typical dielectric constant as a function of light intensity.

The reduce of dielectric constant as increase of Fe_2O_3 concentration is not in line with the increase of applied light intensity. The Fe_2O_3 incorporating glass show different nature as shown in Fig. 6. The light induced on the Fe_2O_3 doped glass is significantly increase by increase of the intensity. Similar phenomena is also shown by Al_2O_3 doped lead silica glass [6].

4. Conclusion

This study concludes that the silica sand can be used as the raw material for the development of borosiliate functional glass. The crystalline state of the produced samples are amorphous. The addition of Fe_2O_3 and light inducement decrease the dielectric properties of borosilicate glass.

Acknowledgments

We would like to thank the higher education on college excellent grants from the directorate general of higher education under the contract no. 9.4.9/UN32.14/LT/2015.

References

[1] Subhadra, M., Kistaiah, P. 2010. *Characterization and Optical Absorption Studies of* VO^{2+}*: Li2O-K2O-Bi2O3-B2O3 glass system.* Journal of Alloys and Compounds 505. Hal 634-639.
[2] Wahyudi., Marzuki A., Cari., Adi P. 2012. *Kajian Indeks Bias Kaca TBZP*. Jurnal Sains Universitas Sebelas Maret.
[3] Lestari, Fifi B., Zainuri, M. 2013. *Pengaruh Variasi Milling Time Pasir Silika Terhadap Sifat Ketahanan Korosi Komposit Pani/*SiO_2 *Pada Plat Baja.* Jurnal Sains dan Seni Pomits Vol 1, No.1.
[4] Shapaan, M., F.M. Ebrahim. 2010. *Structural and electric-dielectric properties of* B_2O_3-Bi_2O_3-Fe_2O_3 *oxide glass.* Journal of Physica B 405. Hal 3217-3222.
[5] Yadav, Avadhesh K., Gautam, Prabhakar Singh. 2014. *Crystallization and dielectric properties of* Fe_2O_3 *doped barium strontium titanate borosilicate glass.* Royal Society of Chemistry. Hal 2819-2826.
[6] Kaur, S, and Singh, K.J., Comparative study of lead borate and silicate glass systems doped with aluminum oxide as gamma-ray shileding materials, IJITEE, 2(5) 2013 pp. 172-175.

IOP Conf. Series: Journal of Physics: Conf. Series **846** (2017) 012008 doi:10.1088/1742-6596/846/1/012008

On Heat Transfer - Stress Analysis of Modified Brick (Reed Filler) Upon Its Production Stage

Kurniati Ornam[1,*], Masykur Kimsan[2], Cadas Teguh Prakasa[2], La Ode Ngkoimani[3], Santi[1]

[1] Halu Oleo University, Architecture Department, 93232 Kendari, Indonesia,
[2] Halu Oleo University, Civil Engineering Department, 93232 Kendari, Indonesia,
[3] Halu Oleo University, Geology Engineering Department, 93232 Kendari, Indonesia

* Email: kurniati.ornam@yahoo.co.id

Abstract. This paper aimed to scrutinize how burning process in modified brick's production impinge on crack as a result of stress differentiation between two consecutive layers of the brick's element. Diffusion engages in burning process of bricks, hence it generates thermal stress on element for different temperature between layers. This research focused on burning process in traditional production ward. Analytical of nonlinear equation and numerical solution, finite difference, were involved to obtain temperature value in each layer, followed by stress calculation. Based on the results, it can be concluded that crack occurs particularly on boundary area, since difussion tends to yield relatively more different value on it. Therefore, certain strategies, that may decrease this differentiation, are required to minimize number of cracks during brick's production.

1. Introduction

Nowadays, bricks are still more preferable building material for housing, especially for wall elements. Moreover, in accordance with environmental issue, utilization of waste and/or recycled material is also gain considerable concern from researchers. Some waste i.e sawdust, reed, sago husk, with their combination with additive material, have been used in brick's production as filler and demonstrated enhanced performance in terms of physical and mechanical properties of bricks [1, 2, 3, 4].

Brick is one of the materials created from burned clay in relatively high temperature. The clay is raw ingredient in the manufacture of bricks that has plasticity and dry shrinkage properties. Plasticity on clay is essential to simplify the process in commencing the manufacture of bricks. should clay be excessively plastic, it has an effect on strength, shrinkage, and outcome number of bricks.

Production of bricks are consisted of preparation, molding, drying, burning, unloading/storaging respectively [5, 6, 7]. The process of burning of bricks should be in balance with the rise in temperature and its speed through several required steps to be considered, i.e evaporation/drying, oxidation stage, full combustion stage, and detention stage. At the stage of full combustion, bricks are burned until cooked and sintering process occurs up to be a solid brick. Temperature of burning process in traditional industries varies between 450 ^{o}C – 920 ^{o}C depended on the nature of the clay.

In the burning process, diffusion ensues, in which the heat flow is going through an object in the form

Content from this work may be used under the terms of the Creative Commons Attribution 3.0 licence. Any further distribution of this work must maintain attribution to the author(s) and the title of the work, journal citation and DOI.
Published under licence by IOP Publishing Ltd

of bricks with dimensions and time. Diffusion process is also related to the strain and stress. Strain occurs due to temperature changes in the structure of the bricks through the process of diffusion combustion. Strain is associated with stress, in which case there will be a tension strain, as a result of elasticity properties of solid materials.

The purpose of this study was to determine the behaviour of sress and strain during burning process of bricks as a result of temperature change.

2. Materials and Method

2.1 Bricks Modelling

As shown on Figure 1, the dimension of bricks in this study followed the standard (5 x 11 x 22 cm) [8, 9, 10, 11, 12, 13, 14]. Brick has to have sharp and rectangle side; raft flat surface, a red color, sonorous sound when it is knocked, have no crack and not easy to be broke. These was the final condition of the brick prior to distribution to market. However, the initial condition of the bricks prior burning was different in terms of humidity. The clafication of brick is following previous research [2,3,4], silty clay with sand.

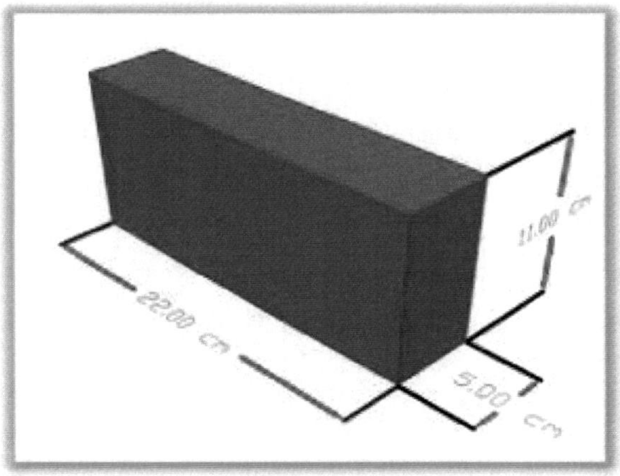

FIGURE 1. Brick's Dimension

2.2 Burning Process

There are 4 (four) stages during burning process as follows. The first stage is the evaporation (drying), i.e. the discharge of water forming the bricks, going up to a temperature of approximately 130 ^{0}C. Next, oxidation stage is where remained of plants (carbon) contained in the clay are burned. This process takes place at a temperature of 200 ^{0}C to 280 ^{0}C. Then, full combustion stage which cooked bricks are sintered to be a solid brick and temperatures vary between 350 ^{0}C and 600 ^{0}C depending on the nature of the clay. Finally, detention phase, where there is detention of temperature for 1-2 hours. In first to third stage, the temperature rise should be in slowly manner, in order to avoid losses on the brick.

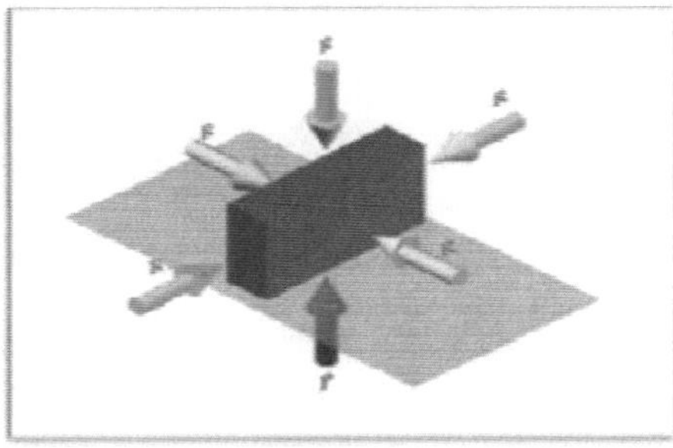

FIGURE 2. Burning Process Modelling and Heat Flow in Brick (*R* is room temperature and *P* is fire temperature)

2.3 Analytical and Numerical Formulation

Analytical solution is the implementation of diffusion equations 1D, Laplace and Poisson 3D, and finite difference methods in the process of burning the bricks to obtain the temperature difference between layers (ΔT). The formulation of diffusion, laplace and poisson are shown respectively in equation 1-3 [15, 16, 17, 18]

$$\frac{dT}{dt} = k\frac{d^2T}{dx^2} \quad (1)$$

$$\frac{d^2T}{dx^2} + \frac{d^2T}{dy^2} + \frac{d^2T}{dz^2} = 0 \quad (2)$$

$$\frac{dT}{dt} = k\left[\frac{d^2T}{dx^2} + \frac{d^2T}{dy^2} + \frac{d^2T}{dz^2}\right] \quad (3)$$

Where *T* is temperature in ^0C; *t* is time in *second*; *x,y,z* (figure 3) are coordinate in catessian space in representing bricks dimension and *k* is thermal conductivity.

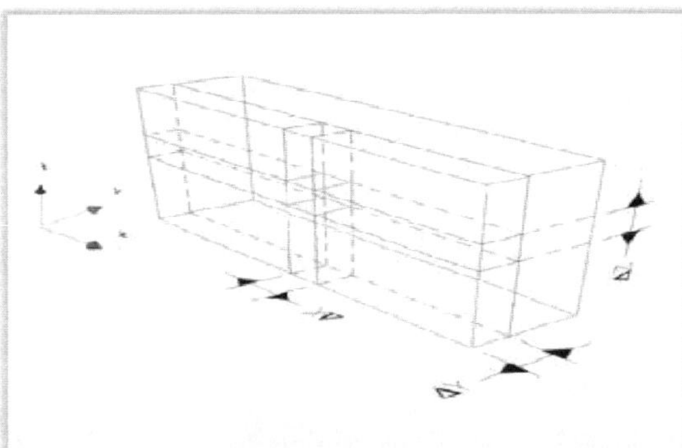

FIGURE 3. Segment of Bricks in Cartessian Coordinate System

2.4 Finite Difference Method

Finite difference method to be applied is the central difference method with *Dirichlet* boundary conditions as follows ($T = u(x, y) = \Delta T$) [19, 20, 21]

$$\frac{d^2T}{dx^2} = \frac{T^n_{i+1,j,k} - 2T^n_{i,j,k} + T^n_{i-1,j,k}}{\Delta x^2} \tag{4}$$

$$\frac{d^2T}{dy^2} = \frac{T^n_{i,j+1,k} - 2T^n_{i,j,k} + T^n_{i,j-1,k}}{\Delta y^2} \tag{5}$$

$$\frac{d^2T}{dz^2} = \frac{T^n_{i,j,k+1} - 2T^n_{i,j,k} + T^n_{i,j,k-1}}{\Delta z^2} \tag{6}$$

$$\frac{dT}{dt} = \frac{T^{n+1}_{i,j,k} - T^n_{i,j,k}}{\Delta t} \tag{7}$$

2.5 Stress-Strain Relationship

An element experiencing temperature difference generates strain and stress. In this study, a simple approach in determining stress in bricks is formulated in equation 8 and 9.

$$\varepsilon_T = \alpha(\Delta T) \tag{8}$$

$$\sigma_T = E_B \times \varepsilon_T \tag{9}$$

Where ε_T is thermal strain, α is thermal coefficient in $^0C^{-1}$, ΔT is temperature difference in ^0C, E_B is modulus of elasticity in *Mpa* and σ_T is thermal stress in *Mpa*.

3. Results and Discussion

3.1 Temperature Change

Figures 4-7 shows temperature change at 4 (four) stages of burning process in 2 cm height. In terms of time, there is no significant difference among first three stages. Temperature change is sharply rising at 4th stage, where burning temperature hit 600 ^0C, e.g. $\Delta T\ (5, 3, 2, t_3)$ = 107.664 ^0C and $\Delta T\ (5, 3, 2, t_4)$ = 216.394^0C. Temperature change is in proportional with burning duration, as can be seen first 20 minutes, then second 20 minutes, followed by 30 minutes and last 50 minutes.

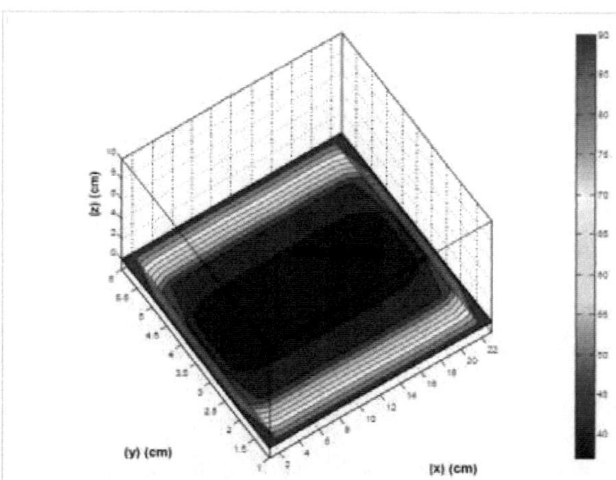

FIGURE 4. Temperature Change (*celcius degree*) in First Stage at z = 2 cm

IOP Conf. Series: Journal of Physics: Conf. Series **846** (2017) 012008 doi:10.1088/1742-6596/846/1/012008

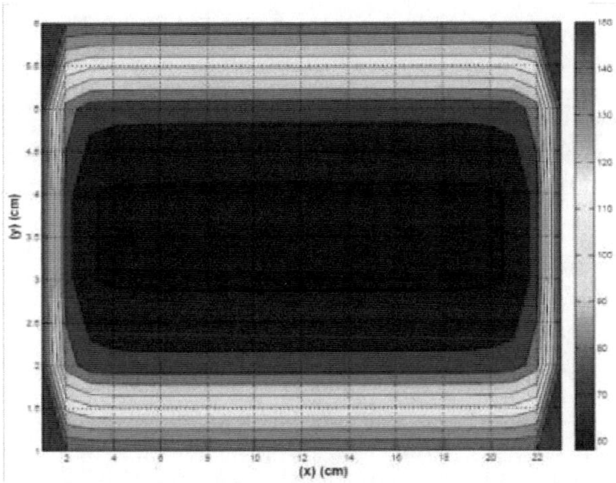

FIGURE 5. Temperature Change (*celcius degree*) in Oxidation Stage at z = 2 cm

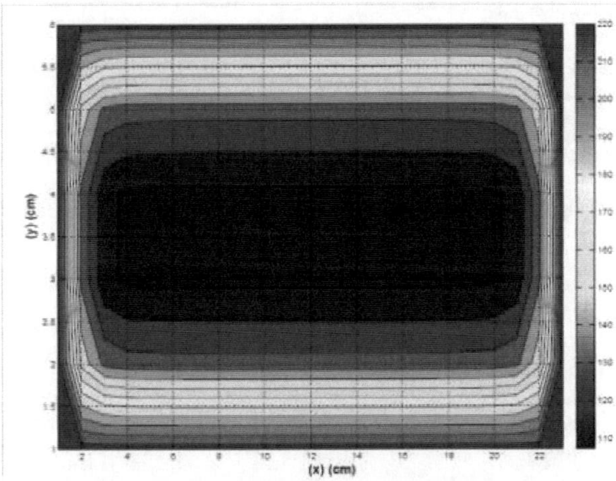

FIGURE 6. Temperature Change (*celcius degree*) in Full Combustion Stage at z = 2 cm

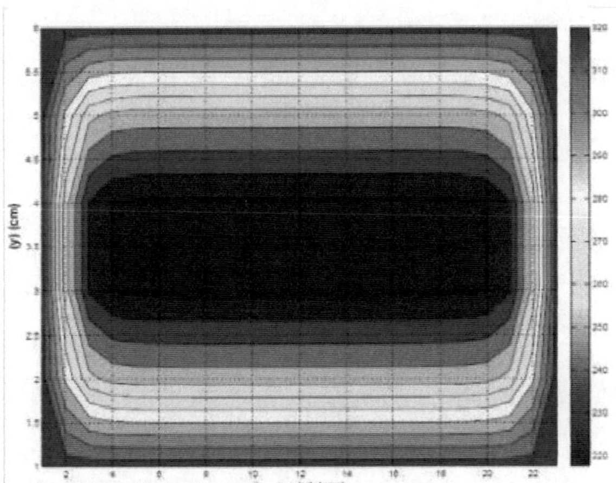

FIGURE 7. Temperature Change (*celcius degree*) in Detention Stage at z = 2 cm

In space point of view, there is considerable temperature difference at start and end point in bricks, e.g. $\Delta T\ (11,\ 2,\ 2,\ t_4) = 253.239\ ^0C$ and $\Delta T\ (11,\ 3,\ 2,\ t_4) = 215.872\ ^0C$. Temperature change in around middle part of bricks share approximately similar value. Moreover, there is about 80 ^0C difference between bottom and top layer of brick's element.

3.2 Stress Distribution

Subsequent to temperature analysis during burning period, stress analysis is performed, as can be seen on Figures 8-9. Figure 8 shows stress analysis on *y-z* plan, and figure 9 shows stress analysis on *x-z* plan. In accordance with temperature analysis, since linear proportional relationship between temperature change and stress, In terms of time, there is no significant difference among first three stages. Stress distribution is sharply rising at 4th stage, e.g. $\sigma\ (5, 3, 2, t_4)$ = 36.6 Mpa and $\sigma\ (5, 3, 2, t_3)$ = 18.1Mpa

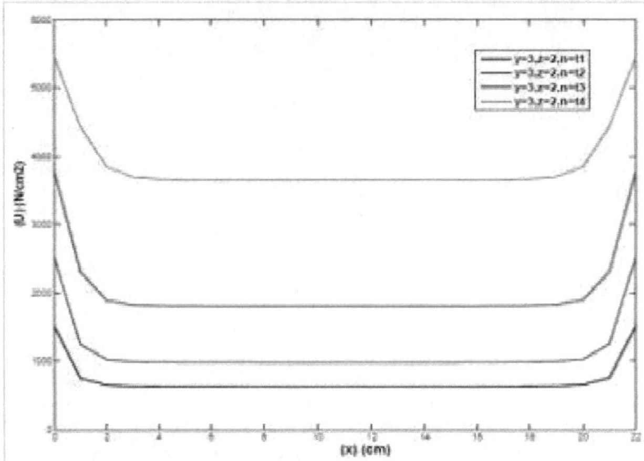

FIGURE 8. Stress Distribution for All Stages at y = 3 cm, z = 2 cm

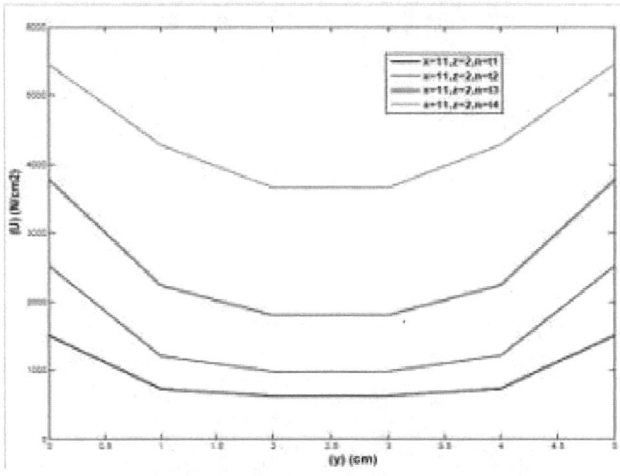

FIGURE 9. Stress Distribution for All Stages at x = 11 cm, z = 2 cm

In terms of position, there is clear difference at start, middle and end point in bricks, e.g. $\sigma\ (11, 2, 2, t_4)$ = 42.8 Mpa and $\sigma\ (11, 2, 6, t_4)$ = 19.8 Mpa. Stress distribution tends to increase at start and end position. This stress distribution may generate crack as if the stress value pass over stress capacity of bricks, both compressive and tensile stress.

4. Conclusion

From the analytical and numerical evaluation, the study has demonstrated that stress distribution in bricks during burning process through 4 (four) stages can be completed by diffusion process, finite difference method, and stress-strain calculation. Space and time are significant variables in establishing temperature change and stress distribution during the burning process. Larger stress in start and end position of bricks may engender crack. This can be avoided by controlling heat transfer inside the bricks.

References

1. J. Gary, B. Sylvain, G. Vincent, J. Philippe, C. S. Ciukaj, " A Homogenised Vibratory Model for Predicting the Acoustic Properties of Hollow Brick Walls", Sound and Vibration, 330, 2011: 3400-3409
2. K. Ornam, M. Kimsan, and E. Cahyono, "Evaluation of Alternative Design of Hollow Brick with Sawdust as Filler for Home-Made Industry," Energy, Environmental and Structural Engineering Series, 32, 2015: 373-376
3. K. Ornam, M. Kimsan, L.O. Ngkoimani, and E. Cahyono, "Physical and Mechanical Properties of Reed Filler Modified Brick," Proceeding s of The Celebes International Conference on Diversity of Walacea's Line, 2015: 362-367
4. K. Ornam, M. Kimsan, "The Effect of Reed as Filler Upon Brick Performance of Building Material," Proceeding of The 2nd Makassar International Conference on Civil Engineering 2015: 257-264
5. E. C. Romão, J. A. Martins, and L. F. M. de Moura, "Three-Dimensional Heat Conduction in Multi-Connected Domains using GFEM with Hexaedrals of 27 Nodes," Mathematics and Computers in Simulation, 6, 2012: 248-256
6. L. Vasek, V. Dolinay, "Simulation Methods for the Heat Distribution Systems," Mathematics and Computers in Simulation, 5, 2011: 439-446
7. M. Skřínská, J. Skřínský, V. Sluka, J. Senčík, S. Malý, G. Tetu, "Mathematical Models for the Prediction of Heat Flux from Fire Balls," WSEAS Transactions on Heat and Mass Transfer, 9, 2014: 243-250.
8. ASTM C216-15, "Standard Specification for Facing Brick (Solid Masonry Units Made from Clay or Shale), USA, 2015
9. ASTM C652-14, "Standard Specification for Hollow Brick (Hollow Masonry Units Made from Clay or Shale), USA, 2014
10. ASTM 67-14, "Standard Test Methods for Sampling and Testing Brick and Structural Clay Tile", USA, 2014
11. ASTM C62-10,"Standard Specification for Building Brick (Solid Masonry Units Made from Clay or Shale), USA, 2010
12. ASTM D 2487-06, "Standard Practice for Clasification of Soils for Engineering Purposes, USA, 2006
13. SNI 15-2094, "Bata Merah Pejal untuk Pasangan Dinding, Balitbang Kimpraswil, Bandung, 2000
14. Yayasan Dana Normalisasi Indonesia NI-10, "Bata Merah sebagai Bahan Bangunan", ed. 2, Bandung, 1973
15. K. Ornam, M. Kimsan, L.O. Ngkoimani, E.Cahyono, " Modelling of Diffussion Process in Porous Bricks", Mathematical and Computational Methods, 1, 2016: 87-91
16. M. G. Adyan, A. Ghaheri, "Solute Transfers Modeling In Layered Porous Media Using Differential Quadrature Method (DQ)," WSEAS Transactions on Heat and Mass Transfer, 9, 2014: 74-83
17. P. Lehner, P. Konečný, P. Ghosh and Q. Tran, "Numerical Analysis of Chloride Diffusion Numerical Analysis of Chloride Diffusion," Mathematics and Computers in Simulation, 8, 2014: 103-106
18. E. Cahyono, Y. Soeharyadi, and Mukhsar, "A Smooth Diffusion Rate Model of Wood Drying: A Simulation Toward More Efficient Process in Industry," Jurnal Teknik Industri, 10, 1, 2008: 1-10
19. F. Mactar, L. Berangere, S. Vincent, "A New Procedure for The Experimental Measurement the Effective Heat Capacity of Wall Elements", Energy and Building, 103, 2015: 62-69
20. M. B. K. Moorthy, T. Kannan, K. Senthilvadivu, "Soret and Dufour Effects on Natural Convection Heat and Mass Transfer Flow past a Horizontal Surface in a Porous Medium with Variable Viscosity," WSEAS Transactions on Heat and Mass Transfer, 8, 4, 2013: 74-83
21. E. Cahyono, L Hamimu, L.O.. Ngkoimani, J Safani, "Interaction of Droplet Diffusions Governed by 2-D Porous Medium Equation," Applied Mathematical Sciences, 8, 127, 2014: 6303 – 6311
22. Chapra, S.C & Canale, R.P, "Numerical Methods for Engineers, 6th ed, Mc Graw-Hill Book Co, New York, 2017

IOP Conf. Series: Journal of Physics: Conf. Series **846** (2017) 012009 doi:10.1088/1742-6596/846/1/012009

Physical Properties of Nyamplung Oil (*Calophyllum inophyllum L*) for Biodiesel Production

Syamsir Dewang[1,*], Suriani[1], Siti Hadriani[1], Bannu, and B. Abdullah[1]

[1]Laboratory of Optical Spectroscopy, Physics Department, Hasanuddin University, Makassar, Indonesia.

*Email: dewang1163@gmail.com

Abstract. Worldwide energy crisis due to the too high of energy consumption causes the people trying to find alternative energy to support energy requirements. The use of energy from environmentally friendly plant-based materials into an effort to assist communities in sufficient of national energy needs. Some processing of Nyamplung (*Calophyllum inophyllum L*) oil production is drying and pressing to produce crude oil. Degumming process is then performed to remove the sap contained in the oil. The next process is to remove free fatty acids (FFA) below 2% that can cause corrosion on the machine when in use. The results performed of the density properties quality to produce oil that appropriate with the international standards by time variation of catalyst. The result was obtained the density value of 0.92108 gr/cm^3 at the time of 3 hours by trans-esterification process, and the best yield value was measured at 98.2% in 2 hours stirring of transesterification.

1. Introduction

The use of oil energy has become a huge need, which is could a trigger of global crisis. Most of oil energies are generally derived from petroleum of fossil material, that causes environmental pollution. Attempts to produce alternative energy derived from plants or vegetable, it is able to develope domestic energy oil especially for environmentally friendly. The contribution of renewable energy to total energy usage is still below 1%[1,2]. Therefore, there should be increased by using renewable energy from plant materials. In Indonesia, the plant of Nyamplung (*Calophyllum inophyllum L*) is one alternative biodiesel feedstock potentially quite large. Biodiesel is commonly produced by triglycerides from the vegetable oils or animal fats. In this research biodiesel is produced from *Calophyllum inophyllum* oil . The advantages of this biodiesel as an alternative fuel to substitute the diesel oil which is high cetane number, environmentally friendly because it contains little of SOx gas, which is a good lubricating power, exhaust emissions are few and relatively clean burning character. In addition to these advantages, the use of biodiesel also provides benefits to machine maintenance [3].

This research observed the biodiesel derived from Nyamplung material. This plant is easy to grow in tropical area especially in Indonesia [4]. The issues faced of vegetable oil research are: How to generate a high yield of oil, reducing the levels of free fatty acids that could produce lathering and inflict the corrosion in machine. Increase the caloric value of fuel or the vegetable oil, and measurement of physical properties such as viscosity, and density, and yield by time variations of catalyst to obtain results in accordance with the standards that have been established nationally and internationally [5]. In this research, the quality of oil will be improved to meet the international standards by varying of catalyst during processing.

Content from this work may be used under the terms of the Creative Commons Attribution 3.0 licence. Any further distribution of this work must maintain attribution to the author(s) and the title of the work, journal citation and DOI.
Published under licence by IOP Publishing Ltd

2. Material and Method

The process of producing Nyamplung seeds into oil biodiesel in this study conducted in several stages, namely: the process of drying and pressing using a pressing machine with strong of 10 tons pressure to remove oil. Furthermore, the screening process or degumming process is to separate the oil from the sap (gum) contained in nyamplung oil consisting of phosphatides, protein, carbohydrate residues, water and resin, and to reduce the free fatty acid.

2.1. Degumming Process

A 500 ml Nyamplung oil is heated on a hot plate to a temperature of 65°C, then added 85% of phosphoric acid by amount of 5% (v/v) from the total of Nyamplung oil, and stirred for one hour. Subsequently, the mixture is allowed to stand for one night, then decanted. Oil obtained from the degumming, then washed with hot water at a temperature of 70°C until the pH of the waste water from the washing process is neutral. The neutralization of oil was heated at 100°C for 15 minutes, then calculated the weight and the levels of free fatty acids (FFA).

2.2 Esterification Process

A number of 50 ml of Nyamplung oil is inserted to a double-mouthed 500 ml of Erlenmeyer flask. Methanol is added with a mole ratio of 20: 1 (w/w) and 10% HCl for oil. Double-mouthed pumpkin mounted to condense the methanol vapor to enter back into the Erlenmeyer. The reaction was performed at 60°C with stirring 300 rpm for 1 hour. After the esterification process is complete, the mixture is put in a test tube, deposited for 8 hours and then measured the levels of FFA on the bottom layer by titrating using NaOH with a concentration of 0.1 N, the indicators used in this titration is PP 0.1%, the indicator as much as 3 drops and titration stopped after the entire color of solution was pink.

2.3 Trans-Esterification Process

A total of 200 ml of oil nyamplung esterification results is poured to the double-mouthed and added methanol at mole ratio of 6: 1 and dissolved by 0.5% NaOH in double mouth flask mounted on the condenser to condense the methanol vapor in order to back into the Erlenmeyer. The reaction is conducted at a temperature of 60° C with stirring 400 rpm for 1, 2 and 3 hours. After the transesterification process is complete, the mixture is inserted in the separating funnel, then deposited for 12 hours. After that the glycerol will settle on the bottom of the separating funnel, so it is easy to be separated. Biodiesel that is formed, then washed with hot water until pH neutral and dried by heating at a temperature of 80° C with vacuum for 20 minutes followed by vacuum drying at a temperature of 90° for 10 minutes.

2.4 Density and Yield Measurement

Measurement of density is using 5ml Pycnometer. The procedure for measuring the density is weighed the empty pycnometer and then filled with biodiesel, if its density is just right volume, pycnometer is closed, and weighing the mass of pycnometer was contained biodiesel. Furthermore, calculate the mass of the fluid that is inserted by subtracting the mass of biodiesel in pycnometer with the mass of empty pycnometer, then count the density of biodiesel by using formulation of $\rho = m / v$. Where ρ is the density of biodiesel in (gr/Cm^3), mass (m) in gram, and volume (v) in cm^3. Furthermore, the Yield (%) measurement is done by value of biodisel (gr) devided by cude oil mass (gr). The value of density and Yield are the two of important physical parameters of biodiesel.

3. Results and Discussion

Nyamplung oil obtained by using the trans-esterification processes to reduce the FFA content until below 2%. If the first esterification process has not FFA less than 2%, then the esterification process repeat again until meet FFA< 2 %. The results of these processes in this experiment can be seen in Table 1.

Table 1. FFA content of Nyamplung oil after Degumming and esterification process.

No	Processing	FFA
1	Degumming	4.08 %
2	Esterification 1	2.18%
3	Esterification 2	1.17%

After second esterification process, we can found that FFA of Nyamplung oil is 1.17% . This is meet the criteria for producing biodiesel.

Table 2 shows the measurement results of mass biodiesel of Nyamplung oil. In this case, the empty pycnometer was adjusted to 11.4439 gram with volume solution of 5 mL.

Table 2. Measuring of sample biodiesel

No	Biodiesel Solutions	Mass of Pycnometer and Biodiesel (gr)	Mass of Biodiesel (gr)
1	Aquades as control	16.4807	5.0308
2	Sample 1	16.0394	4.5955
3	Sample 2	16.0413	4.5974
4	Sample 3	16.0493	4.6054

Sample 1 is biodiesel at trans-esterification processing time of 1 hour
Sample 2 is biodiesel at trans-esterification processing time of 2 hours
Sample 3 is biodiesel at trans-esterification processing time of 3 hours

The density (ρ) is calculate by using equation,

$$\rho = \frac{m}{v} \qquad (1)$$

where m is mass, v is volume and aquades as a control. The aquades density is $\rho_{aquades} = \frac{5,0308}{5} = 1,00736\ gr/cm^3$. While, density of biodiesel from trans-esterification processing time of 1 hour is calculated as follow:

$$\rho_1 = \frac{4,5955}{5} = 0,9191\ gr/cm^3 \qquad (2)$$

The density of biodiesel resulted from trans-esterification processing time of 2 and 3 hours is shown in Figure 1.

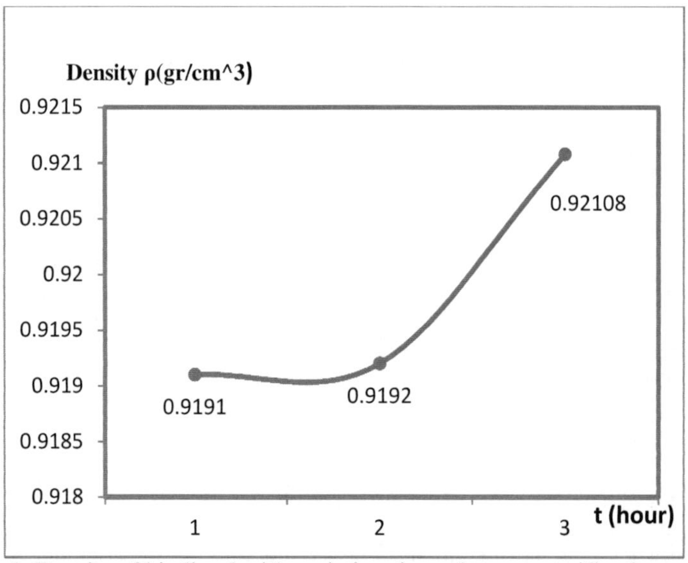

Figure 1. Density of biodiesel with variation time of trans-esterification process.

From Figure 1 above we can see that the highest density was 0.92108 gr/cm^3 (trans-esterification time of 3 hours). For trans-esterification time of 1 and 2 hours, the density obtained almost same. This reveals that time of transesterification process can affect the properties of biodiesel. However, the density biodiesel obtained in this research exceed than the standard for biodiesel (ISO standards) which has density from 0.85 to 0.89 g / cm^3.

IOP Conf. Series: Journal of Physics: Conf. Series **846** (2017) 012009 doi:10.1088/1742-6596/846/1/012009

The results also suggest that time of trans-esterification process should not exceed than one hour to meet the standard. Furthermore yield of trans-esterification process was measured. It found that the highest yield obtained at trans-esterification time of 2 hours which is 98.2%. The yield for 1 and 3 hour are 94.8% and 97.7% respectively. This result suggests that the most effective time of transesterification processing of Nyamplung biodiesel is 2 hours .

4. Conclusion

Experiment for produce biodiesel from biooil of Namplung was successfully performed. The free fatty acid (FFA) can be reduced to 1.17% after two step esterification. This biooil was used in trans-esterification process to measure the density of Nyamplung biodiesel. There found that the highest yield is 98.2% from 2 hours in trans-esterification. These results indicate the most effective time to produce biodiesel at the condition. It suggests that Nyamplung is good for biodiesel raw material.

Acknowledgments
The writer wishes many thanks to Rector of Hasanuddin University, and Faculty Dean of Mathematics and Natural Sciences, Hasanuddin University of their sponsoship to attend the ICTAP 2015 Halu Oleo Univesity, Kendari.

References
[1] M. Mittelbach, P. Tritthartand H. Junek. 1985. *Diesel Fuel Derived from Vegetable Oils, II: Emission Tests Using Rape Oil Methyl Ester*. EnergyAgricul. 4:207–215.
[2] Ozgul, S dan S. Turkay. 2002. *Variable Affecting the Yield of Methyl Ester derived from in situ Esterification of Rice Bran Oil*. JAOCS, 76:611-614.
[3] B.K. Venkanna, and Ventakaramana, R.C. 2009. *Production and Optimization from Callophylum Inophyllum Linn Oil (Honne Oil) – A Three stage method*. Bioresource Technology 100: 5122 – 5125.
[4] B. Freedman, E.H. Prydeand T. L. Mounts.1984.*Variable Affecting the Yield of Fatty Ester from Transesterified Vegetable Oil*.JAOCS, 61:1638-1643.
[5] J.B Friday, and D. Okano. 2006. *Species Profiles for Pacific Island Agroforestry Calophy llluminophyllum.*
[6] A.C Dweek, and T, Meadows, 2002. *Tamanu (Callophylum inophyllum) the Africa, Asia Polynesian and Pasific Panacea.* Int J. Cos. Sci,24:1-8.
[7]. Hwai Chyuan Ong, H.H. Masjuki, T.M.I Mahlia, A.S. Silitonga, W.T. Chong, K.Y. Leong. *Optimization of Biodiesel Production and Engine Performance from High Free Fatty Acid Calophyllum inophyllum Oil in Cl Diesel Engine*. Journal Energy Conversion and Management, Volume 81, Pages 30-40, (2014),
[8]. Jahirul M.I., Brown J.R, Senadeera W, Ashwath N, Laing C, Leski-Taylor J, Rasul M.G. *Optimisation of Bio-Oil Extraction Process from Beauty Leaf (Calophyllum inophyllum) Oil Seed as a Second Generation Biodiesel Source*. Journal Procedia Engineering, Volume 56, pages 619-624, (2013)

Application of Local Adsorbant From Southeast Sulawesi Clay Immobilized *Saccharomyces Cerevisiae* Bread's Yeast Biomass for Adsorption Of Mn(Ii) Metal Ion

Halimahtussaddiyah R*, Mashuni, Budiarni
Department of Chemistry, Faculty of Mathematics and Natural Sciences, Halu Oleo University, Kendari, Sulawesi Tenggara, 93231

*Email : halimahhalimah124@yahoo.co.id

Abstract. Southeast Sulawesi has a great stock of clay. It is probably to use as a source of adsorbent. The adsorbent capacity of clay can be largered with teratment using bread's yeast as biomass. At this research, study of analysis adsorption of Mn(II) metal ion on clay immobilized *Saccharomyces cerevisiae* bread's yeast biomass adsorbent has been conducted. The aims of this research were to determine the effects of contact time, pH and concentration of Mn(II) metal ion and to determine the adsorption capacity of clay immobilized *S. cerevisiae* biomass for adsorbtion of Mn(II) metal ion. Activated clay was synthesized by reaction of clay with $KMnO_4$, H_2SO_4 and HCl. *S. cerevisiae* biomass was result by bread's yeast mashed. Immobilization of *S. cerevisiae* biomass into clay was done by mixing of ratio of *S. cerevisiae* bread's yeast biomass and clay equal to 1:3 (mass of biomassa : mass of clay). The adsorption capacity was determined by using Freundlich and Langmuir adsorption isoterms. The results of FTIR spectrums showed that the functional groups of clay immobilized *S. cerevisiae* biomass were Si-OH (wave number 1643 cm^{-1}), Si-O-Si (wave number 1033 cm^{-1}), N-H (wave number 2337 cm^{-1}), O-H (wave number 3441cm^{-1}), and C-H (wave number 2931 cm^{-1}). The result of adsorption capacity from Mn(II) metal ion of contact time optimum 120 minutes , pH optimun at 7 and concentration optimum 50 mg/L were 1,816 mg/g; 0,509 mg/g and 2,624mg/g respectively.The adsorption capacity of Mn(II) metal ion with ratio 1:3 (biomass : clay) was 0,1045 mg/g. Type of isothermal adsorption followed the Freunlich adsorption.

1. Introduction

Advances in science and technology followed by a very rapid industrial development give a lot of positive and negative impacts for life. The negative impact is produced pollutants that interfere the environment, particularly the aquatic environment such as the spread of heavy metals into the environment at concentrations above the threshold include manganese. Manganese (Mn) are included in one of the essential micro heavy metals needed by plants in relatively small quantities of up to 50 ppm [1].Clay is inorganic compound composing soil, containing active sites such as silanol (Si-OH), a siloxane (Si-O-Si), aluminol (Al-OH) and iron oxides which can adsorb harmful metal ions in the environment [2]. Southeast Sulawesi is rich in clay minerals by the number of clays 537,000 million m^3. Others interesting properties of clay are have a large specific surface area, multi-layered structure, a Bronsted and Lewis acid, have high mechanical and chemical stability [3]. Application this clay for porous ceramics for various applications were also reported previously [4]. The porous ceramic were produced by mixing clay and ash of sago waste from local farmer and processing by using conventional heating as well as microwaves.

Another method which has been developed to reduce the heavy metals in environment, especially in industrial waste is biosorption. Biosorption is most promising technologies in wastewater treatment. The method using living or dead cells biomass to absorb metal. Process of metal absorption by biomass is a combination of passive accumulation processes. The process also does not depend on the biomass

Content from this work may be used under the terms of the Creative Commons Attribution 3.0 licence. Any further distribution of this work must maintain attribution to the author(s) and the title of the work, journal citation and DOI.
Published under licence by IOP Publishing Ltd

metabolic processes. Biosorption involves chemical and physical adsorption process, ion exchange, coordination interactions, complexation, chelatization, and microprecipitation [5].

Biomass from bread's yeast has been widely studied related to its potential as bioacumulator and biosorbent of heavy metals. Adsorption of Cr^{3+} metal ions have been conducted, using biomass adsorbent of bread's yeast immobilized on montmorillonite clay with maximum absorption capacity 5.025 m/g [6]. The use of biomass from bread's yeasts, caused material percentage of cell wall as a source of high metal binding. The cell wall consists primarily of polysaccharides, lipids and proteins which have functional groups such as carboxylic, hydroxyl, and amino which can bind with metal ions [7]. Other materials than clay for ceramics from local biomass such as sago wastes and rice ash husk [8-10] as well as commercial powder such alumina have been also reported in detail previously [11-13].

In this study, biomass of bread's bread's yeast immobilized on clay. The advantage of bread's yeast biomass after immobilized on clay was created of material which not easily to degraded, with a relatively large size so it is more efficient to use as an adsorbent of heavy metal ion.

2. Research Method

2.1 Materials

The materials used in this study include bread's yeast of fermipan brand, clay, standard solution of Mn(II) 1000 ppm (E. Merck), sulfuric acid (H2SO4 95%, E. Merck), hydrochloric acid (HCl 37%, E. Merck), nitric acid (HNO3 56%, E. Merck), solid potassium permanganate (KMnO4, E. Merck), pH indicator paper (Macherey-Nagel), Whatman filter paper (cat. No. 42), aquabidest, aquadest, and tissue paper.

2.2 Research Procedure

2.2.1 Adsorbent manufacture from clay, synthesized of clay immobilized by bread's yeast and characterization by FTIR.

Clay samples were sampled from Amesiu village of Pondidaha District of Southeast Sulawesi province at a depth ranging 10-20 cm. Clay has been cleaned of dirt, dried, pulverized and then sieved with a 200 mesh sieve. Subsequently, 200 grams of clay mixed with KMnO4 0.5M while stirring with a stirrer for 4 hours at temperature of 80ºC. The result was filtered and the sediment was washed repeatedly until the washing water reached a neutral pH, then heated in an oven for 12 hours at temperature of 80ºC. Clay results of KMnO4 0.5M treatment then added H_2SO_4 6M while stirring with a stirrer for 4 hours at temperature of 80ºC. The results were filtered and the precipitant washed repeatedly until the washing water reached a neutral pH and then heated in an oven for 12 hours at a temperature of 80ºC. Furthermore, clay results of KMnO4 0.5M and H_2SO_4 6M treatment were added HCl 6M while stirring with a stirrer for 3 hours at temperature of 80ºC and then washed repeatedly until the washing water reached a neutral pH, then heated in an oven for 12 hours at temperature of 80ºC.

Immobilization of bread's yeast biomass on clay made by mixing the clay and bread's yeast that has powdered using variations of ratio 1: 3 (mass of yeast: mass of clay) [14]. Analysis of samples with spectrophotometer fourrier Transform Infra Red (FTIR) [15].

2.2.2 Effect of Contact Time on Manganese Metal Ion Adsorption [6]

A total of 40 mL solution of Mn(II) metal ion with a concentration of 50 ppm were mixed with 0.5 gram of clay adsorbent immobilized by *S. cerevisiae* biomass. The contact time varied successively 5, 10, 30, 60, 90 and 120 minutes. The filtrate separated by filtered using Whatman filter paper and analyzed by Atomic Absorption Spectrophotometer (AAS) to determine the concentration of the Mn(II) metal ion which were not adsorbed.

Immobilisasi biomassa ragi roti pada lempung dilakukan dengan pencampuran antara lempung dan ragi roti yang telah dihaluskanmenggunakan variasi perbandingan 1:3 (gram ragi : gram lempung)[7]. Analisis sampel dengan Spektrofotometer *Fourrier Transform Infra Red* (FTIR) [8].

2.2.3 Effect of pH of solution on the Adsorption of Manganese Metal Ion [6]

Solution of Mn(II) 10 ppm put into 6 glasses of erlenmeyer each 40 mL, which was then the pH of the solution in the glasses varied on 3, 4, 5, 6, 7 and 8 with the addition of HCl and NaOH. Furthermore, the Mn(II) metal solution with different pH were contacted with an adsorbent. The filtrate was filtered using Whatman filter paper and then analyzed using Atomic Absorption Spectrophotometer

(AAS) to determine the metal ion were not adsorbed by the adsorbent.

2.2.4 Effect of Mn(II) Metal Ion concentration on Manganese Metal Ion Adsorption [6]

A total of 40 mL solution of Mn(II) metal ion with concentrations of 5, 10, 15, 20, 50 and 75 mg/L were mixed with 0.5 gram of adsorbent clay immobilized by S. *cerevisiae* biomass. The filtrate separated by filtered using Whatman filter paper and analyzed by Atomic Absorption Spectrophotometer (AAS) to determine the concentration of the Mn(II) metal ion which were not adsorbed.

3. Results and Discussion

FTIR analysis were used to identify the main functional groups of biomass of bread's yeast immobilized on clay. Fig. 1 shows the cluster of biomass NH at wave number 2337 cm^{-1} and wave number 3441 cm^{-1} in carboxylic OH group from biomass. In addition, there were also an absorption band of Si-OH group at wave number 1643 cm^{-1} and CH absorption band at wave number 2931 cm^{-1}, while group of Si-O-Si appears at wave number 1033 cm^{-1}. The spectrums shows there has been a hydrogen bond between H atoms in the NH group, CH and OH from biomass with O atoms in Si-OH group and Si-O-Si of clay.

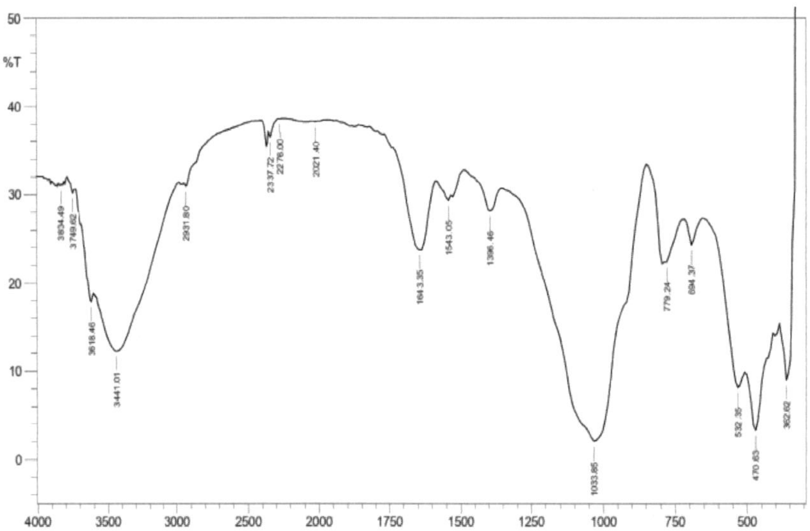

Fig. 1. The spectrum of biomass which immobilized on clay

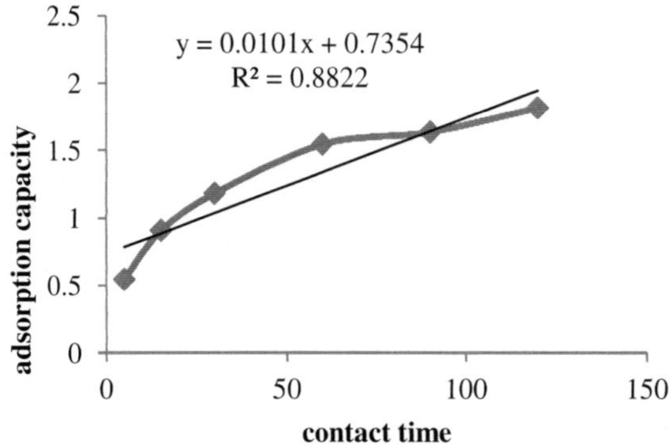

Fig. 2. Graphic of Effect of Contact Time of Mn(II) Metal Solution (mg/L) on Mn(II) Metal Ion Adsorption capacity by Bread's Yeast Biomass immobilized Clay (adsorbent weight = 0.5 gram)

IOP Conf. Series: Journal of Physics: Conf. Series **846** (2017) 012010 doi:10.1088/1742-6596/846/1/012010

3.1 Effect of Contact Time on Mn(II) Metal Ion Adsorption

Effect of contact time on the adsorption of Mn(II) metal ion by yeast biomass adsorbent immobilized clay are presented in **Fig. 2** and **Table 1**. Based on **Fig. 2** and **Table 1**, increased of Mn(II) metal ion adsorption by adsorbent yeast biomass immobilized by clay with increasing of contact time. At the contact time of 5 minutes the adsorption capacity 0.546 mg/g. By increasing the contact time to 10, 30, 60, 90 and 120 minutes, then the adsorption capacity of the adsorbent for the adsorbate were respectively 0.909; 1,182; 1,546; 1.636 and 1.816 mg/g. The increasing of the adsorption capacity because of the longer time contacting the more likely the process of diffusion and annealing of the Mn(II) metal ion in the adsorbent better.

Table 1. Comparison of Adsorption Capacity on Ssome Contact Time Variations

Contact Time (minute)	Adsorption Capacity mg/gram
5	0,546
15	0,909
30	1,182
60	1,546
90	1,636
120	1,816

3.2 Effect of pH on Mn(II) Metal Ion Adsorption

Effect of pH on the adsorption of Mn(II) metal ion by yeast biomass adsorbent immobilized clay are presented in **Fig. 3** and **Table 2**.

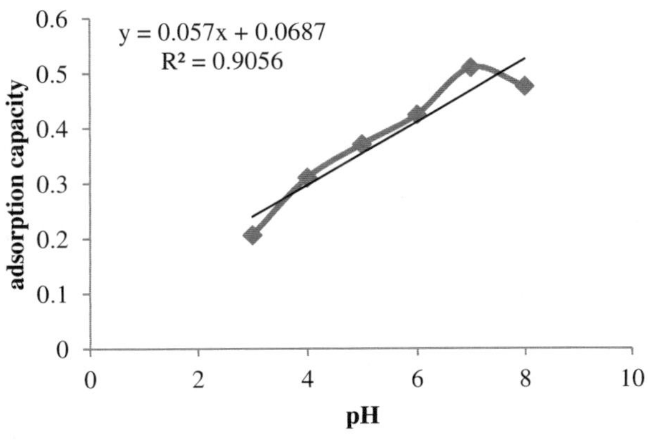

Fig. 3. Graphic of Effect of Mn(II) Metal Ion pH of solution (mg/L) on Adsorption Capacity of Mn(II) Metal Ion by Bread's yeast Biomass Immobilized Clay (adsorbent weight = 0.5 gram)

Table 2. Comparison of Adsorption Capacity on Some pH of Solution Variations

pH of Solution	Adsorption Capacity mg/gram
3	0,206
4	0,310
5	0,370
6	0,424
7	0,509
8	0,475

IOP Conf. Series: Journal of Physics: Conf. Series **846** (2017) 012010 doi:10.1088/1742-6596/846/1/012010

Based on Fig. 3 and Table 2, an increased in the adsorption capacity along with increasing of pH of solution from pH 3 to 6 by the adsorption capacity successively 0.206; 0.310; 0.370 and 0.424 mg/gram to achieve optimum absorption at pH 7 i.e. 0,509 mg/g. Decrease of adsorption capacity occurs at a pH of 8 i.e. 0.475 mg/g.

The absorption of the Mn(II) metal ion by bread's yeast biomass is strongly influenced by the pH of the solution. Absorption efficiency of Mn(II) metal ion increased from pH 3 to pH 6 to achieve maximum absorption at pH 7. This is because at low pH, the surface of the cell wall of biomass protonated or positively charged, so the adsorption of metal that occurs is very small, because carboxylate group tend to be in a neutral form with the reaction:

$$R\text{-}COO^- + H^+ \longrightarrow R\text{-}COOH \qquad (1)$$

At high pH, the surface of the cell wall biomass negatively charged, so the adsorption of metal becomes larger. The higher pH will also make more group R-COO biomass that can act as ligand in complex formation also more and more much, by the reaction:

$$R\text{-}COOH \longrightarrow R\text{-}COO^- + H^+ \qquad (2)$$

The presence of negative charge will cause interactions between positively charged metal with active sites on the surface of the negatively charged cell walls. At the same time, the surface ligands will compete with OH^- in binding metal cations, that will lead to an increase in metal adsorption by biomass. At pH 8 shows the pattern of absorption of the Mn(II) metal ion decreased. These are because the pH of the number of OH ions more, so that OH^- ions formed tend to bind to the Mn(II) metal ion to form hydroxide $Mn(OH)_2$.

3.3 Effect of Mn(II) Metal Ion solution Concentration on Mn(II) Metal Ion Adsorption

Effect of concentration on the adsorption of Mn(II) metal ion by biomass yeast immobilized clay are presented in Fig. 4 and Table 3.

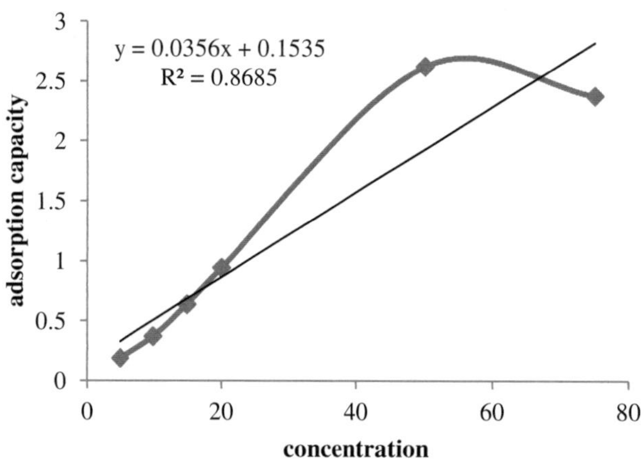

Fig. 4. Graphic of Effect of Mn(II) Metal Solution concentrations (mg/L) on Adsorption capacity of Mn(II) Metal Ion by Bread's yeast biomass Immobilized Clay (adsorbent weight = 0.5 gram)

Table 3.Comparison of Adsorption Capacity on Some Solution Variations

Concentration (mg/L)	Adsorption Capacity mg/gram
5	0,191
10	0,373
15	0,639
20	0,943
50	2,624
75	2,377

Increasing of Adsorption capacity also occurred in several variations of the Mn(II) metal ion concentration. Adsorption starting from a concentration of 5 mg/L with a higher adsorption capacity equal to 0.191 mg/g and continued to increase until the concentration equal to 50 mg/L i.e. 2,624 mg/g and at a concentration of 75 mg/L adsorption capacity descend to 2.377. In Fig. 4 and Table 3, it is clear that the state of the active site of the fourth adsorbent on each addition concentration were not saturated due to continued improvement of the adsorption of Mn(II) metal ion or in other words all of the active sites on the clay is considered not saturated up to a concentration 50 mg/L. However, at a concentration of 75 mg/L occur decreasing of adsorption capacity. This shows that at a concentration of 75 mg/L of active sites on the fourth adsorbent has saturated. On Oscik [16], Langmuir explained that the surface of the adsorbent viz bread's yeast biomass immobilized clay has a number of active sites which proportional to the surface area of the adsorbent. At each active sites is only one molecule can be adsorbed. In the state of active sites which has not been saturated with adsorbate of the Mn(II) metal ion, then increasing the concentration of Mn(II) metal ion which contacted with bread's yeast biomass immobilized clay whose numbers will still produce uptake of Mn(II) metal ion which increases linearly. When the active site has been saturated with adsorbate, then increasing concentrations of Mn(II) metal ion in further will not increase the amount of Mn(II) metal ion which is absorbed.

Freundich's Isothermal adsorption curve, Mn(II) metal ion by the yeast biomass adsorbent immobilized on clay made by plotting log to log qe to log Ce, where qe is the amount of adsorbed adsorbate and Ce is the concentration of the adsorbate in the adsorption equilibrium are presented in Fig. 5.

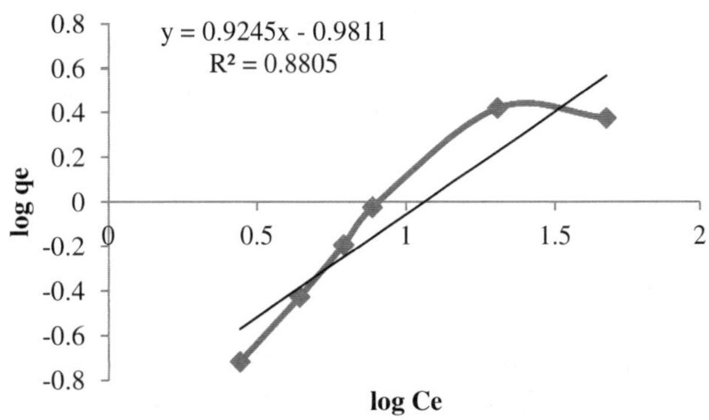

Fig. 5. Graphic of Freundlich adsorption isotherm of Mn(II) Metal Ion by yeast Biomass adsorbent Immobilized on Clay.

Langmuir adsorption isothermal curve Mn(II) metal ion by yeast biomass adsorbent immobilized on clay made by plotting Ce/qe on Ce where qe is the amount of adsorbed adsorbate and Ce is the concentration of the adsorbate in the adsorption equilibrium are presented in **Fig. 6**.

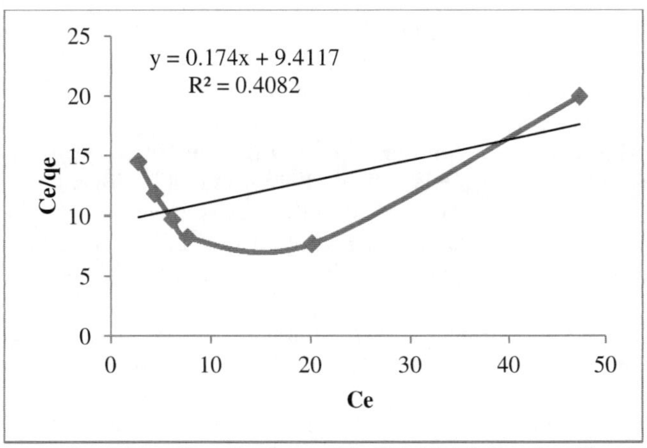

Fig. 6. Graphic of Langmuir Adsorption Isotherm Mn(II) Metal Ion by Yeast Biomass Adsorbent Immobilized on Clay.

Based on Fig. 5 and Fig. 6 above, the linear regression values obtained in Freundlich equation for yeast biomass adsorbent immobilized on clay i.e. 0.8426 while the value of the linear regression of Langmuir equation for the adsorbent obtained 0.5351. Therefore, it can be stated that the type of metal ion adsorption of Mn(II) by the adsorbent tends to follow Freundlich isotherm linearity for linear regression values for Freundlich isotherm is greater than the value of the linear regression on the Langmuir isotherm, thus determining the adsorption capacity used Freundlich equation. Freundlich equation obtained on the adsorption capacity (K_F) for yeast biomass adsorbent immobilized on clay equal to 0.1045 mg/g. In this study, yeast biomass adsorbent immobilized on clay which adsorbed Mn(II) metal ion categorized as physical adsorption [17].

4. Conclusion

The adsorption capacity of the Mn(II) metal ion obtained optimum contact time in 120 minutes, the optimum pH 7 and the concentration of Mn(II) 50 mg/L, the adsorption capcity equal to 2,624 mg/gram. Based on Freundlich adsorption isotherm equation, the known value of the adsorption capacity of yeast biomass immobilized clay equal to 0.1045 mg/gram.

References

1. Lusiana, Muchyar, M. Najib. 2000. Unsur-unsur Hara Esensial. Lambung Mangkurat University, Banjarbaru. [in Indonesian]
2. Tan, 1995,Dasar-Dasar Kimia Tanah, Gadjah Mada University Press, Yogyakarta. [in Indonesian]
3. Bhattacharyya, K.G. and Gupta, S.S., (2008b), Kaolinite and Montmorillonite as Adsorbents for Fe(III),Co(II) and Ni(II) in a Aqueous Medium, Applied Clay Science, 41, pp. 1–9.
4. H. Aripin, S. Mitsudo, I. N. Sudiana, N. Jumsiah, I. Rahmatia, B. Sunendar, L.Nurdiwijayanto, S. Mitsudo, S.Sabchevski, Preparation of Porous Ceramic with Controllable Additive and Firing Temperature, Advanced Materials Research, Vol. 277 (2011) pp. 151-158
5. Herrero, R., B. Cordero, P. Lodeiro, & C. Rey-Castro. 2002. Interactions of Cadmium(II) and protons with dead biomass of marine algae Fucus sp. Departamento de Química Física e Enxeñería Química I. Coruña, Spanyol.
6. Muslimin, 2010, Pemanfaatan Lempung Montmorilonit Terammobilisasi Biomassa Ragi Roti Untuk Mengadsorpsi Ion Logam Cr(III),Minithesis, Haluoleo University, Kendari. [in Indonesian]
7. Hughes, M.N., and Poole, R.K., 1990, Metal and Microorganism, Chapman and Hall, London.
8. I. N. Sudiana, S. Mitsudo, T. Nishiwaki, P. E. Susilowati, L. Lestari, M. Z. Firihu, H. Aripin, Synthesis and Characterization of Microwave on Sintered Silica Xerogel Produced from Rice Husk Ash, Journal of Physics: Conference Series Vol. 739 No. 1, (2016) 012059
9. Balcu, I., Macarie, C.A., Segneanu, A.E. dan Pop, R.O. 2011. Combined Microwave-Acid Pretreatment of the Biomass. Progress in Biomass and Bioenergy Production.
10. Dehani, F. R., Argo, B. D. dan Yuliningsih, R. 2013. Utilization of Microwave Irradiation to Maximize Pretreatment Process Degradation of Lignin Jeremi Paddy. Jurnal Biopress Komoditas Tropis Vol.1 No. 1, 175-182 [In Indonesia].
11. W. H. Sutton, Microwave processing of Ceramic Materials, Microwave Solutions for Ceramic Engineers, American Ceramic Society, (2005), 35-65.
12. M. Z. Firihu., I. N. Sudiana, S. Mitsudo, Microwaves Enhanced Sintering Mechanisms in Alumina Ceramic Sintering Experiments, Contemporary Engineering Sciences, Vol. 9, 2016, 5, 237 – 247
13. I. N. Sudiana, S. Mitsudo, M. Z. Firihu, Effect of Initial Green Samples on Mechanical Properties of Alumina Ceramic , Contemporary Engineering Sciences, Vol. 9, 2016, no. 12, 595-602.
14. I. N. Sudiana, S. Mitsudo, M. Z. Firihu, H. Aripin, Effect of High-Frequency Microwave on Micro Hardness of Alumina Ceramic, Material Science Forum , Vo. 872, 2016, pp.114-117.
15. Fahmiati, Mashuni, Alimin, dan Amiruddin, 2007, Pengembangan Teknologi Pemisahan Logam Berat dengan Memanfaatkan Biomassa S. cerevisiae dari Ragi yang Diimmobilisasi pada Lempung sebagai Adsorben, Report of Compete Grant Research, FMIPA, Haluoleo University, Kendari. [in Indonesian]
16. Darmayani, S. 2010, Karakterisasi Lempung Montmorilonit Immobilized Ragi Roti Menggunakan Analisis BET, FTIR dan SEM, Minithesis, Haluoleo University, Kendari. [in Indonesian].
17. Oscik, K., 1982, Adsorption, John Wiley, New York.
18. Al-Duri. B., 1995, A Review in Equiliberium in Single and Multicomponen Liquid Adsorption System, Review in Chemical Engineering, 11, p.p. 101-143

IOP Conf. Series: Journal of Physics: Conf. Series **846** (2017) 012011 doi:10.1088/1742-6596/846/1/012011

Microwave-Assisted Hydrolysis of Chitosan from Shrimp Shell Waste for Glucosammine Hydrochlorid Production

Ahmad Zaeni[1,*], **Endang Safitri**[2], **Badrotul Fuadah**[2], **I Nyoman Sudiana**[2]
[1]Department of Chemistry, Faculty of Mathematic and Natural Science, Halu Oleo University, Kampus Bumi Tridharma Anduonohu, Kendari 93232, Indonesia
[2]Department of Physics, Faculty of Mathematic and Natural Science, Halu Oleo University, Kampus Bumi Tridharma Anduonohu, Kendari 93232, Indonesia

* E-mail: ahmad.zaeni@gmail.com

Abstract. Chitin is the most widespread renewable natural sources following cellulose as the main source of chitosan. Chitin is isolated from crustacean waste and shrimp shells. Chitosan is derived from chitin throuhgt demineralisation, deproteination, decolorisation and deacetylation process using chemicals such as sodium hydroxide, hydrogen chloride and acetone. Glucosamine hydrochloride (GlcN-Cl) can be produced by hydrolysis of chitosan by using hydrogen chloride. During deacetylation and hydrolysis the solution is heated by hotplate or furnace. In this paper we use microwave instead of hotplate for production chitosan and GlcN-Cl. The research investigates effect of microwaves to amount of rendemen and their property. The chitosan was characterized its moisture content, solubility, and degree of deacetylation (DDA). Whereas the glucosammine hydrochloride characterized its functional groups using FTIR and crystallization by using X-Ray Difraction (XRD). The experimental results show that the use of microwave energy on deacetilation of chitosan and hydrolisis processes can decrease time consuming and reactant concentration during production. the DDA value obtained was very high from 70 to 85%. The results also show that microwaves meet chitosan and GlcN-Cl standards.

1. Introduction

Osteoarthritis (OA) is one of the leading causes of disability (disability) in the elderly population in the world. OA is the fifth leading cause of disability in the population of developed countries, and the ninth highest cause in developing countries [1]. The main cause of disability (disability) in Australia and Asia is cancer. After cancer, many disability caused by lower back pain and OA with a prevalence of about 15% of the total cases. The results show in the world 9.6% of men and 18.0% of women aged over 60 have symptoms of OA. 80% of them due to OA causing limitation of movement, and 25% can not perform major activities of daily living [2]. WHO also estimates that in 2020 OA will be one of the causes of physical disability in humans after rheumatoid arthritis, osteoporosis and lower back pain.

The prevalence of OA continues to increase along with the increase in population, age, and risk factors such as obesity [1]. Indonesia is the 4th with the number of elderly people (elderly) the most after China, India and the United States [3]. Approximately 56.7% of patients in the outpatient clinic of the Department of Rheumatology (Cipto Mangunkusumo Hospital) has been diagnosed is one one variant OA. Osteoarthritis patients is very haunted by the pain caused by degradation and joint inflammation that occurs [4], so the treatment is focused on reducing pain by administration of nonsteroidal antiinfla-mmatory drug (NSAID) in the long term. However, this drug side effects if used long term. The use of NSAIDs for a long time, especially in older people reportedly turned out to cause a lot of side effects, such as gastrointestinal disorders, impaired liver function,

Content from this work may be used under the terms of the Creative Commons Attribution 3.0 licence. Any further distribution of this work must maintain attribution to the author(s) and the title of the work, journal citation and DOI.
Published under licence by IOP Publishing Ltd

impaired kidney function and so on [4]. Dietary supplements glucosamine and chondroitin sulfate has been recommended as a safe and effective option for preventing and treating symptoms of OA [5].

Glucosamine is a natural compound found in the human body is composed of glucose and amino acid glutamine, other than that glucosamine is a key element of glycosamino-glycans (GAGs) in the cartilage of cartilage and synovial fluid. The function of glucosamine in the body is to produce synovial fluid that serves as a lubricant to the cartilage, so that the movement of the bones to be good. Shortage of synovial fluid in the body will cause joint disorders, such as stiff joints so that movement will result in OA disease. Giving oral glucosamine can help the production of synovial fluid so as to prevent and cure of disease [5]. Until now, glucosamine has not been produced in Indonesia that are imported from other countries with a fairly high. Research to produce glucosamine have been done by some researchers. Islam et al. (2011) have managed to hydrolyze glucosamine hydrochloride amounted to 63.45% of the chitin shells of shrimp[6]. Afridiana (2011) also managed to extract glucosamine hydrochloride from chitin shells of shrimp through chemical hydrolysis method with a concentration of 45.64%. Hydrolysis is safe for human consumption [5].

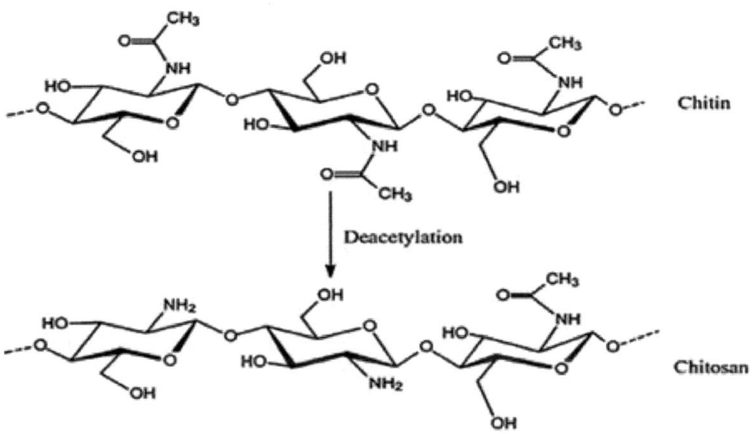

Figure 1. Deacetylation chitin to chitossan

Glucosamine can be obtained from the hydrolysis of chitin. Chitin is an organic material in animal groups molluscs, crustaceans, insects, arthropods and in the cell walls of plants low grade. The content of chitin in the skin of crustaceans such as shrimp shells 20-40%, -35% 15 shell crabs, and squid cartilage 97.20%. Chitin contained in squid cartilage, usually conjugated with a protein. Therefore, in this study glucosamine hydrochloride will produced from shrimp shell waste.

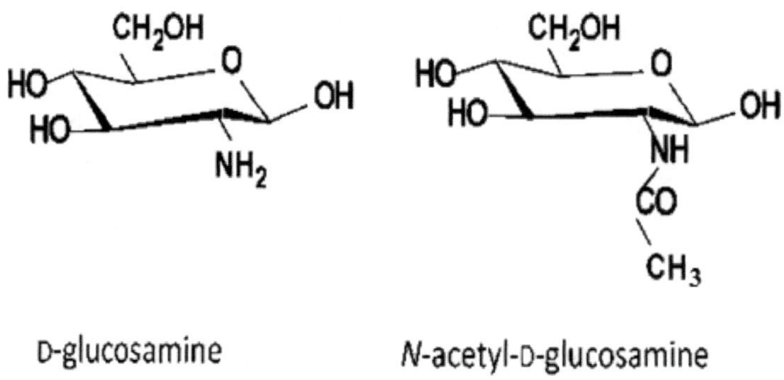

Figure 2. Hydrolysis product of chitosan and chitin

One of main process to produce Glucosamine Hydrochloride is hydrolysis. Conventionally, hydrolysis is performed by soaked in a chemical solution (i.e. hydrochloric acid) with certain concentrations and heated by using hot plate or electric furnace at temperature 80-100 °C. Recently, the use of microwave energy to heat material or to speed up chemical reaction as a substitute of conventional heating has been reported. Some peculiar results were found such faster densification of ceramics[7-9], improve microstructure and mechanical of ceramics [10-14], drying and chemical reaction [15-17], synthesizing organic materials [18-20]. In this study, we applied this microwave energy on deacetylation and hydrolysis. The experimental results were then compared to the results of conventional ones.

2. Material and Method

The shrimp shells were collected from local fisheries in Kendari, Indonesia. Shrimp shell waste was washed and dried in the sun at room temperature before processing.The composition of the shell parts was determined by Thermogravimetry. The degree of deacetylation (DDA) of the chitin, chitosan, and glucosamine were determined by Fourier Tansform Infrared Spectroscopy (FTIR). Crystallinity was determined by X-Ray Difraction (XRD). All chemicals used in this study was bought from SIGMA-ALDRICH. The procedure of preparation of glucosamine from exoskeleton of shrimp is follow the procedure on Mojarrad and Arifin [21-22].

2.1 Demineralisation

50 grams of shrimp shell dissolved in hydrochloric acid of 1.0 M with a ratio of 1:10 (sample: solvent), then stirred with a stirrer for 1 hour at a temperature of 75°C. After filtered and washed using distilled water until neutral pH, then dried in an oven at 60 ^{o}C.

2.2 Deproteination

Deproteination (removal of protein) is done by a solution of NaOH 3.5% with a ratio of 1:10 (sample: solvent), then stirred with a stirrer for 2 hours at a temperature of 65°C. Then the samples were filtered and washed with distilled water until pH neutral and dried at 60°C

2.3 Decolorisation

Decolorisation can be done by solution of 0,315% NaOCl for 5 minute or extracted by acetone for several hours. The resulted then washed with water and dried in oven at 60 ^{o}C.

2.4 Deacetylation

Deacetylation performed 50 % sodium hydrokhlorid solution in water, after that continue by heating wich microwave for several minute.

2.5 Hydrolysis

Two gram chitosan and HCl was prepared for glucosamine production. Hydrolysis performed by two methods, microwave and conventional. By using microwave with varying exposure time from 5 to 20 minutes. Conventional hydrolysis was performed by using hot plate at temperature of 90°C for 90 minutes. Samples were soaked in a solution of hydrochloric acid at a ratio 1: 9 (w / v). HCl solution was varied at 20%, 25%, 32% and 37%. The use of microwave in these process are not only in the deacetylation process but also in the hydrolysis of this produced chitosan.

3. Results and Discussion

Main exoskeleton of shrimp components are protein, mineral, and chitin. After characterization, we found the composition as shown in Table 1. to remove the proteins and minerals do deproteinization and demineralization .Deproteinization was done by soaking the samples in NaOH 3,5% solution for 2 hours at temperature 90°C.While demineralization was performed by using HCl 1,0 M at ambient temperature for 1 hour. The use of microwave in the deacetylation of chitin and hydrolysis of chitosan decreasing heating temperature and time significantly.

Table 1. Characterization of shrimp shell

No.	Material	Percentage
1	Protein	29,5 %
2	Mineral	50,75%
3	Chitin	19,75%

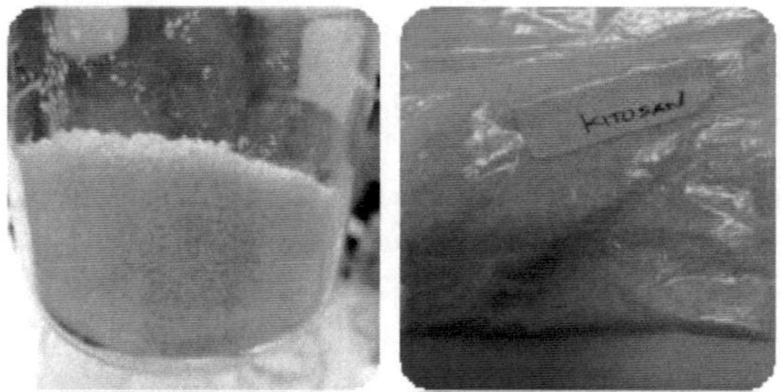

Figure 3. Chitin and chitosan produced from exoskeleton of shrimp

Analysis of functional groups to identify functional groups and calculates the degree of deacetylation (DD) was performed by using FTIR (Fourier Transform Infra Red). FTIR spectra of chitosan showed absorption at 3441.87 cm wave number region-1 (O-H stetching), 1658.78 (C = O amide) as shown in Fig. 4 and 5. Absorption at wave number 1658.78 cm-1 (amide peak) appear due to chitosan is yet deasetilation overall. Quality of chitosan can be seen also from the large percent degree of deacetylation (DD). The existence of chitin into chitosan supported by data degree of deacetylation of chitosan which is calculated using the Baseline Method according to Khan *et.al* (2002), is 61.78% [23-24]. According Hargono *et al.* (2008) [25], standard degree of deacetylation of the chitosan from 60 to 100% so that the chitosan found in this experiment meet the standard. The higher DD means the more pure chitosan.

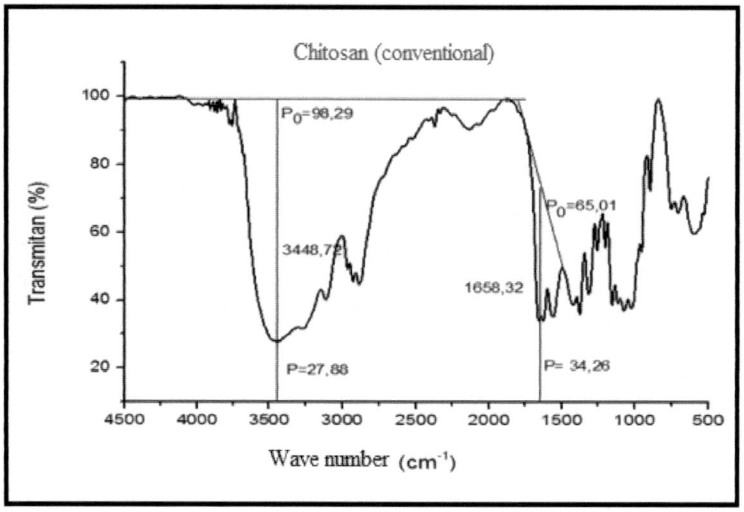

Figure 4. FTIR Spectrum of chitosan produced by hot plate (conventional)

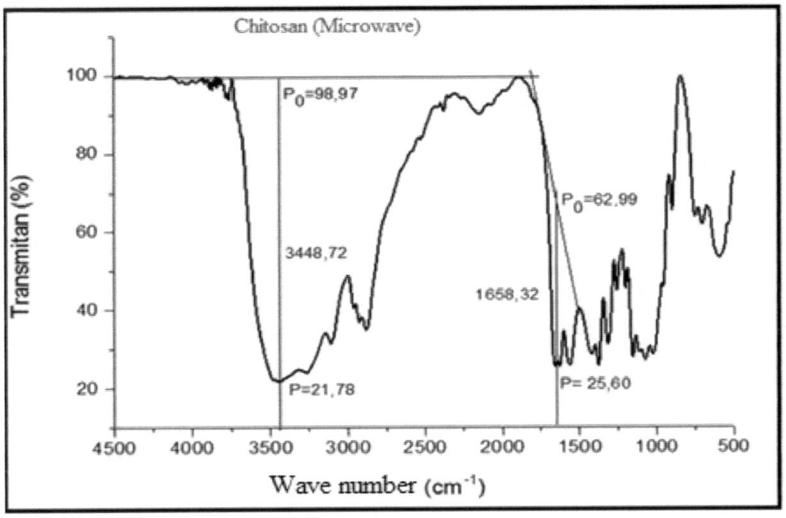

Figure 5. FTIR spectrum of chitosan produced by microwave

Figure 6. show FTIR spectrum of Glucosamine hydrochloride(GlcN-Cl) hydrolysis by using microwave for 15 minutes. The spectrum shows that an absorption at wave number of 3348.42 cm^{-1} as a result of vibration of -OH group and followed by absorption at wave number 3292.49 cm^{-1} derived from stretching vibration of NH amines Fig. 6. Brugnerotto (2001) explained that the monomer GlcN-HCl will show O-H functional group at 3350 cm^{-1} whereas polymer GlcN-HCl will show O-H functional group closer to 3450 cm^{-1}[26]. From the Fig. 6. indicate that the GlcN-HCl found in this experiment is monomer.

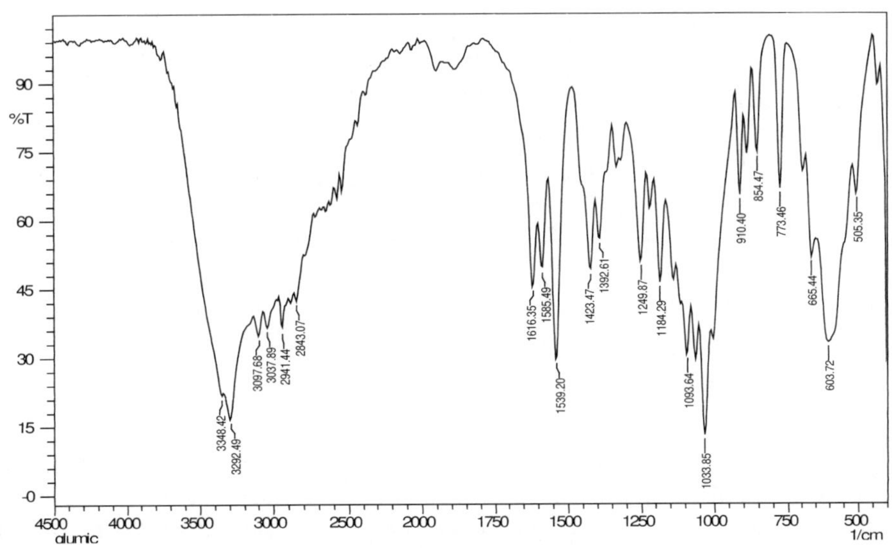

Figure 6. FTIR spectrum of Glucosamine hydrochloride(GlcN-Cl) hydrolysis by using microwave

Figure 6. show FTIR spectrum of Glucosamine hydrochloride(GlcN-Cl) hydrolysis by using by chemical methods with heated by hot plate for 90 minutes. The spectrum shows similarities with the microwave one for 15 minutes. It indicates that microwave accelerate the process of hydrolysis of GlcN-Cl. In addition with shorter time, we can found the almost same yields than that of result in hydrolisis by using conventional method as shown in Figure 8.. It can be seen that the yield obtained in the variation of microwave heating time 15, 11, 7,and 3 minutes give yield of 30.14%, 38.25%, 6.8% and 6.35% respectively. While by using conventional for 90 minutes we found yield of 51,01 %.

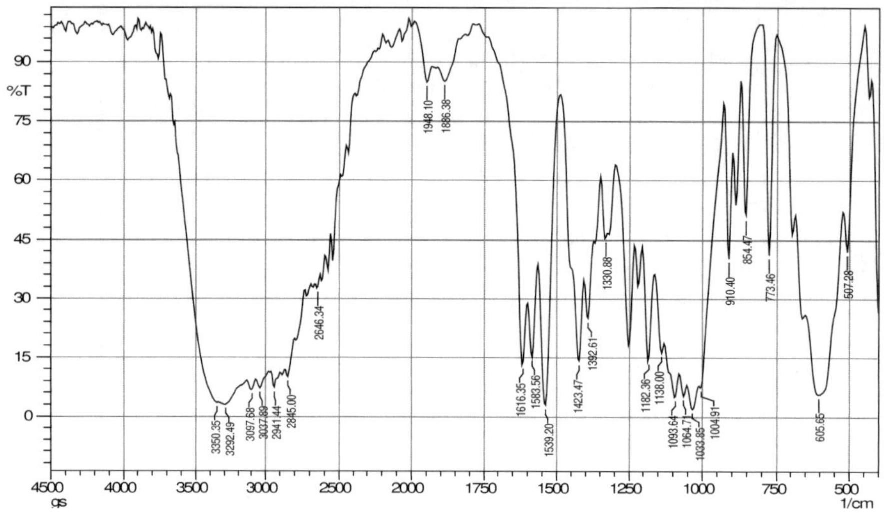

Figure 7. FTIR spectrum of Glucosamine hydrochloride (GlcN-Cl) hydrolysis by using hot plate

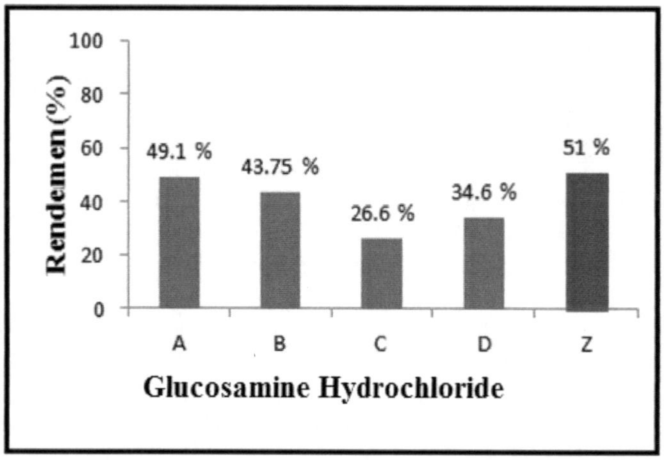

Figure 8. Glucosamine hydrochloride's rendemen from hydrolisis of chitosan (microwave : (A) 3 minutes; (B) 7 minutes; (C) 11 minutes; (D) 15 minutes; and (Z) conventional (hotplate)

4. Conclusion

Application of microwave energy for deacetylation of chitin and hydrolysis of chitosan process for glucosammine hydrochlorid production was successfully performed. Experimental results show that microwave decreased heating temperature and heatingtime significantly. It indicates that microwave accelerates these processes. FTIR results showed that absorption bands of glucosamine hydrochloride meet the standard of GlcN-Cl. These results suggested that microwave is appropriate technology for production of GlcN-Cl.

Acknowledgments

This work has been partially financial supported by the Directorate of Higher Education, at the Ministry of Research Technology and Higher Education, Republic of Indonesia and University of Halu Oleo.

References

1. S. Tanna,2014. Osteoarthritis: Opportunities to Address Pharmaceutical Gaps. WHO: Priority Medicines for Europe and the World.
2. _, 2014. World Health Statistics, http://www.who.int/gho/publications/worldhealthstatistics/
3. Ernawati, 2012, Pembuatan Glukosamin Hidroklorida (GlcN HCl) dengan Metode Autoklaf, Skripsi, Institut Pertanian Bogor, Bogor.
4. N. Kertia,Danang, 2011. Effect of the combination of curcuminoid and essential oil on the serum ureum and creatinine level of patients with osteoarthritis, Majalah Farmasi Indonesia, 22(3), pp. 151 – 157. [in Indonesian]
5. N. Afridiana, 2011. Recovery Glukosamin Hidroklorida dari Cangkang Udang Melalui Hidrolisis Kimiawi sebagai Bahan Sediaan Suplemen Osteoarthritis, Skripsi, Institut Pertanian Bogor. Bogor. [in Indonesian]
6. Islam, M. M., Masum, S. M., Rahman and Shaikh, 2011, Preparation of Glucosamine Hydrochloride from Indigenous Shrimp Processing Waste, Bangladesh Journal of Scientific and Industrial Research, **46(3)**, 376-377.
7. W. H. Sutton, Microwave processing of Ceramic Materials, Microwave Solutions for Ceramic Engineers, American Ceramic Society, (2005), 35-65.
8. I.N. Sudiana, Ryo Ito, S. Inagaki, K. Kuwayama, K. Sako, S. Mitsudo, Densification of Alumina Ceramics Sintered by Using Sub-millimeter Wave Gyrotron, J. Infrared, Millimeter, and Terahertz Waves. 34 (2013), pp. 627-638.
9. M. Zamrun F., I. N. Sudiana, S. Mitsudo, Microwaves Enhanced Sintering Mechanisms in Alumina Ceramic Sintering Experiments, Contemporary Engineering Sciences, Vol. 9, 2016, 5, 237 – 247
10. S. Mitsudo, S. Inagaki, I.N. Sudiana, K. Kuwayama, Grain Growth in Millimeter Wave Sintered Alumina Ceramics , Advanced Materials Research, Vol.789 (2013), pp. 279-282
11. I.N., Sudiana, S. Mitsudo, K. Sako, S. Inagaki, L. O. Ngkoimani, I. Usman, H. Aripin, The microwave effects on the properties of alumina at high frequencies of microwave sintering, American Institute of Physics (AIP) Conference Proceeding, **1719**, 030014 (2016); doi: 10.1063/1.4943709. .
12. I.N. Sudiana, S. Mitsudo, T. Nishiwaki, P. E. Susilowati, L. Lestari, M Zamrun Firihu, H. Aripin, Effect of Microwave Radiation on the Properties of Sintered Oxide Ceramics, Contemporary Engineering Sciences, Vol. 8 No. 34, (2015), 1607-1615
13. I. N. Sudiana, S. Mitsudo, M. Z. Firihu, Effect of initial green samples on mechanical properties of alumina ceramic, Contemporary Engineering Sciences, Vol. 9, 2016, no. 12, 595-602
14. I. N. Sudiana, S. Mitsudo, M. Z. Firihu, Effect of Initial Green Samples on Mechanical Properties of Alumina Ceramic, Contemporary Engineering Sciences, Vol. 9, 2016, no. 12, 595-602.
15. M. Z. Firihu, I.N. Sudiana, 2.45 GHz microwave drying of cocoa bean , ARPN Journal of Engineering and Applied Sciences Vol. 12 No. 19, 2016, pp. 11595-11598.
16. Nikolic, S., Mojovic, L., Rakin, M., Pejin,D. dan Savic, D. 2008. A Microwave-Assisted Liquefaction as a Pretreatment for the Bioethanol Production by the Simultaneous Saccharification and Fermentation of Corn meal. CI and CEO. Vol 14(4), 231-234.
17. H. Aripin, S. Mitsudo, I. N. Sudiana, N. Jumsiah, I. Rahmatia, B. Sunendar, L.Nurdiwijayanto, S. Mitsudo, S.Sabchevski, Preparation of Porous Ceramic with Controllable Additive and Firing Temperature, Advanced Materials Research, Vol. 277 (2011) pp. 151-158.
18. Kumar P, Barrett D.M., Delwiche M.J. dan Stroeve, P. 2009. Methods for Pretreatment of Lignocellulosic Biomass for Efficient Hydrolysis and Biofuel Production, Industrial and Engineering Chemistry Research Vol. 48 (8), pp 3713–3729
19. I. N. Sudiana, S. Mitsudo, T. Nishiwaki, P. E. Susilowati, L. Lestari, M. Z. Firihu, H. Aripin, Synthesis and Characterization of Microwave on Sintered Silica Xerogel Produced from Rice Husk Ash, Journal of Physics: Conference Series Vol. 739 No. 1, (2016) 012059.
20. Y. Wu, Zhang, C., Liu, Y., Fu,Z., Dai, B. dan Yin, D. 2012. Biomass Char Sulfonic Acids (BC-SO3H)-Catalyzed Hydrolysis of Bamboo under Microwave Irradiation. Bioresources. 7(4), 5950-5959.
21. J. S. Mojarrad, Mahboob, N., Valizadeh, H., Ansarin, M., and Bourbour, S., 2007, Preparation of Glucosamine from Exoskeleton of Shrimp and Predicting Production by Response Surface Metodhology, Journal of Agricultural and Chemistry, **55,** 2245
22. Z. Arifin, dan Dady Irawan., 2015, Microwave-Assisted Deasecylation of Chitin from Shrimp Shells, Prosiding Seminar Nasional Teknik Kimia "Kejuangan", Pengembangan Teknologi Kimia untuk Pengolahan Sumber Daya Alam Indonesia, Yogyakarta [In Indonesian].

23. Khan T.A. Peh K.K, dan Hung S.C.2001, reporting Degree of Deacetylation Value of Chitosan: The Influence Analytical Methods. J Pharm Pharmaceut Sci. vol 5. No 3 205-212.
24. H.M.S. Kingston, H.J. Haswell, 1997. Microwave-Enhanced Chemistry: Fundamental, Sample Preparation, and Applications, American Chemical Society.
25. A. Hargono, and S. Indro, 2008. Pembuatan Kitosan dari Limbah Cangkang Udang serta Aplikasinya dalam Mereduksi Kolesterol Lemak Kambing, Reaktor, 12(1). 53-57.[In Indonesian]
26. J. Brugnerotto ., 2001. An Infrared Investigation Interaction with Chitin and Chitosan Characterization, Polymer 42: 3569-3580.

IOP Conf. Series: Journal of Physics: Conf. Series **846** (2017) 012012 doi:10.1088/1742-6596/846/1/012012

Characterization of Briquette from the Corncob Charcoal and Sago Stem Alloys

Lina Lestari[1*], Viska Inda Variani[2], I Nyoman Sudiana[1], Dewi Purnama Sari[2], Wa Ode Sitti Ilmawati[1], Erzam Sahaluddin Hasan[2]

[1]Department of Physics,Faculty of Mathematics and Natural Sciences, Halu Oleo University, Kendari, Indonesia
[2]Physics Laboratory, Faculty of Mathematics and Natural Sciences, Halu Oleo University, Kendari, Indonesia

*E-mail: linalestari68@yahoo.com

Abstract. The briquettes fabricated from charcoal of corncob (*zea mays,L*) and sago stem (*metroxilon sago rottb*) have been produced and characterized. The samples were prepared step by step carefully. The charcoal powder filtered by strainer with mesh size of 70-80 to get the homogeneous particle size. Briquettes are made by mixing corncob charcoal powder, sago stem charcoal and sago adhesive with a mass ratio of 4:5:1, 4.5: 4.5: 1, 5:4:1. The materials are mixed with hot water and stirred to get homogeneous blend. Then they are compacted by pressure of 34.66kg/cm^2, 69.32kg/cm2, and 103.98kg/cm^2 to form a cylindrical shape with diameter of 4 cm. The cylindrical briquettes then were dried at temperature of 60°C for 48 hours. After dried, the samples where then characterized their density and water, ash, volatile matter, fixed carbon contents. The burning rate, combustion temperature, and ignition time were also determined. The experimental results show that the briquettes have average densities from 0.602 to 0.717gr/cm^3. The density increase with the increasing of forming pressure. The increasing of pressure also result in the decreasing of moisture content from 2.669% to 0.842%. The ash content is found from 3.459% to 8.766%. Volatile matter and fixed carbon are varies from 13.658% and 21.168% and 67.667% to 80.758% respectively. The lowest burning rate is 0.0898gr/s and the optimum burning temperature is 499.2^{o}C with the lowest ignition time of 1.58 minutes. These briquette's parameters agree wit the quality standard of industrial briquette.

1. Introduction

Processing and utilization of biomass resources are being developed as more valuable material [1-2] as well as as alternatives to conventional energy source [3-4]. The biomass is a solid waste and it is commonly used as a fuel source. Biomass energy can be used as an alternative energy source to replace fossil fuels. The advantage of biomass is it does not contain sulfur and also did not caused the air pollution. It also improves the efficiency of utilization of forest and agriculture resources [5]. There are abundant biomass from agricultural wastes in Indonesia especially in southeast Sulawesi such as corncobs and sago stems. The corncob waste contains celluloses and lignins about 44.9% and 33.3%, respectively, while sago stem has celluloses of about 57% and lignins of 38% [6]. These facts indicate that the corncobs and sago stems can be used as material for producing the charcoal briquette. The charcoal briquette charcoal is charcoal which is made by changing the shape, the size and the density by applied a certain pressing to the mixture of charcoal powder with an adhesive.

Content from this work may be used under the terms of the Creative Commons Attribution 3.0 licence. Any further distribution of this work must maintain attribution to the author(s) and the title of the work, journal citation and DOI.
Published under licence by IOP Publishing Ltd

The manufacturing of briquette alloy of corncob charcoal and sago stem charcoal need an appropriate method to improve the briquette quality. Commonly, the production of briquettes are using conventional method, especially for drying is using electric furnace or fire. Recently, the use of microwave radiation or hot gases as a substitute for the conventional heating has been reported for processing materials. The applications have been reported in sintering ceramic [7-10], synthesizing organic materials [11-13] as well as speeding up a chemical reaction and drying [14-15]. The theoretical analysis were also reported [16-17]. Microwave radiation is believed to be able to increase the rate of a chemical reaction and may improve the microstructure of briquette and then result in increase their caloric energy. The use this technology for briquette is performing in our research group and will be reported soon in separated papers.

Chemically, sago stem contains 28% of amylose and 72% of amylopectin and can be used as adhesive [3]. Sago starch is potential as an adhesive of charcoal briquettes of the mixtures of charcoal corncob and sago stem. The ratio of the mass of charcoal corncob and sago stem is 9:1 which gives the optimum composition of charcoal briquette [18]. Briquette quality also depends on the particle size and pressure. Charcoal briquettes of corncobs with adhesive from sago is produced with the size of 80 mesh, and the pressure of 94,22kg/cm^2. The resulting briquettes have a caloric value of 7607,4707 cal/g, the ash content of 5.3042%, moisture content of 2.2492% and the density of 0,7419gr/cm^3 [19]. The charcoal briquette of sago stem made with adhesive with composition of 9:1, particle size of 70 mesh, and pressure of 94,22kg/cm^2 produce the high burning temperature of 428,2^{o}C [20]. In this research, we study the effect of material composition and forming pressure to quality of briquette of corncob and sago stem alloy. The samples will characterize carefully such as their density, water content, ash content, volatile matter, fixed carbon, the burning rate, burning temperature, and burning time. By varying the composition there is expected to obtain the best composition of the charcoal briquettes for their optimum quality.

2. Material and Method

2.1 Briquette Fabrication

The material used in this research are the corncob wasted *(Zea mays,L)* and sago stem wasted (*Metroxilon sagu Rottb*). The corncob wasted is taken from Tikep, Muna, Indonesia. Sago stem waste is collected from Konda district of Konawe Selatan, Indonesia. Before carbonation, corncob and sago stem waste is dried in the oven with temperature of 110°C for 120 minutes. The corncob is carbonated by using furnace for 60 minutes to achieve temperature's carbonation of 499°C and sago stem is furnace about 160 minutes to reach the temperature of 530°C. The charcoal of corncob and sago stem are then mashed to obtain the charcoal powder. The charcoal powder is then filtered by strainer with size of 80 mesh. The residue is the filtered again with the strainer of 70 mesh size. The resulted powder is then used as ingredient for making the charcoal briquette. The charcoal powder of corncob and sago stem are mixed with the adhesive from sago starch. The mass ratio composition are 4:5:1, 4.5:4.5:1, 5:4:1, respectively for total mass of the mixed material of 10g. The mixed material is then added with 2 ml of water and stirred until homogeneous.

The mixed samples with these composition were then formed to cylindrical shape with diameter of 4 cm with hole of 0.8 cm in the center. They were compressed with the pressure of 34.66kg/cm^2, 69.32kg/cm^2,and 103.98kg/cm^2, respectively. The resulted briquettes were then dried in oven at 60°C for 48 hours.

2.2 Characterization of briquettes

After fabrication, the briquettes were then characterized. About one gram of sample heated in the oven with the temperature of 105°C until the constant mass (no evaporation) is reached. The briquette was then cooled before weighted. The moisture content is calculated as the ratio of evaporated water's mass compared with sample's mass before heating. The volatile matter is determined in the same way but the sample is heated at temperature of 750 ^{o}C for 7 minutes. Volatile matter is evaporated mass during heating but water content. The ash content of the briquette is determined by burning the one gram briquette sample till all the sample become ash. The ash was then cooled and then weighted. The ash content is determined as presentation of the resulted ashes compared with the original sample. Fixed carbon is the carbon in the sample while the density is the mass ratio per unit volume of the sample. Burning testing is performed in the open space at temperature 25°C. This is to obtain the ignition time, burning rate and the maximum temperature of the briquettes. The temperature were measured using infrared thermometer.

3. Results and Discussion

3.1 Density

The density of the briquette made from the charcoal of corncobs and sago stem alloys with different compositions is presented in Figure 1. The blue and the red lines are the density for the composition of charcoal corncob, the charcoal of sago stem and the adhesive of sago starch with ratio of 4 : 5: 1 and 4.5 : 4.5 : 1, respectively. While the green line is for the ratio of 5:4:1. It is shown that the density is increased with the increasing of the pressure. The briquette's density is from 0.602 g/cm^3 to 0.717g/cm^3. The highest average density is found at the composition 4.5:4.5:1. The pressure applied during forming was not only to create material bonding and distance between atoms shorter but also to spread the adhesive thoroughly to whole charcoal body. The briquette with forming by suitable pressure will have a high density and high mechanical strength. However, if we applied not suitable pressure during fabrication, it may make briquette more fragile [20]. According the British standard, the value of the briquette density is more than 0.48g/cm^3 while the Japanese standard is lower than 1.2 g/cm^3 [21]. Based on this, the resulted briquette made from the charcoal of corncobs and sago stem alloys satisfy these standard.

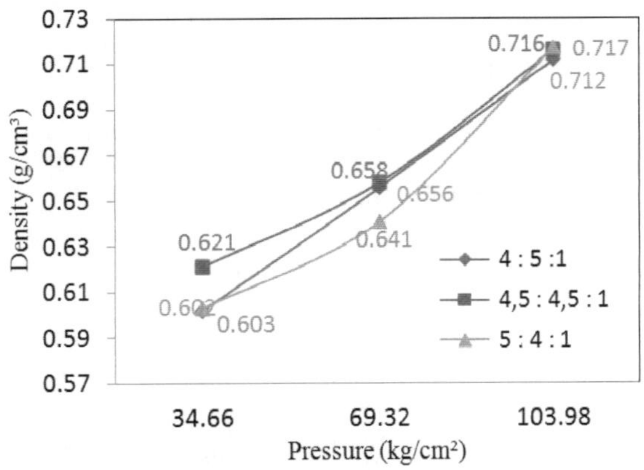

FIGURE 1. Density of briquette made from corncobs and sago stem alloys with different composition ratio.

3.2 Moisture content

Moisture content of charcoal from corncob and sago stem alloys with different composition is given in Figure 2. The blue and the red lines are the density for the composition of charcoal corncob, the charcoal of sago stem and the adhesive of sago starch with ratio of 4:5:1 and 4.5:4.5:1, respectively. The green line is for the ratio of 5:4:1. From the figure, we can see that the moisture content is decreased with the increment of the pressure. The moisture content of the briquettes are from 0.842% to 2.669%. The lowest average moisture content is obtained at composition of 5:4:1. Generally, the moisture content of the briquette is expected to be low because of the lower its moisture content the higher its burning power. The standard value of briquette moisture content is less than 3.6% for British standard and less than 8% for Japan standard [8]. Our briquettes meet to these standards.

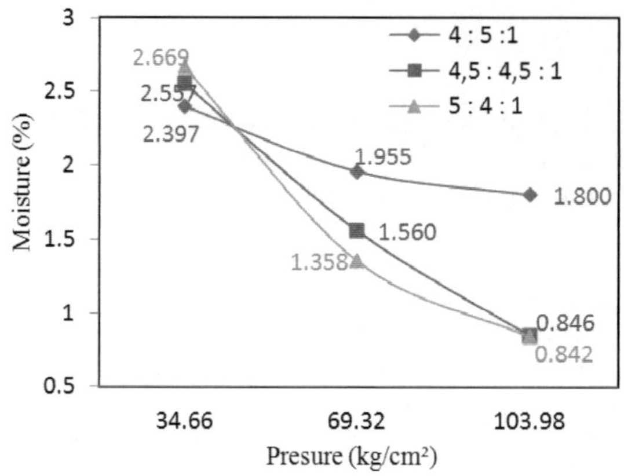

FIGURE 2. Moisture content of briquette made from corncob and sago stem alloys with different composition ratio.

3.3 Ash, volatile matter, and fixed carbon contents

The ash, volatile matter and fixed carbon contents of the briquettes made from the charcoal of corncob and sago stem alloys and adhesive for different composition ratio are presented in Table 1. From Table 1, we see that the highest fixed carbon content is the briquette made from the charcoal of corncob and sago stem alloys together with composition of 4.5:4.5:1. We also found that the charcoal of corncob contributed to the highest ash as well as volatile matter contents. According to the Japanese standard, the values for ash, volatile matters and the fixed carbon contents of briquettes are from 3% – 6%, 15% - 30%, and 60% – 80 %, respectively [21]. Althought the values for ash content exceed the standard value, but for volatile matter and fixed carbon contents are in good agreement with the standard.

TABLE 1. The ash content, volatile matter,and fixed carbon briquettes

Composition Ratio	Ash Content (%)	Volatile Matter (%)	Fixed Carbon (%)
4:5:1	6.884	19.785	71.375
4.5:4.5:1	6.757	18.720	72.963
5:4:1	7.521	20.984	70.136

3.4 Burning rate, ignition time and maximum temperature

Burning rate is the rate of briquette's mass combusted in air at certain time. Figure 3 shown the average burning rate of the briquettes of three composition ratios. The blue and the red lines are for the composition of charcoal corncob, the charcoal of sago stem and the adhesive of sago starch of 4:5:1 and 4.5:4.5:1, respectively. While the green line is for the ratio of 5:4:1.The lowest burning rate of the briquette is found for the composition ratio of 5:4:1 while the highest one is for the composition ration of 4:5:1.

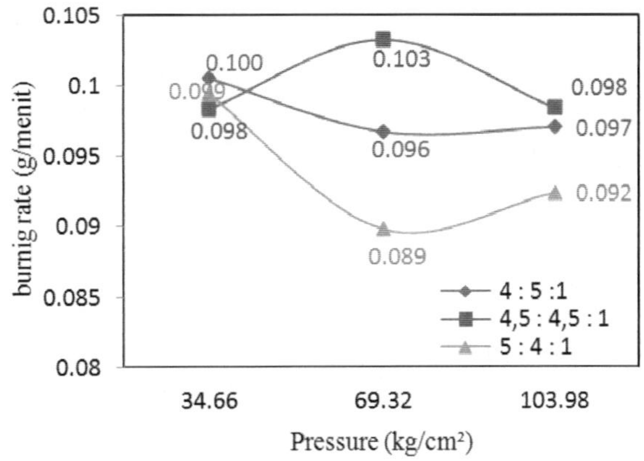

FIGURE 3. Burning rate of briquette made from corncob and sago stem alloys and adhisive with different composition ratio.

The ignition time is time needed by the briquette for starting to blame. It determines how combustible the briquette. From Table 2, we found that the lowest iginition time is 1.58 minutes for briquette with the composition ratio of 5:4:1 and the the forming pressure of 103.98kg/cm^3. The highest temperature is 499.2°C for briquette with the composition ratio of 4.5:4.5:1 and the pressure of 34.66kg/cm^3.

TABLE 2. Ignition time and the maximum temperature of briquette

Composition Ratio	Pressure (kg/cm^2)	Ignition Time (menit)	Maximum Temperature (°C)
4:5:1	34.66	4.30	495.8
	69.32	3.20	477.2
	103.98	3.15	480.8
4.5:4.5:1	34.66	2.50	461.2
	69.32	2.29	478.6
	103.98	1.58	491.6
5:4:1	34.66	4.80	499.2
	69.32	4.10	488.4
	103.98	3.30	474.4

4. Conclusion

The fabrication the briquette from chorcoal of corncob and sago stem alloys was successfully performed. For adhesive, the sago strach was applied. The composition of corncob charcoal and sago stem charcoal alloys and an adhesive are 4:5:1, 4.5:4.5:1, and 5:4:1, rescpectively. And the corresponding forming pressure are 34.66 kg/cm^2, 69.32 kg/cm^2, and 103.98 kg/cm^2. The briquete's characteristics are meet to the international standard for industrial briquette. The briquette with composition ratio of 5:4:1 and forming pressure of 34.66kg/cm^2 give the maximum temperature of 499.2 °C.

Acknowledgments
The work was is supported by the Ministry of Research, Technology and Higher Education of the Republic of Indonesia through PPT Research Grant. We also acknowledge support from Halu Oleo University.

References

1. H. Aripin, S. Mitsudo, I. N. Sudiana, N. Jumsiah, I. Rahmatia, B. Sunendar, L.Nurdiwijayanto, S. Mitsudo, S.Sabchevski, Preparation of Porous Ceramic with Controllable Additive and Firing Temperature, Advanced Materials Research, Vol. 277 (2011) pp. 151-158.

2. I. N. Sudiana, S. Mitsudo, T. Nishiwaki, P. E. Susilowati, L. Lestari, M. Z. Firihu, H. Aripin, Synthesis and Characterization of Microwave on Sintered Silica Xerogel Produced from Rice Husk Ash, Journal of Physics: Conference Series Vol. 739 No. 1, (2016) 012059.
3. P. Agung, Utilization of Biogas as Alternative Energy. Fakultas Teknik, Universitas Gajah Mada.Yogyakarta. 2008. [In Indonesian]
4. P. Kumar, Barrett D.M., Delwiche M.J. dan Stroeve, P. 2009. Methods for Pretreatment of Lignocellulosic Biomass for Efficient Hydrolysis and Biofuel Production, Industrial and Engineering Chemistry Research Vol. 48 (8), pp 3713–3729
5. W. Syafi'i, . Forests as Future Energy. Kompas, 15 April 2003. [In Indonesian]
6. Warisno. 2003. Hybrid Corn Farming. Canisius, Yogyakarta. [In Indonesian]
7. W. H. Sutton, Microwave processing of Ceramic Materials, Microwave Solutions for Ceramic Engineers, American Ceramic Society, (2005), 35-65.
8. I.N. Sudiana, Ryo Ito, S. Inagaki, K. Kuwayama, K. Sako, S. Mitsudo, Densification of Alumina Ceramics Sintered by Using Sub-millimeter Wave Gyrotron, J. Infrared, Millimeter, and Terahertz Waves. 34 (2013), 627-638.
9. I. N. Sudiana, S. Mitsudo, M. Z. Firihu, Effect of Initial Green Samples on Mechanical Properties of Alumina Ceramic , Contemporary Engineering Sciences, Vol. 9, 2016, no. 12, 595-602
10. S. Mitsudo, S. Inagaki, I.N. Sudiana, K. Kuwayama, Grain Growth in Millimeter Wave Sintered Alumina Ceramics , Advanced Materials Research, Vol.789 (2013), pp. 279-282.
11. Y. Wu, Zhang, C., Liu, Y., Fu,Z., Dai, B. dan Yin, D. 2012. Biomass Char Sulfonic Acids (BC-SO3H)-Catalyzed Hydrolysis of Bamboo under Microwave Irradiation. Bioresources. 7(4), 5950-5959.
12. H. Aripin, S. Mitsudo, I.N. Sudiana, T. Saito, S. Sabchevski, Structure Formation of a Double Sintered Nanocrystalline Silica Xerogel Converted From Sago Waste Ash, Transactions of the Indian Ceramic Society, DOI: 10.1080/0371750X.2014.980850 (2015)
13. M. Z. Firihu, I.N. Sudiana, 2.45 GHz microwave drying of cocoa bean , ARPN Journal of Engineering and Applied Sciences Vol. 12 No. 19, 2016, pp. 11595-11598.
14. I.N., Sudiana, S. Mitsudo, K. Sako, S. Inagaki, L. O. Ngkoimani, I. Usman, H. Aripin, The microwave effects on the properties of alumina at high frequencies of microwave sintering, American Institute of Physics (AIP) Conference Proceeding, **1719**, 030014 (2016); doi: 10.1063/1.4943709.
15. M. Zamrun F., I. N. Sudiana, S. Mitsudo, Microwaves Enhanced Sintering Mechanisms in Alumina Ceramic Sintering Experiments, Contemporary Engineering Sciences, Vol. 9, 2016, 5, 237 – 247
16. I.N. Sudiana, S. Mitsudo, T. Nishiwaki, P. E. Susilowati, L. Lestari, M Zamrun Firihu, H. Aripin, Effect of Microwave Radiation on the Properties of Sintered Oxide Ceramics, Contemporary Engineering Sciences, Vol. 8 No. 34, (2015), 1607-1615
17. L. Lestari, Aripin,Yanti, Zainudin,Sukmawati,Marliani. 2010. Adhesive Material Effect on the Quality of Corn Cob Charcoal Briquette. Jurnal Aplikasi Fisika, 6, 2, pp. 93-96. [In Indonesian]
18. L. Lestari, M. Dula, L. Agusu, Zainudin, A. Maskur, H. Alrasyid. 2011. Optimization of Material Composition, Particle Size and Pressure in Charcoal Briquette -Corn Cob Alloy. Jurnal Aplikasi Fisika, 7, 2, pp. 88-91. [In Indonesian]
19. L. Lestari, E.S Hasan, V.I. Variani. 2014. Optimization of Particle Size, Composition and Pressure in Charcoal Briquette Sago Stem Alloy. Jurnal Aplikasi Fisika,10, 1,pp. 27-30. [In Indonesian]
20. Poespowati, 2009. Effect of compression pressure and adhesives type to Coconut Shell Charcoal Briquettes from Palm (Elaeis Quinensis Jacq). TIN FATETA IPB, Bogor. [In Indonesian]
21. Mujiono, 2009. Economic and Technic Aspect Analysis of Biobriket Charcoal Sawdust and Rice Husk. Universitas Muhammadiyah, Surakarta. 2009. [In Indonesian]

IOP Conf. Series: Journal of Physics: Conf. Series **846** (2017) 012013 doi:10.1088/1742-6596/846/1/012013

Fast Microwave-assisted Pretreatment for Bioconversion of Sawdust Lignocellulose to Glucose

I Nyoman Sudiana[1,*], Seitaro Mitsudo[2], Prima Endang Susilowati[3], Desak Ketut Sutiari[1], Made Widana Arsana[4], Muhammad Zamrun Firihu[1], La Ode Ngkoimani[5], La Aba[1], Erzam Sahaluddin Hasan[1], Edi Cahyono[6], Svilen Sabchevski[7], Haji Aripin[8] , Komang Gde Suastika[9]

[1]Department of Physics, Faculty of Mathematic and Natural Science, Halu Oleo University, Kampus Bumi Tridharma Anduonohu, Kendari 93232, Indonesia
[2]Research Center for Development of Far-Infrared Region, University of Fukui, 3-9-1 Bunkyo, Fukui-shi 910-8507, Japan
[3]Department of Chemistry, Faculty of Mathematic and Natural Science, Halu Oleo University, Kampus Bumi Tridharma Anduonohu, Kendari 93232, Indonesia
[4]Department of Agrotechnology, Faculty of Agriculture, Halu Oleo University, Kampus Bumi Tridharma Anduonohu, Kendari 93232, Indonesia
[5]Department of Geophysics, Faculty of Earth Sciences and Technology, Halu Oleo University, Kampus Bumi Tridharma Anduonohu, Kendari 93232, Indonesia
[6]Department of Mathematics, Faculty of Mathematic and Natural Science, Halu Oleo University, Kampus Bumi Tridharma Anduonohu, Kendari 93232, Indonesia
[7]Lab. Plasma Physics and Engineering, Institute of Electronics of Bulgarian Academy of Sciences, Bulgaria
[8]Center for Material Processing and Renewable Energy, Faculty of Learning Teacher and Education Science, Siliwangi University, Jl. Siliwangi 24 Tasikmalaya 46115, West Java, Indonesia
[9] Department of Physics Education, Palangka Raya University, Indonesia, 73112

*E-mail: sudiana75@yahoo.com

Abstract. A preliminary study of application microwave energy for bioconversion of cellulosic sawdust to glucose was performed. The effects of the microwave were compared to those of the conventional method for each solvent. It was expected that a broader mechanism responsible for the microwave effects on the chemical processes, especially the pretreatment on the hydrolysis of cellulose can be explained. Reagents used were an acid (HCl), an alkali (NaOH), and distilled water (H_2O). The experimental results showed that the microwave-assisted pretreatment on the lignocellulosic sawdust faster than by using conventional heating (hotplate). Moreover by using microwave a higher glucose content compared to the conventional method was found. With microwave during hydrolisis, high temperatures and high reagent concentrations were not required. Pretreatment with a microwave at 800 Watt and solvent NaOH 22,50 mg/mL at a temperature of 120°C appeared to be most efficient found in this experiment. These results indicate that microwave effective for bioconversion of cellulosic sawdust to glucose. The microstructure evaluation by using SEM and XRD should be performed to understand more detail the effect especially on their cellulosic structural evolution.

Content from this work may be used under the terms of the Creative Commons Attribution 3.0 licence. Any further distribution of this work must maintain attribution to the author(s) and the title of the work, journal citation and DOI.
Published under licence by IOP Publishing Ltd

IOP Conf. Series: Journal of Physics: Conf. Series **846** (2017) 012013 doi:10.1088/1742-6596/846/1/012013

1. Introduction

Renewable biomass resources and appropriate energy technology are being developed as alternatives to conventional fuels. Bio-ethanol or bio-hydrogen that can be obtained via the fermentation of starchy sugar is one of the biomass-based energies that can be a renewable and friendly energy source [1]. However, such a use of starch will compete with the fulfillment for food demands. Therefore, the second generation of bio-ethanol by using cellulose containing organic wastes, such as sago wastes, saw dusts, banana trunks, etc. has been developed [2]. Upon the production of the second generation of bio-ethanol, it requires a pretreatment in order to gain cellulose with a low content of lignin and of hemicelluloses. The cellulose is hydrolyzed to form glucose which is then fermented to produce bio-ethanol [3]. The pretreatment can be done by using a physical method (milling), a physical-chemical method (auto hydrolysis, hydrothermolysis, alkali oxidation), chemical method (application of alkali, mixture of acids), biological method (fermentation, addition of enzymes) and heating or combination of the methods above [4]. The use of cellulosic wastes for the production of bio-ethanol still encounters some problems including low amount of product and high cost production. The problems are brought about by the structure of cellulose being covered with lignin that makes it hard to hydrolyze. Therefore, to produce glucose from wood it is required an optimal technology especially at the pretreatment stage [5].Novianti, *et.al.* studied the use of a conventional heating and reported that the hydrolysis using sulfuric acid (50%) for 2 h on saw dust (8:1, v/w) produced the highest sugar content (43,72%) [6]. In addition, a waste of coconut bunches has also been tested to produce glucose via hydrolysis using HCl (12%) at a temperature of 90-100^{0}C for 30 minutes and resulted in glucose of 12,795 ppm. From the used conventional method, it was found that the produced glucose was still not optimal.

Recently, the use of microwave radiation as a substitute for the conventional heating has been reported for sintering ceramic [7-12], synthesizing organic materials [13-15] or speeding up a chemical reaction and drying [16-17]. Microwave radiation is believed to be able to increase the rate of a chemical reaction. The chemical reaction principle with a microwave oven is to make use of the ability of molecules to move and dipole in the materials or conduction ions in a substance that play a role in changing the electromagnetic energy to heat [18-19]. In a conventional method, the thermal energy is transferred to the materials via convection, conduction, or heat radiation from the material surface. On the other hand, the microwave energy is directly transferred to the materials via the interaction between molecules and electromagnetic fields.The *in situ* energy conversion is very interesting for chemical applications and material processing [19]. The heating with a microwave oven depends more on the molecular characteristics and the condition of reactions compared to a conventional heating .

A number of reports states that microwave can speed up a reaction based on the speed and uniformity of the heat transfer produced by the microwave. Microwave irradiation has an effect on the content of cellulose, hemicellulose and lignin. It can increase the percentage of cellulose by reducing the lignin and hemicellulose content. A pretreatment with a microwave oven at a frequency of 2.45 GHz and with a wattage of 950 Watt with varied times and rice straw powder sizes showed a highly significant increase in cellulose content and a highly significant reduction in lignin and hemicelluloses content [2]. Combining microwave pretreatment with NaOH can increase the glucose content and reduce the lignin and hemicellulose content [21]. A pretreatment with a microwave oven combined with NaOH increased the cellulose content of banana stems compared to other treatments. This is because NaOH can interact with the phenolic lignocelluloses and cause the lignocellulose bonds to break [3-4]. Microwave radiation occurring during the pretreatment at an NaOH concentration of 3 M with a duration of exposure of 40 minutes had an effect on the reduction in the lignin and hemicellulose content and an increase in the cellulose content of sugarcane wastes [1]. Pretreatment with NaOH resulted in a higher sugar content compared to Na_2CO_3 and $Ca(OH)_2$.

Acids and alkalis also significantly increase glucose content. Microwave pretreatment with an acid has advantages at the reaction conditions such as corrosion, high temperatures, and economically cheaper. If the concentration of acid/alkali is increased, the glucose content also increases [21-22] reported that pretreatment with microwave combined with distilled water, NaOH and H_2SO_4 at a temperature ranging from 130 to 200°C on miscanthus showed H_2SO_4 resulted in maximum glucose content at a temperature of 180°C. At 200°C, however, the glucose content significantly decreased. The mechanism of how a microwave increases the chemical reaction has yet to elucidate. A number of authors has proposed a theory as for the presence of ponderomotive force or the emergence of new hot spots in the reacting materials during the radiation process by the microwave. In the current study, the effects of microwave energy on the degradation of saw dust celluloses were observed. Those effects were compared to the effects of the conventional heating. The degradation was expected to be able to break down the hemicellulosic bonds, reduce the lignin content, and convert crystal structures of the celluloses to simple sugars [20-21]. In addition, it was expected to find a broader explanation of the microwave effects on the chemical processes, especially on the pretreatment for cellulose hydrolysis.

2. Material and Method

2.1 Sample Preparation

Samples of 'Kuma' sawdusts were collected in Kendari, Indonesia. The saw dusts were uniformly dried in the sun for ± 10 h, then sieved with a 20-mess sieve. The solvent pH of 5 was achieved by adding HCl, a pH of 7 by adding distilled water, and a pH of 9 by adding NaOH.

2.2 Hydrolisis and characterization

The study was arranged of varied pH2 and varied microwave wattages. The sample pH consisted of 3 levels- pH 5, pH 7 and pH 9, while the microwave wattages consisted of 400, 560 and 800 Watt. Measurements were conducted with a heating period of 15 minutes for each sample.
Sample hydrolysis was done by using HCl, distilled water, and NaOH. Saw dusts measuring 20 mesh weighing 20 g were thoroughly mixed with 200 mL of a solvent. The samples were heated with a microwave oven having a wattage of 400 Watt for 15 minutes. Data collections were repeated by using a 560 Watt and 800 Watt microwave. After being heated, the samples were cooled down at a room temperature, then sieved with sieve paper to separate the liquid from the saw dusts. The glucose content was determined by using the DNS (3,5-dinitrosalicylic acid) method (BIO BASIC INC). The residues of sieving were dried in the oven at a temperature of 45°C until its water content became zero. Sample hydrolysis without heating was conducted by dissolving 20 g of saw dusts into the solvent at pH of 5, 7 and 9, consecutively. The samples were thoroughly stirred, kept still for 15 minutes, and sieved by sieve paper. The liquid resulting from the sieving was measured for its glucose content by using the DNS method. The sieving remnant was dried in the oven at a temperature of 45°C. Sample hydrolysis using a conventional heating was done by heating bath water, with a treatment on each solvent pH 5, pH 7 and pH 9 for 15 minutes, after which the samples were cooled down and then sieved to separate the remnant from the liquid. Microstructure of samples then characterized by using Scanning electron microscope (SEM) and X-Ray Diffraction (XRD).

3. Results and Discussion

3.1 Effects of Microwave Energy on the Reaction using an Acid

NaOH was used in the current study to find a solvent with a pH of 9. Figure 1 shows that the glucose content increased with wattage in microwave-NaOH pretreatment. The duration of energy exposure used was 15 minutes, while the tested microwave power were 400, 560 and 800 Watt. The results showed that the use of NaOH with microwave assisted energy exposure could increase glucose content from 7,36 mg/mL (without heating) to 16,12 mg/mL (400 Watt), 20,10 mg/mL (560 Watt) and 22,50 mg/mL (800 Watt). In agreement with Dawson, *et.al.*, NaOH degrades the lignin that covers cellulose and hemicellulose by breaking down the lignin per se. The structural damage of the lignin and hemicellulosic bonds could increase the glucose content of the saw dusts [23]. In NaOH-containing saw dusts, the energy could spread through ionic conduction where the movements of the ions produce energy and increase rates of collision that converts kinetic energy to heat.

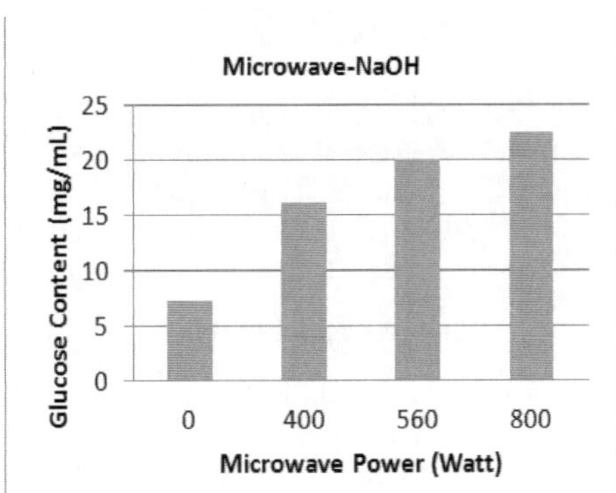

Figure 1 Glucose content vs. microwave power in microwave-NaOH pretreatment

An increase in microwave wattage and temperature could increase the energy and temperature received by any heated material. This causes a faster reaction. That cellulose was hydrolyzed relatively faster at a higher temperature. The interaction between microwaves and the materials brings about the reduced number of hemicelluloses that bind celluloses. This is in agreement with Wu *et al.* (2008) that microwave radiation cause the occurrence of physical explosions on the microfibers which then result in the disintegration of structures that are hard to degrade such as lignin that protect cellulose [24]. In addition, the electromagnetic fields used in the NaOH-microwave solvent could produce chemical-physical effects that can speed up the disintegration of the crystal structures in celluloses. Therefore, the celluloses found in pretreatment with NaOH-microwave combination produced a higher cellulose content than the cellulose content before being given such a pretreatment.

3.2 Effects of Microwave Energi on the Reaction using H_2O

According to Lidstroem, *et al.* , H_2O is a polar solvent in which dipolar polarization occurs due to the interaction among electrical field components [25]. When a microwave energy is exposed to a material, polar molecules will follow the electrical fields and adapt themselves to the field phase. However, group of intra polar molecules will undergo inertia and cannot follow the fields. This causes particles to move randomly and such a random interaction will produce hot spots. He also reported that heating with a microwave could increase the rate of a chemical reaction and increase glucose content because of a higher number of disintegrated lignin and hemicelluloses with the microwave heating.

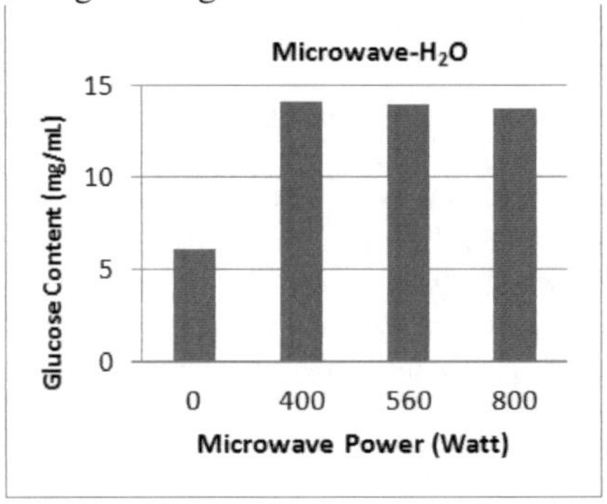

Figure 2 Glucose content vs. microwave power in microwave-H_2O pretreatment

Figure 3 shows effect of microwave power on glucose content for various solvents. Of the three solvents, the highest glucose content (22.50 mg/mL) was found in NaOH at a wattage of 800 W. There was no significant change in the glucose content with HCl and H_2O despite an increase in microwave power. This is concomitant with previous results that pretreatment with microwave heating using different types of solvents resulted in different glucose contents, where NaOH is found to produce the highest glucose content [24-25].

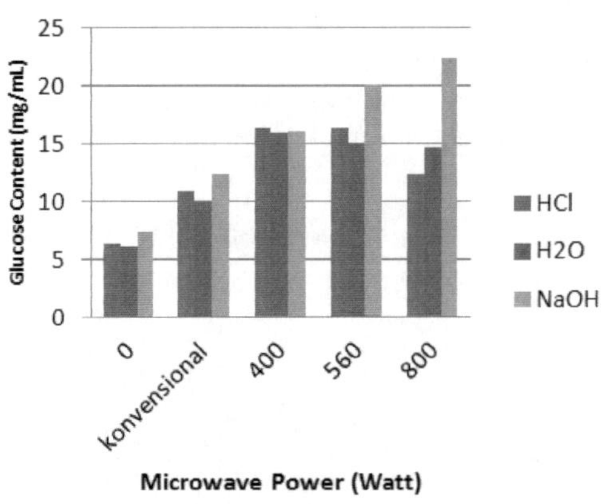

Figure 3 Glucose content vs. microwave power in various solvents pretreatment

An SEM test was used to study the microstructures of saw dusts before and after the pretreatment. The preliminary results indicates that the abundance of bonds of the intact celluloses before treatment and a great number of bonds disappear after being given a microwaves pretreatment. The details of the microstructure evolution after pretreatment in this investigations will be reported in separated paper.The heat energy of the microwave is produced from a direct interaction between the polar regions of the saw dust biomass and the oscillation of the electromagnetic fields. Microwave irradiation produces an efficient internal heat that can result in a direct and uniform heating on the mixed reactions. The reflection and diffusion on the local barrier cause the occurrence of hot spots and superheating effects as reported previously [7-10, 26-28]. There were also reported that ionic conduction (crystalline) and non-ionic conduction (amorph) occur in celluloses. The structure of hydrogen bonds in the cellulosic crystalline can cause the movements of protons under the electromagnetic fields at suitable conditions. Therefore, the cellulosic crystalline can act as an absorber for the microwave active energy, which can increase biomass decompositions. During this process, lignin/hemicelluloses undergoes disintegration, the effect of microwave absorbance increases and causes the biomass to degrade. This is in agreement with the work of Alvira et al. that pretreatment could break down the lignin that blankets hemicellulose and cellulose, and to reduce the crystal structure of the cellulose [29].

4. Conclusion

The use of microwave for the pretreatment of cellulose hydrolysis was successfully conducted. The reactants used were an acid (HCl), an alkali (NaOH), and distilled water (H_2O). The use of microwave produced a higher glucose content compared to conventional method, especially in the combination with the alkali (NaOH). This indicates that microwave oven is effective for a pretreatment in the conversion of lignocellulose in saw dusts to glucose. The SEM images show the differences in the microstructures of the cellulose surfaces. Microwave-assisted alkali pretreatment showed a reduced number of bonds as a result of a more severe degradation on the microstructures of the cellulose surface compared to other treatments. This is in concomitant with the XRD test where an increase of crystalline index was found from the saw dusts. This result indicates that microwave technology is very promising for converting sawdust lignocellulose to glucose.

Acknowledgments
This work has been partially supported by the Directorate of Higher Education, at the Ministry of Research Technology and Higher Education, Republic of Indonesia and University of Halu Oleo, Kendari.

References
[1] Nikolic, S., Mojovic, L., Rakin, M., Pejin,D. dan Savic, D. 2008. A Microwave-Assisted Liquefaction as a Pretreatment for the Bioethanol Production by the Simultaneous Saccharification and Fermentation of Corn meal. CI and CEO. Vol 14(4), 231-234.
[2] Dehani, F. R., Argo, B. D. dan Yuliningsih, R. 2013. Utilization of Microwave Irradiation to Maximize Pretreatment Process Degradation of Lignin Jeremi Paddy. Jurnal Biopress Komoditas Tropis Vol.1 No. 1, 175-182 [In Indonesia].
[3] Jingjing, L. 2011. Isolation of Lignin from Wood. Saimaa University of Applied Sciences Faculty of Technology.
[4] Kumar P, Barrett D.M., Delwiche M.J. dan Stroeve, P. 2009. Methods for Pretreatment of Lignocellulosic Biomass for Efficient Hydrolysis and Biofuel Production, Industrial and Engineering Chemistry Research Vol. 48 (8), pp 3713–3729
[5] Thostenson, E.T. dan Chou, T.W. 1999. Microwave processing : Fundamental and Applications, Elsevier Composites Applied Science and Manufacturing: Part A 30, 1055-1071.
[6] Novianti, Mappiratu dan Musafira. 2013. Utilization Sawdust Waste for Production Bioethanol using yeast cells. J. of Natural Science, Vol. 2. No. 3, 9-19 [in Indonesia]
[7] W. H. Sutton, Microwave processing of Ceramic Materials, Microwave Solutions for Ceramic Engineers, American Ceramic Society, (2005), 35-65.
[8] I.N. Sudiana, Ryo Ito, S. Inagaki, K. Kuwayama, K. Sako, S. Mitsudo, Densification of Alumina Ceramics Sintered by Using Sub-millimeter Wave Gyrotron, J. Infrared, Millimeter, and Terahertz Waves. 34 (2013), 627-638.
[9] I. N. Sudiana, S. Mitsudo, M. Z. Firihu, Effect of Initial Green Samples on Mechanical Properties of Alumina Ceramic , Contemporary Engineering Sciences, Vol. 9, 2016, no. 12, 595-602
[10] S. Mitsudo, S. Inagaki, I.N. Sudiana, K. Kuwayama, Grain Growth in Millimeter Wave Sintered Alumina Ceramics , Advanced Materials Research, Vol.789 (2013), pp. 279-282.
[11] M. Z. Firihu, I.N. Sudiana, Properties of 2.45 GHz microwave sintered S_iO_2 from rice husk ash and

Al_2O_3, ARPN Journal of Engineering and Applied Sciences Vol. 12 No. 19, 2016, pp. 11595-11598

[12] H. Aripin, S. Mitsudo, I. N. Sudiana, N. Jumsiah, I. Rahmatia, B. Sunendar, L.Nurdiwijayanto, S. Mitsudo, S.Sabchevski, Preparation of Porous Ceramic with Controllable Additive and Firing Temperature, Advanced Materials Research, Vol. 277 (2011) pp. 151-158

[13] M. Zamrun F., I. N. Sudiana, S. Mitsudo, Microwaves Enhanced Sintering Mechanisms in Alumina Ceramic Sintering Experiments, Contemporary Engineering Sciences, Vol. 9, 2016, 5, 237 – 247

[14] I.N. Sudiana, S. Mitsudo, T. Nishiwaki, P. E. Susilowati, L. Lestari, M Zamrun Firihu, H. Aripin, Effect of Microwave Radiation on the Properties of Sintered Oxide Ceramics, Contemporary Engineering Sciences, Vol. 8 No. 34, (2015), 1607-1615

[15] H. Aripin, S. Mitsudo, I.N. Sudiana, T. Saito, S. Sabchevski, Structure Formation of a Double Sintered Nanocrystalline Silica Xerogel Converted From Sago Waste Ash, Transactions of the Indian Ceramic Society, DOI: 10.1080/0371750X.2014.980850 (2015)

[16] Wu, Y., Zhang, C., Liu, Y., Fu,Z., Dai, B. dan Yin, D. 2012. Biomass Char Sulfonic Acids (BC-SO3H)-Catalyzed Hydrolysis of Bamboo under Microwave Irradiation. Bioresources. 7(4), 5950-5959.

[17] M. Z. Firihu, I.N. Sudiana, 2.45 GHz microwave drying of cocoa bean , ARPN Journal of Engineering and Applied Sciences Vol. 12 No. 19, 2016, pp. 11595-11598

[18] H.M.S. Kingston, H.J. Haswell, 1997. Microwave-Enhanced Chemistry: Fundamental, Sample Preparation, and Applications, American Chemical Society.

[19] Taylor, M. Atri., S.S. dan Minhas, S. 2005. Developments in microwave chemistry. Evalueserve.

[20] Stuerga, D. 2006. Microwaves in Organic Synthesis, Second Edition. Wiley-VCH.

[21] Singh, R., Tiwari, S., Srivastava, M. dan Shukla, A. 2014. Microwave Assisted Alkali Pretreatment of Rice Straw for Enhancing Enzymatic Digestibility. Hindawi Publishing Corporation Journal of Energy, Vol. 2014, pp. 1-7. DOI: http://dx.doi.org/10.1155/2014/483813

[22] Balcu, I., Macarie, C.A., Segneanu, A.E. dan Pop, R.O. 2011. Combined Microwave-Acid Pretreatment of the Biomass. Progress in Biomass and Bioenergy Production.

[23] L. Dawson and R. Boopathy, 2008. Cellulosic Ethanol Production From Sugarcane Bagasse Without Enzymatic Saccharification, Bio Resources 3(2), pp. 452-460

[24] Y. Wu, Zhang, C., Liu, Y., Fu,Z., Dai, B. dan Yin, D. 2012. Biomass Char Sulfonic Acids (BC-SO3H)-Catalyzed Hydrolysis of Bamboo under Microwave Irradiation. Bioresources. 7(4), 5950-5959.

[25] Lidstroem, P., Tierney, J., Wathey, B. dan Westman, J. 2001. Microwave Assisted Organic Syntesis-a Review. Tetrahedron. Vol. 57, pp. 9225-9283.

[26] I.N. Sudiana, S. Mitsudo, K. Sako, S. Inagaki, L. O. Ngkoimani, I. Usman, H. Aripin, The microwave effects on the properties of alumina at high frequencies of microwave sintering, American Institute of Physics (AIP) Conference Proceeding, **1719**, 030014 (2016); doi: 10.1063/1.4943709

[27] I. N. Sudiana, S. Mitsudo, M. Z. Firihu, H. Aripin, Effect of High-Frequency Microwave on Micro Hardness of Alumina Ceramic, Material Science Forum , Vo. 872, 2016, pp.114-117.

[28] H. Aripin, S. Mitsudo**,** I Nyoman Sudiana, Bambang Sunendar , Svilen Sabchevski, Structural Characterization of a Glass Ceramic Developed from TiO_2 and a Novel Material-Silica Xerogel Converted from Sago Waste Ash, Material Science Forum Vol. 872, 2016, pp.81-86.

[29] P. Alvira, E. Tomás-Pejó., M. Ballesteros, and M.J. Negro, 2010. Pretreatment Technologies for an Efficient Bioethanol Production Process Based on Enzymatic Hydrolysis: A review. Bioresource Technology, 101(13), pp. 4851–4861.

Air Pollutants Minimalization of Pollutant Absorber with Condensation System

Yayat Ruhiat*, Firmanul Catur Wibowo, Yuvita Oktarisa

Dept. of Physics Education, Faculty of Teacher Trainings and Education
University of Sultan Ageng Tirtayasa

Email: yruhiat@untirta.ac.id

Abstract. Industrial development has implications for pollution, one of it is air pollution. The amount of air pollutants emitted from industrial depend on several factors which are capacity of its fuel, high chimneys and atmospheric stability. To minimize pollutants emitted from industries is created a tool called Pollutant Absorber (PA) with a condensing system. Research & Development with the approach of Design for Production was used as methodology in making PA. To test the function of PA, the simulation had been done by using the data on industrial emissions Cilegon industrial area. The simulation results in 15 years period showed that the PA was able to minimize the pollutant emissions of SO2 by 38% NOx by 37% and dust by 64%. Differences in the absorption of pollutants shows the weakness of particle separation process in the separator. This condition happen because the condensation process is less optimal during the absorption and separation in the separator.

1. Introduction

The development of industrial gives implication to several aspects, which are the speed of air pollution, odor, vibration, noise, and water pollution. Based on the result studies in some industry areas, air pollutant increase significantly. In 1960-2010 the concentration of Corbon Dioxide pollutant rised from 320-390 ppm, along this period the rate of CO_2's concentartion increased 20,5 % . By this situatiion the emition of CO_2 from industrial will be increased 0,4 % every year. The kind of air pollutants which are produced by industry depend on the category of fuel used. The bigger capacity of fuel, the higher chimney used [1]. The height of the chimney used by industry affect to the separation's distance of its emission [2]. The higher of the chimney used, the further distance of separation of pollutant, which emitted by industry [3]. Meanwhile, the capacity of pollutants emitted depends on the stability of atmosphere condition on that area. Area with unstable atmosphere condition (A), the concentration of pollutant emitted is high, otherwise if the area is on stable atmosphere (E), the concentration of pollutant emitted is low [4]. The large size of air pollutants emitted by industry in various stabilities of atmosphere forces an effort to minimize the emission; one way, which can be done is by doing modification of chimney.

Decreasing polution by modification of chimney was used by adding Electricity Cataly Oxidation (ECO) SO_2 and electricity coagulation. Some research about pollutant absorbing have been done, for instance Pollutant detection by absorbption using mie scattering and topographic targets as retroreflectors [5], Effects on air pollutant removal by plant absorption and absortion [6], Environmental Samples Analysis by atomic absorption spectometry and inductively couple plasma-optical emission spectroscopy [7], and Light with AERONET measuremnets [8]. Most of the previous

Content from this work may be used under the terms of the Creative Commons Attribution 3.0 licence. Any further distribution of this work must maintain attribution to the author(s) and the title of the work, journal citation and DOI.
Published under licence by IOP Publishing Ltd

reseach focus on reaction, the speed of component's disolacement from gas to liquid (absortion), and efficiency of gas control (absortion) which have been done by using tools for contaminent in the form of gas (absorber). Furthermore, model-related absorption of pollutants insome atmospheric conditions have been conducted by several researchers, including: the effect on the recovery model of the flow of CO_2 in the absorption of CO_2 by K_2CO_3 solution in a packed column with a non-isothermal conditions [9]. Furthermore, research regarding on apsortion pollutant model have been done by several researcher, for example the effect of flowing model on CO_2 recovery on the absortion of CO_2 in the methyldiethanolamine which promotor piperazine in coloumn packed [10]. After that, a new approach to estimating surface facility costs for shale gas development [11. Based on several work system tools, which already exist, Pollutant Absorber (PA) with condensation system was made to reduce air pollutant. PA is one of control technic used to absorb the pollutant before flow out from the chimney, to maximum the efficiency, this technic working based on condensation system. The different between PA with previous tools can be seen from the function of its, PA can absorb and separate pollutant based on different fuel. All of the pollutant are absorbed and after that separated by using separator, and finally these pollutants absorbed.

2. Material and Method

The methodology of this research is Research & Development technique with Design For Production (DFP) approach. The research methodology using a DFP approach using stratified system perspective of manufacturing activities to guide the product development process [12]. Tools used in this research are *softwarae screen3* dan *WrPlot.* To get information about the concentrate of the pollutant which emitted, data about industrial emition is needed, meanwhile to get information about the distance of pollutant separation, climate's data is needed. The process of working reseach can be seen from Figure 1.

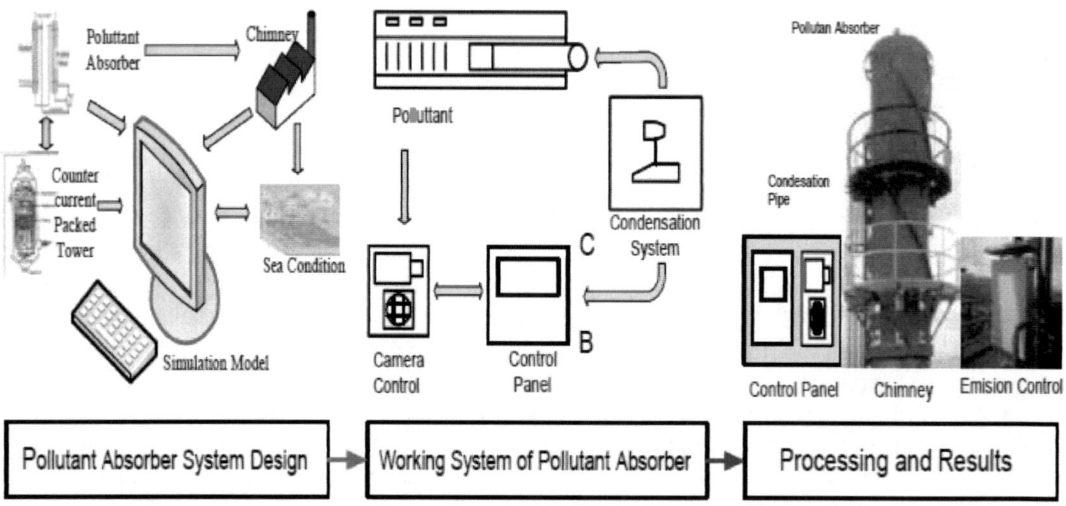

Figure 1. Design and control of minimization of air pollutant

Based on Fig. 1 above, the design and simulation tool based on the study of industrial chimneys and pollutant absorber mounted on the chimney. Based on the research approach, the treatment of the pollutants carried out in several stages, as shown in Figure 2. Based on these images, pollutants coming out of the chimney is absorbed by the filter accommodated on chamber1, then these pollutants are separated on chamber2, then any kind of pollutants are stored in chamber3, The flow of pollutants in each sub appliance, detected using an instrument of control. To keep the flow of particles in each pipe is made by condensing system. PA working system absorbs all kinds of pollutants coming out of the chimney. Pollutants were arrested filter absorber accommodated in the chamber through a capillary tube. To maintain, modify, and change the smoke particles to separate into charged ions created by the condensing system to maintain the energy through the process enthalpy.

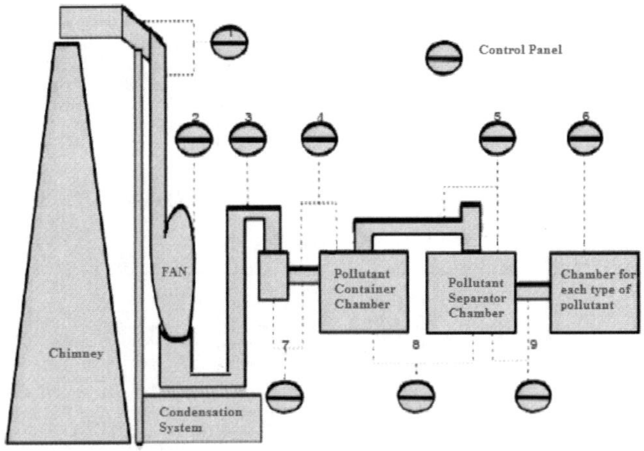

Figure 2. Reduction of air pollutants control design

3. Results and Discussion

PA is a tool that serves to absorb all kinds of pollutants emitted from industrial chimneys. In absorbing pollutants emitted, this tool works with the absorption process. The absorption process is the process of absorption of chemical pollutants. The process of absorption in a packed column is controlled by diffusion with fast chemical reaction in the liquid film, while the in tray columns controlled by physical absorption [13]. Pollutants from chimney absorbed by using absorber, the tool works by three processes namely Absorption, Separation and Adsorption (ASA). To test the function of the tool, simulation using data on industrial emissions industrial estate in Cilegon and climate data started Year 2005-2015 of Meteorology and Geophysics (BMKG) Serang.

To analyze the concentration of pollutants emitted every industry use SCREEN3 software. On the software required input data on climate, namely: atmospheric stability and wind speed. Based on climate data from BMKG Serang over a period of 10 years atmospheric stability obtained in Cilegon is currently on the stability of the A to E with an average wind speed of 2.4 m / sec. Using data from industrial emissions in Cilegon industrial area, as well as the climatic conditions in the region, obtained by the difference between the industry by using a control technique of absorption with no control of the absorption of pollutants. Results SCREEN3 software running distance and pollutant concentrations in two different situations that shown in Figure 3 below. Based on the picture, it appears there is a big difference between the air pollutant concentrations of industries that use control techniques by using techniques that do not control. Using data from atmospheric stability and wind speed are the same, obtained by two different patterns. Industries that do not use a control technique of absorption (Figure 3A) the concentration of air pollutants at first small, but the concentration is on the rise. The maximum pollutant concentrations occur at a distance of 7000 meters, then decomposed by the wind pollutant concentrations resulting in decreased concentration. Meanwhile, for industries that use control techniques absorption of air pollutants (Figure 3B) concentrations of air pollutants emitted steadily declining. However, at a distance of 0-500 meters there is increased concentration of air pollutants, it is possible PA mounted on a chimney is not working optimally. Another thing that happened from the look of the picture above, running SCREEN3 done with the assumption of new industry operates. Furthermore, the distribution range of pollutants, but depending on atmospheric stability also depends on wind speed and height of the chimney. The greater the wind speed, the concentration of pollutants around the chimney increases. This is because polluters will more quickly reach certain locations because of the wind. Additionally, strong winds will be faster deflect the current polluters moved up so that the trajectory of the movement of pollutants will be closer to the ground surface [4, 14].

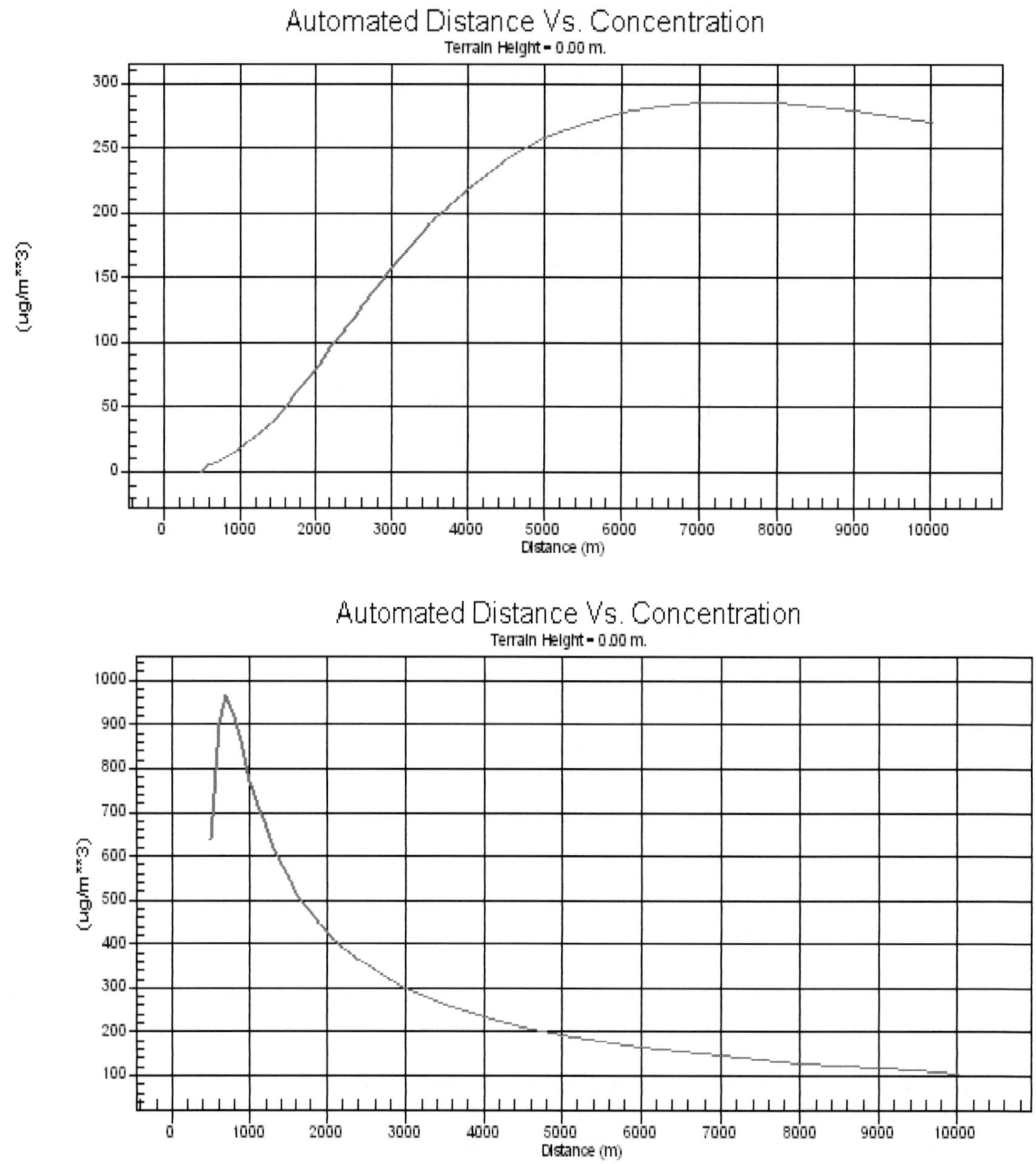

(a) Without control (upper) (b) Using the techniques of control (bottom)

Figure 3. Comparison between the distance and the concentration air pollutants

To test the reliability of the control apparatus functions pollutant PA validated by using the location as a case study of PT Chandra Asri in Cilegon industrial area. The industry is used as an indicator of measurement, because the industry is already using a chimney with absorption control techniques. The results of the analysis of pollutant concentrations emitted by the PA was then compared with the results of the measurement. This was done to test the reliability of the control techniques PA absorb air pollutants, the results of the comparison between models and measurements, is shown in Figure 4. Based on these images, pollutants emitted from industries first large concentration, but after going through the process of ASA pollutant concentrations to be reduced , the high concentration of pollutants in the beginning of the simulation, because the use of environmental boundary conditions assuming an open chimney.

Figure 4. Model results once there absorbent

Furthermore, to determine the function of the PA to absorb pollutants, simulations conducted with data input discharge emission measurement results in Cilegon Industrial Zone in the period from January to July 2014. The simulation was conducted to determine the effectiveness of PA installation in minimizing the emissions of air pollutants. Additional information on thesesimulations, will predict the concentration of air pollutants emitted from industries. The type of air pollutant simulated using PA namely: sulfur dioxide (SO2), sodium oxide (NOx) and dust. These pollutants, would have predicted concentration in a span of 30 years, from 2014 to 2043 year. The prediction of pollutant emission is shown in Figure 5.

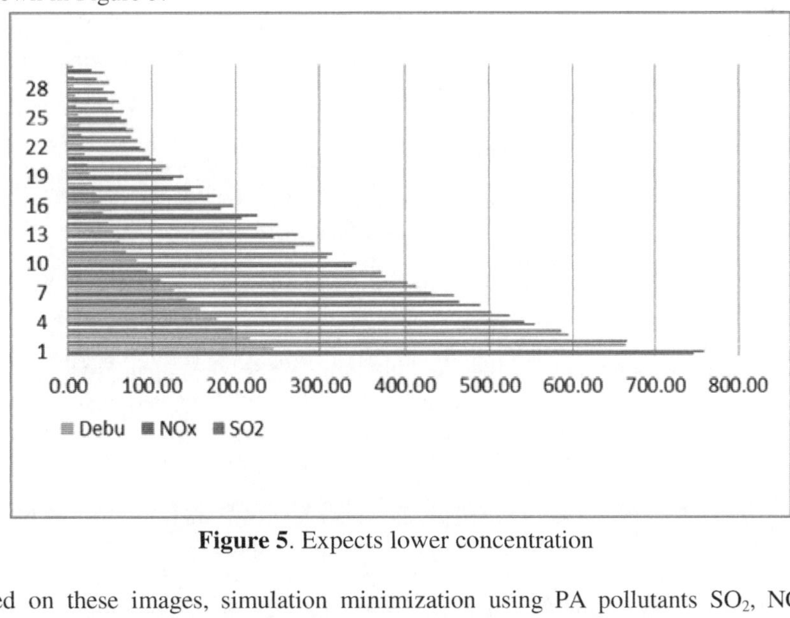

Figure 5. Expects lower concentration

Based on these images, simulation minimization using PA pollutants SO_2, NOx and dust every year cuts. SO_2 pollutant emissions at the time of measurement of 2014 amounted to 747.00 ug / m3 in a span of 15 years PA is able memnimalisasi pollutant emissions of SO_2 by 38%. Meanwhile, when the NOx pollutant emissions measurements in the same year amounted to 758.73 ug / m3 during that time, the PA is able to minimize NOx by 37%. However, for the pollutant emissions of dust initially at 244.44 ug / m3 in the same time frame, the PA is able to minimize by 64%. Based on simulation results, it appears that the pollutant concentrations which do not react with other elements, more rapidly absorbed by the PA. PA tool mounted on the chimney industry would be effective if it is used to absorb the dust compared to other pollutants. PA has not optimlanya work function in absorbing all kinds of air pollutants, shows the ineffectiveness of the separator functions to separate the types of pollutants. Weak particle separation process, one possible cause less optimal condensation process on the separator.

The simulation results PA installation in the chimney industry showed a reduction in pollutant emissions. Their means of PA be the solution to minimize pollutant emissions from industry, it will encourage the industry to an area. Every year all industries increased, for industrial goods metals and base metals rose between 15.40 to 15.67% [15]. By successfully and installation of PA on every chimney industry, assumed the environment inside and outside the industry to be clean.

4. Conclusion

PA to the condensing system is a tool to absorb all kinds of pollutants that emitted from industrial chimneys. The tool works by three processes namely Absorption, Separation and Adsorption (ASA). In absorbing pollutants emitted, this tool works with the absorption process. The results of the simulation work function tool, PA is able to minimize the pollutants emitted from industries. In the period of 15 years, PA able to minimize the pollutants SO_2 by 38%, NOx by 37% and dust by 64%. The difference in minimizing pollutant concentrations showed the weakness of the process of particle separation in the separator. This condition happen because the condensation process is less optimal during the absorption and separation in the separator.
Therefore, condensation system for the absorption and separation of contaminants need to continue to be detected by the instrument controls.

Acknowledgments
The work was financially supported by Reseacrh and Comuunity servise of Sultan Ageng Tirtayasa University

References
1. Frooqnia, A., Verdín, C. T., & Sepehrnoori, K. (2016). Inference of Rock Pressure- Production Properties from Gas-Oil Production Logs. *Journal of Natural Gas Science and Engineering*, 124-143, Volume 36, Part A.
2. Hajra, B. (2014). A Review of Some Recent Studies on Buoyancy Driven Flows in an Urban Environment. *International Journal of Atmospheric Sciences*, 1-15, Volume 2014.
3. Park, Y. K., & Chang Wu, S. (2012). Two-dimension Numerical Simulation of Stack Flue Gas Dispersion. *Journal of the Korean Oil Chemists' Society*, 33-39, Volume 29.
4. Ruhiat, Y. (2009). *Model Prediksi Distribusi Laju Penyebaran Pencemar SO2 dan Debu dari Kawasan Industri Cilegon.* Bogor: IPB.
5. Byer, R., & Garbuny, M. (1973). Pollutant Detection by Absorption Using Mie Scattering and Topographic Targets as Retroreflectors. *Applied Optics* , 1-10, Volume 12 No 7.
6. Fujii, S., Cha, H., Kagi, N., Miyamura, H., & Kim, Y. S. (2004). Effects on air pollutant removal by plant absorption and adsorption. *Building and Environment*, 105-112, Volume 40.
7. Popescu, I., Stihi, C., Cimpoca, G., Dima, G., Vlaicu, G., Gheboianu, A., et al. (2009). Environmental Samples Analysis by Atomic Absorption Spectrometry (AAS) and Inductively Coupled Plasma-Optical Emission Spectroscopy (ICP-AES). *Romamnia Journal Physic*, 1-6, Volume 54.
8. Chin, M., Diehl, T., Dubovik, O., Eck, T., Holben, B., Sinyuk, A., et al. (2009). Light absorption by pollution, dust, and biomass burning aerosols: a global model study and evaluation with AERONET measurements. *Annales Geophysicae*, 3439-3464.
9. Kusnarjo, Kuswandi, Susianto, & Altway, A. (2009). Pengaruh Model Aliran terhadap Recovery CO2 pada Absoprsi Gas CO2 oleh Larutan K2CO3 Didalam Packed Column dengan Kondisi Non-Isothermal. *Reaktor*, 154-160, Volume 12 No3.
10. Wiguna, G. A., Nazaruddin, N., Susianto, & Altway, A. (2014). Model Simulasi Absorpsi Gas CO2 dalam Larutan Methyldiethanolamine (MDEA) berpromotor Piperazine (PZ) dalam Packed Column. *Jurnal Teknik Pomits* , 1-6, Volume 2 No. 1.
11. Zhu, L., & Yuan, J. (2016). A new approach to estimating surface facility costs for shale gas development. *Journal of Natural Gas Science and Engineering*, 202-212, Volume 36.

12. Herman, J., & Chincholkar, M. (2002). Reducing Throughput Time During Product Design. *Journal of Manufacturing System*, 1-8, Volume 20 No. 6.
13. Gary, Rochele, T., & Dang, H. (2003). *Carbon Dioxide Absorption Rate and Solubility in Monoethanolamine/Piperazine/Water.* Austin: University of Texas .
14. Ruhiat, Y., Wibowo, F. C., Hasani, A., Kuswantoro, M., & Fahrizal, R. (2016). Spatial and Temporal Model Development for Analysis the Impact of Concentration of Pollutant and Emission to the Air Quality. *Advance Science Letters*, 1-4, Volume 22 No 12.
15. Badan Pusat Statistik. (2016). *Badan Pusat Statistik Indonesia.* Jakarta: BPS.

IOP Conf. Series: Journal of Physics: Conf. Series **846** (2017) 012015 doi:10.1088/1742-6596/846/1/012015

Estimation of Coal Distribution in Tawanga Village, East Kolaka Regency, Southeast Sulawesi by Using DC Resistivity Method

Laode Ngkoimani*[1], Laode Saleh Isa[1], Jahidin[1], Syamsurizal[2], Suryawan Asfar[3], Andi Makkawaru[4]

[1]Department of Geophysical Engineering, Universitas Halu Oleo, Kendari
[2]Physics Department, Universitas Halu Oleo, Kendari
[3]Department of Geological Engineering, Universitas Halu Oleo, Kendari
[4]Energy and Mineral Resource Agency, Southeast Sulawesi

Email: laodem@yahoo.com

Abstract. The purpose of this study is to determine coal distribution and plating in Tawanga Village, Eastern Kolaka Southeast Sulawesi. The measurements were performed in 3 lines which length of every line is 32 m, 64 m, and 80 m by using Geoelectrical method. The data then analyzed by using Res2Dinv software to find out resistivity and rock layer structure. By using this software revealed that resistivity for between 0.0913 and 438 Ωm, 0.552 and 344 Ωm, 0.898 and 271 Ωm for the first line, second line and third line ranging repectively. Based on these results then coal distribution was estimated. That is found that they are around 130-438 Ωm for the first line, 137-344 Ωm for the second line, and 120-271 Ωm for the third line. They distributes at 0,5 - 12,4 meter from northwest to southeast.

1. Introduction

1.1 Background

Tawanga village is located in Eastern Kolaka Southeast Sulawesi Province. Based on geological consideration, there are three rock units in Tawanga [1-2] such as schist units consisting of mica and clorite schist, coal unit, and alluvial unit. Schist unit distribution takes place for 89,80% of all study area, locates in eastern and western part and separated by alluvial unit (Qal). Coal distribution in coal unit locates river area whose wide is around 1,03% of all area and whose thickness of coal is around 22 m based on measurement of A-B geological section. While alluvial unit distribution is about 10,20% and located in middle part [1-2]. However, the detail coal distribution data in this area was not avaliable yet.The distribution is important for mining plan.

Investigation for grounding and lightning prevention usually by using soil electrical resistivity. This method is important to get parameters that reflects physical properties of soil and subsurface structure [3-6]. For measuring vertical alteration of soil resistivity, a Vertical Electrical Sounding (VES) method is commonly applied [7]. There are three configurations of electrode that are applied in VES profiling i.e Wenner, Schlumberger, and dipole-dipole array [8]. Among them, the Wenner array is the more simple and easier to applied [9]. Several previous reports, the application of the Wenner arrays method has been published. It successfully applied to determine soil resistivity and to estimate soil content within 1,50 m from the surface sandy soil [6], detecting the encroachment of saline water bodies into soil [10-12], and for monitoring reclamation of saline soils and salinity in the irrigation projects [13-16]. In addition, the new theory is developed for improving this method, explaining the effects of soil properties such as salt concentration and water content on soil electrical conductivity[17-19]. This method will be used in this investigation. By using this method, coal distribution and plating in Tawanga Village, Eastern Kolaka Southeast Sulawesi will be determined.

Content from this work may be used under the terms of the Creative Commons Attribution 3.0 licence. Any further distribution of this work must maintain attribution to the author(s) and the title of the work, journal citation and DOI.
Published under licence by IOP Publishing Ltd

IOP Conf. Series: Journal of Physics: Conf. Series **846** (2017) 012015 doi:10.1088/1742-6596/846/1/012015

1.2 Geological Consideration

Figure 1 shows the Tawanga Geological Map. Generally, rock formations in this area are arranged severals rock types such as mekongga formation, Tokala formation, Meluhu formation, ultramafic igneous, Pompangeo complex, Langkawa formation, Boepinang formation and alluvial. Based on study of the its geological history, this region was started at pre Permo-carbon age. They are establishing sedimentary rocks and limestone where settled in the inner part of neritic marine. After then the rocks acquired elevation and transformed to Mekongga metamorohic and Palezoikum marble rocks. In Permo-Trias ages, granite infiltrated metamorphic rocks. Meluhu and Tokala formation were precipitated inconsistently on metamorphic rocks in the middle Trias until the end of Trias age.

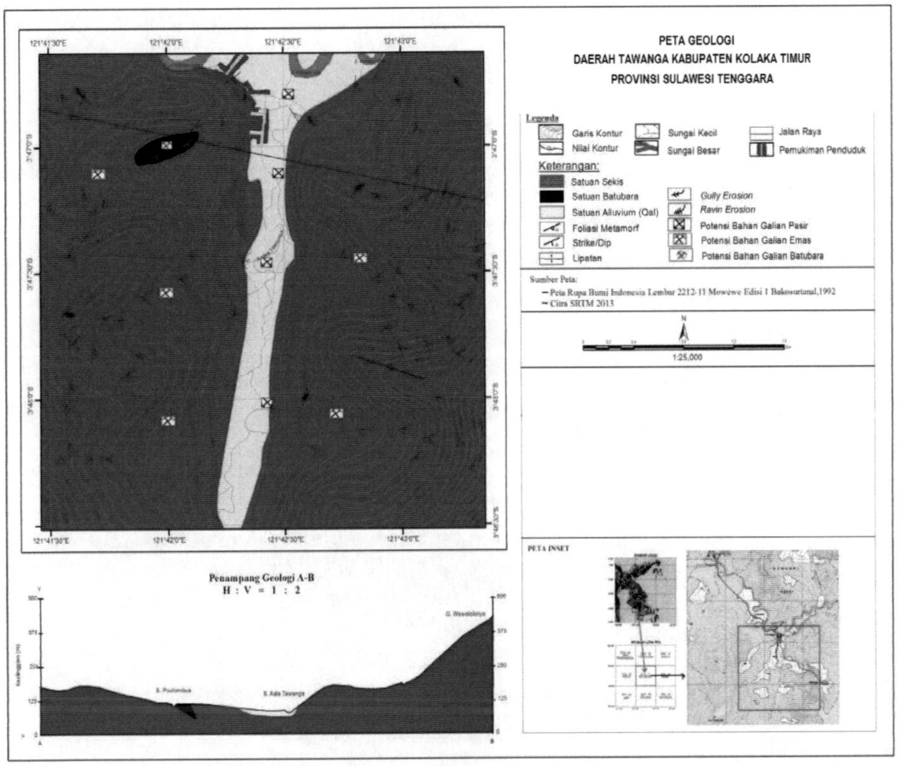

Figure 1. Tawanga Geological Map

Tawanga exists on Southeast Peninsula of Sulawesi so that its geological structure similar to Sulawesi Southeast Peninsula's geological structure which dominated by f Northwest - Southeast drift fault. A fragment of continent in Southeast Peninsula of Sulawesi was named as the continental fragment of Southeast Sulawesi [1-2]. It goes along Northwest- Southeast with length about 205 km and width about 97,5 km. It lay mostly in Sulawesi Southeast Peninsula region.

2. Method

The Tawanga village, Uluiwoi district, East Kolaka Regency, Southeast Sulawesi geographically lay between $3^0 46'39''$ S and $3^0 48'26''$ S and between $121^0 41'30''$ N and $121^0 43'20''$ N. Before data acquisition, we were find out the condition of the area and then geoelectrical experiment method by using Geores Wenner Array device was performed. Four electrodes were injected into the soil surface along a straight at predetermined distances (a) (Fig. 2). The electrical potential of current flow was measured between a pair of inner electrodes (P_1 and P_2 in Fig. 2) while electrical current flowed was measured between a pair of outer electrodes (C_1 and C_2 in Fig. 2). The resistivity measurements performed for Wenner array consist of three lines. The first line namely #TW1 has 32 meter in length with electrode space about 2 meter. The second line namey #TW2 has 64 meter in length with electrode space about 4 meter, and the third line namely #TW3 has 80 meter in length with electrode space about 5 meter.

IOP Conf. Series: Journal of Physics: Conf. Series **846** (2017) 012015 doi:10.1088/1742-6596/846/1/012015

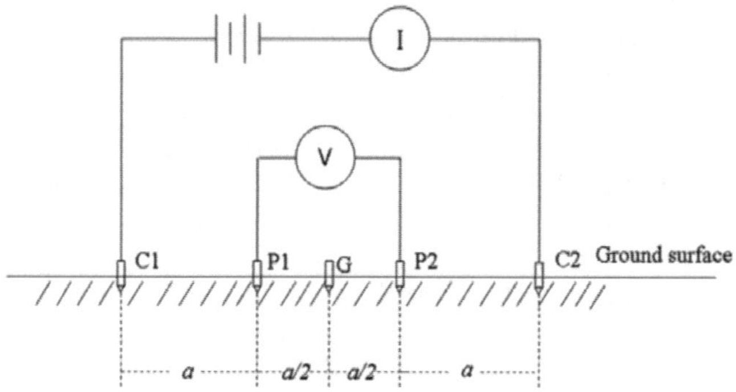

Figure 2. Schematic Diagram of Wenner Array [20]

The potential is determined by using following Equation :

$$V = -\int_0^{\infty} \frac{\rho_o I}{2\pi r_o} dr_o = \frac{\rho_o I}{2\pi r_o} \qquad (1)$$

Data processing was carried out by using Res2dinv software to find out the material resistivity variation in subsurface.

3. Results and Discussion

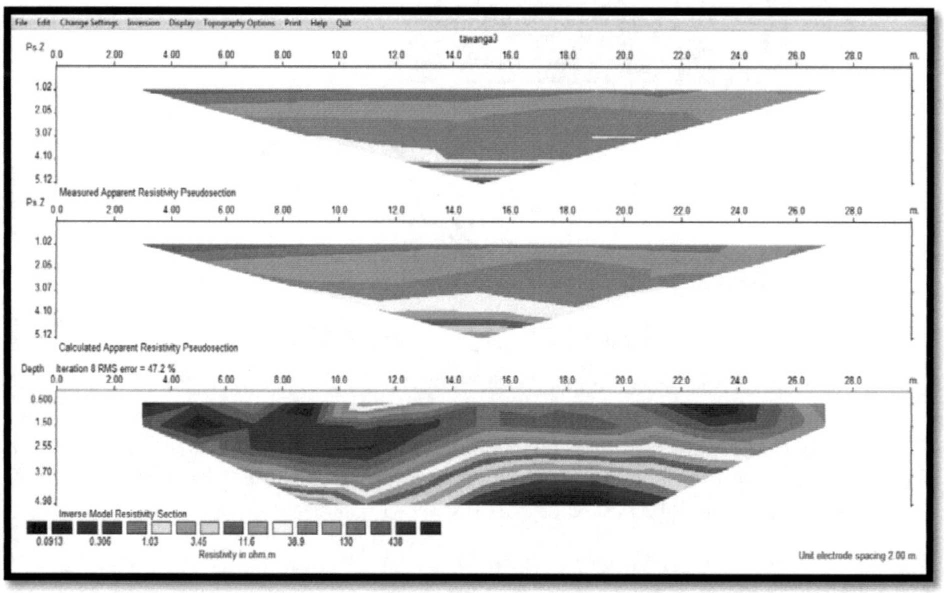

Figure 3. Distribution of Resistivity Value for The First Line (TW1)

Figure 3 shows the contour of the resistivity of TW1 line. The variation of resistivity is from 0.0913 Ωm to 438 Ωm in 0.5 - 4.98 m from surface. Interpretation result of Figure 3 revealed that the distribution of coal resistivity is around 130-438 Ωm. The coal distribution exists in line 3 to 27 m and about 0.5 - 4 m under surface. While the outcrop of coal distribution is 50 - 65 cm in thickness and tends to Northwest - Southeast. The values of resistivity that vary from 0.0913 to 1.03 Ωm are indicated as wet clay, whereas the material with resistivity from 1.03 to 38.9 Ωm are indicated as clay as shown in previous line (TW1).

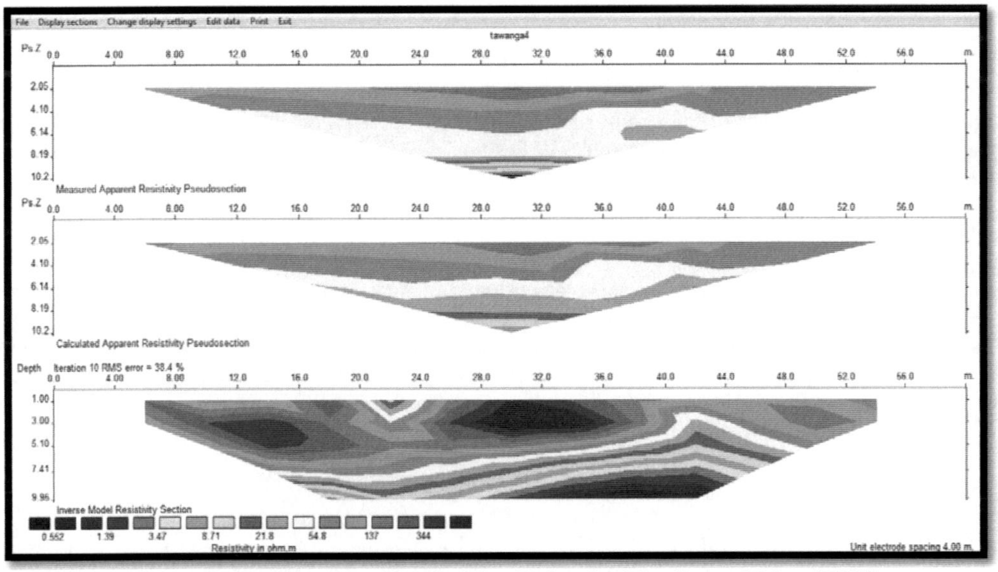

Figure 4. Distribution of Resistivity Value for The Second Line (TW2)

The result of TW2 data processing is shown in Figure 4. It can be seen that resistivity variation is found in interval from 0.552 Ωm to 344 Ωm and between 1 m and 9.95 m in depth. From Fig. 4 can be estimated the coal resistivity distribution. They are about 137-344 Ωm lie at 6 m to 54 m from Nortwest to Southeast and about 1.2 - 8 m undersurface. From this data processing also we found that clays exist at 20 to 24 m, whereas outcrop of coal found in 40 - 45 cm from surface. From the reference data we can conclude that material with resistivity 0,552 to 1,39 Ωm are indicated as wet clay and material with resistivity 1,39 to 54,8 Ωm are indicated as clay.

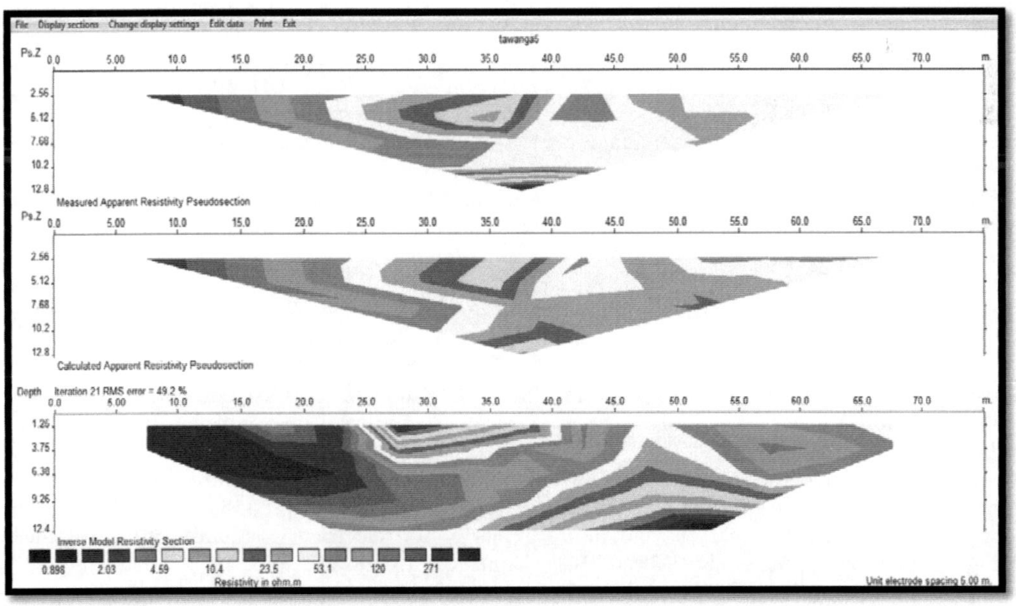

Figure 5. A distribution of The Resistivity Value for The Third Line (TW5)

According to TW3 data processing results (Fig. 5) shows that the resistivity is found from 0.898 Ωm to 271 Ωm in 1.25 m to 12.4 m undersurface. Based on Figure 5, the estimation of coal distribution was performed. There are found that distribution of coal resistivity is in range 120 - 271 Ωm at 1.25 m and 12.4 m tends from Northwest to Southeast . The coal distribution is estimated up to 75 m, laterally, however at 25 - 40 m and 46 - 50 m exists clays. The material resistivities with values from 0,898 to 2,03 Ωm are indicated as wet clay and as clay for resistivity of 2,03 to 53,1 Ωm.

Verma *et al.* have described in detail about coal resistivity anomaly variation using Wenner array. They found that the resistivity values of coal are from 120 to 860 Ωm. They also reported that coal resistivity anomaly variation determined by using 1/2 Schlumberger in range 120 Ωm to 1140 Ωm. Singh *et al.* (2004) estimated coal layer in north part of Jharia in depth of 10 - 31 m from surface. They found that the highest resistivity value in range 989 Ωm to 1632 Ωm. Both of these studies are agree with resistivity values found in Tawanga village. It indicated that Tawanga village has potential coal.

4. Conclusion

The investigation of coal distribution and plating in Tawanga Village, Eastern Kolaka Southeast Sulawesi was successfully performed. The measurements were performed in 3 lines which length of every line is 32 m, 64 m, and 80 m by using Geoelectrical method. A Res2Dinv software was applied for find out resistivity and rock layer structure. The experiment results revealed that resistivity lie between 130-438 Ωm for the first line, 137-344 Ωm for the second line, and 120-271 Ωm for the third line at 0,5 - 12,4 meter from northwest to southeast. The results indicates that Tawanga village has potential coal for mining.

Acknowledgement
We would like to thank Directorate P2M-Ditjen DIKTI for helping in funding this research under the scheme of Penelitian Riset Fundamental in year 2015.

References

[1] Surono. 2013. Geologi Lengan Tenggara Sulawesi. Badan Geologi Kementrian Energi dan Sumber Daya Mineral : Bandung.
[2] Surono., Hartono, U. 2013. Geologi Sulawesi. LIPI Press : Jakarta
[3] Sakamoto, S., 2001. Seasonal and diurnal variation of upper soil resistivity in the Cerro Chascón science preserve. Alma Memo 369, 1–7.
[4] Sakamoto, S., Sekiguchi, T., 2001. Spatial distribution of near-surface soil resistivity in the Cerro Chascón science preserve. Alma Memo 346, 1–9.
[5] Samouëlian, A., Cousin, I., Tabbagh, A., Bruand, A., Richard, G., 2005. Electrical resistivity survey in soil science: a review. Soil & Tillage Research 83, 173–193
[6] Zhu J, Kang H.Z and Gonda, Y. 2007, Application of Wenner Configuration to Estimate Soil Water Content in Pine Plantations on Sandy Land, Pedosphere 17(6) : 801-812.
[8] Rucker, D.F., Glaser, D.R., Osborne, T., Maehl, W.C., 2009. Electrical resistivity characterization of a reclaimed gold mine to delineate acid rock drainage pathways. Mine Water and the Environment 28, 146–157.
[7] Nejad, H.T., 2009. Geoelectric investigation of the aquifer characteristics and groundwater potential in Behbahan azad university Farm, Khuzestan Province, Iran. Journal of Applied Sciences 9, 3691–3698.
[10] Rhoades, J. D. and Ingvalson, R. D. 1971. Determining salinity in field soils with soil resistance measurements. Soil Science Society of America Proceedings. 35: 54-60.
[11] Rhoades, J. D., Manteghi, N. A., Shouse, P. J. and Alves, W. J. 1989b. Estimation of soil salinity from saturated soil-electrical conductivity. Soil Science Society of America Journal. 53: 428-433.
[12] Rhoades, J. D. and Oster, J. D. 1986. Solute content. In Klute, A. (ed.) Methods of Soil Analysis. Part 1. Physical and mineralogical methods. America Society of Agronomy, Madison. pp. 985-1006.
[13] Corwin, D. L. and Lesch, S. M. 2003. Application of soil electrical conductivity to precision agriculture: Theory, principles, and guidelines. Agronomy Journal. 95: 455-471.
[14] Corwin, D.L., 2008. Past, present and future trends of soil electrical conductivity measurement using geophysical methods. In: Allred, B.J., Daniels, J.J., Ehsani, M.R. (Eds.), Handbook of Agricultural Geophysics. CRC Press, Boca Raton, pp. 17–44
[15] Corwin, D.L., Lesch, S.M., 2003. Application of soil electrical conductivity to precision agriculture: theory, principles, and guidelines. Agronomy Journal 95, 455–471.
[16] Mori, Y., Hopmans, J. W., Mortensen, A. P. and Kluitenberg, G. J. 2003. Multi-functional heat pulse probe for the simultaneous measurement of soil water content, solute concentration, and heat transport parameters. Vadose ZoneJournal. 2: 561-571.
[17] Rhoades, J. D.,Waggoner, B. L., Shouse, P. J. and Alves, W. J. 1989c. Determination of soil salinity from soil and soil-paste electrical conductivity: Sensitivity analysis of models. Soil Science Society of America Journal. 53: 1368-1374.

[18] Hendrickx, J. M. H. and Kachanoski, R. G. 2002. Nonintrusive electromagnetic induction. In Dane, J. H. and Topp, G.C. (eds.) Methods of Soil Analysis. Part 4. Physical Methods. SSSA Book Series No. 5. Soil Science Society of America, Madison. pp. 1036-1297.

[19] Kachanoski, R. G., Hendrickx, J. M. H. and DeJong, E. 2002. Section 3.1.3.9. Electromagnetic induction. In Dane, J. H. and Topp, G. C. (eds.) Methods of Soil Analysis. Part 4. Physical Methods. SSSA Book Series No. 5. Soil Science Society of America, Madison. pp. 497-545.

[20] L. Song, Jiaojun Zhu, Qiaoling Yan, Hongzhang Kang, 2012, Estimation of groundwater levels with vertical electrical sounding in the semiarid area of South Keerqin sandy aquifer, China, Journal of Applied Geophysics 83 (2012) 11–18

[21] Verma, R.K., and Bhuin, N.C. 1979. Use of Electrical Resistivity Methods for Study Of Coal Seams in Parts of The Jharia Coalfield, India. Geoexploration, 17 :163-176

[22] Singh, K. K.K., Singh, K. B., Lokhande, and Prakash, A. 2004. Multieletrode Resistivity Imaging Technique for the Study of Coal Seam. Journal of Scientific and Industrial Research 63 : 927-930

IOP Conf. Series: Journal of Physics: Conf. Series **846** (2017) 012016 doi:10.1088/1742-6596/846/1/012016

Near-Surface Fault Structures of the Seulimuem Segment Based on Electrical Resistivity Model

Nazli Ismail[1, 3,*], Muhammad Yanis[2], Syafrizal Idris[2], Faisal Abdullah[1], Bukhari Hanafiah[1]

[1)] Laboratory of Geophysics, [2)]Department of Physics, Faculty of Mathematics and Natural Sciences, Syiah Kuala University,Jl. Syech Abdurrauf, Kopelma Darussalam, Banda Aceh, 23111, Indonesia.
[3)] Tsunami and Disasters Mitigation Research Center (TDMRC), Syiah Kuala University

* Email: nazli.ismail@unsyiah.ac.id

Abstract. The Great Sumatran Fault (GSF) system is arc-parallel strike-slip fault system along the volcanic front related to the oblique subduction of the oceanic Indo-Australian plate. Large earthquakes along the southern GSF since 1892 have been reported, but the Seulimuem segment at the northernmost Sumatran has not produced large earthquakes in the past 100 years. The 200-km-long segment is considered to be a seismic gap. Detailed geological study of the fault and thus its surface trace locations, late Quaternary slip rate, and rupture history are urgently needed for earthquake disaster mitigation in the future. However, finding a suitable area for paleoseismic trenching is an obstacle when the fault traces are not clearly shown on the surface. We have conducted geoelectrical measurement in Lamtamot area of Aceh Besar District in order to locate the fault line for paleoseismic excavation. Apparent resistivity data were collected along 40 m profile parallel to the planned trenching site. The 2D electrical resistivity model provided evidence of some resistivity anomalies by high lateral contrast. This anomaly almost coincides with the topographic scarp which is modified by agriculture on the surface at the northern part of Lamtamot. The steep dipping electrical contrast may correspond to a fault. However, the model does not resolve well evidences from minor faults that can be related to the presence of surface ruptures. A near fault paleoseismic investigation requires trenching across the fault in order to detect and analyze the geological record of the past large earthquakes along the Seulimuem segment.

1. Introduction

Aceh Province is the northern part of Sumatra passed by the Great Sumatran Fault (GSF) System. This active fault extends along 1900 km from the southern part of Sumatra to Andaman Sea. The fault is split up in several segments [1]. Most of the segments have shown their seismic activities within the last century with magnitudes above 6.0 and 7.7, except for the Aceh and Seulimuem segments in the northern part of Sumatra [2]. The seismic gap behave along these two segments for more long time could be suspected to be a silent enemy for thousands people in Aceh Province.

The slip rate of the two northern segments of the GSF are predicted 34 mm per year [3] [4] [2]. An intensive research of GPS geodetic observation crossing the Seulimuem segment suggest that the accumulated seismic moment in 170 years correspond to an earthquake of magnitude 7 [5]. Recently, in 2013 there were two events occurred close to the Seulimuem segment, i.e. Tangse (October 22nd) and Bener Meriah (July 2nd), with magnitude 5.6 and 6.2, respectively. Both events are considered small, but they caused many victims and heavy damaged.

Content from this work may be used under the terms of the Creative Commons Attribution 3.0 licence. Any further distribution of this work must maintain attribution to the author(s) and the title of the work, journal citation and DOI.
Published under licence by IOP Publishing Ltd

In order to reduce risks affected by such disasters in the future, it is important to study earthquake history that has been occurred along the fault segment. By understanding the geologic structures of the fault, when the fault last displaced can be estimated and the magnitude of the earthquake that generated by the previous displacement can be predicted. By this so called paleoseismic study, it can also investigate periodicity of large earthquakes occurred along the fault.

Trenching is a common technique for investigating an active fault with a surface rupture, carefully excavating a series of trenches to reveal the structure and magnitude of the fault displacement. However, very few paleoseismic studies have been done on land in Aceh. Most of research on paleoseismic and paleotsunami studies have been reported form the coastal land in Aceh Province. The studies mostly focused on prehistoric earthquakes generated from the subduction area beneath the Indian Ocean [6] [7].

Since the Seulimuem segment is considered to be seismic gaps. An intensive geological study of the fault and thus its surface trace locations, late Quaternary slip rate, and rupture history are needed urgently. However, successful of paleosesimic investigation is influenced strongly by a proper area selection for trenching. Geophysical methods are believed to be fast and cheap method that can be used for locating the trenching site.

Based on field and geomorphic observation, we have selected Lamtamot, District of Aceh Besar for candidate paleosesimic excavation site (Figure 1). Some geological features indicated as fault traces are found in Lamtamot. At the southern part of the village we found fault trace outcrop cut perpendicular the Seulimuem segment. About 500 meter to the north from this outcrop, we also found a 7.6 meters height wall expected as fault scarp directed parallel to the Seulimuem Segment (see inset (B) in Figure 1). Although the fault features are clear, both of the sites are situated at topographic area that causes some difficulties for further excavation.

At the northern part, topography of the area is almost flat, where it is suitable for trenching site. However indication of fault line is still uncertain. Prior to the excavation, we have done geoelectrical measurement for near surface mapping.

Figure 1. Area of study in Lamtamot village.

2. Method

Area of study is paddy field of Lamtamot village situated at southern of Seulawah Agam volcano District of Aceh Besar (about 100 km to the east from Banda Aceh). The area is mostly covered by fluvial sediment, however about 50 centimeters of the uppermost layer is considered as disturbed layer by agricultural activities.

Two profiles in east-west direction crossing perpendicularly to the suspected fault line were selected for geoelectrical data measurement (Figure 1). Length of each profile is 42 meters with electrodes spacing 1.5 meters. The Wenner-Schlumberger electrodes configuration were selected for data acquisition [8]. The data collection was carried out with the help of multi-electrode electrical resistivity imaging system.

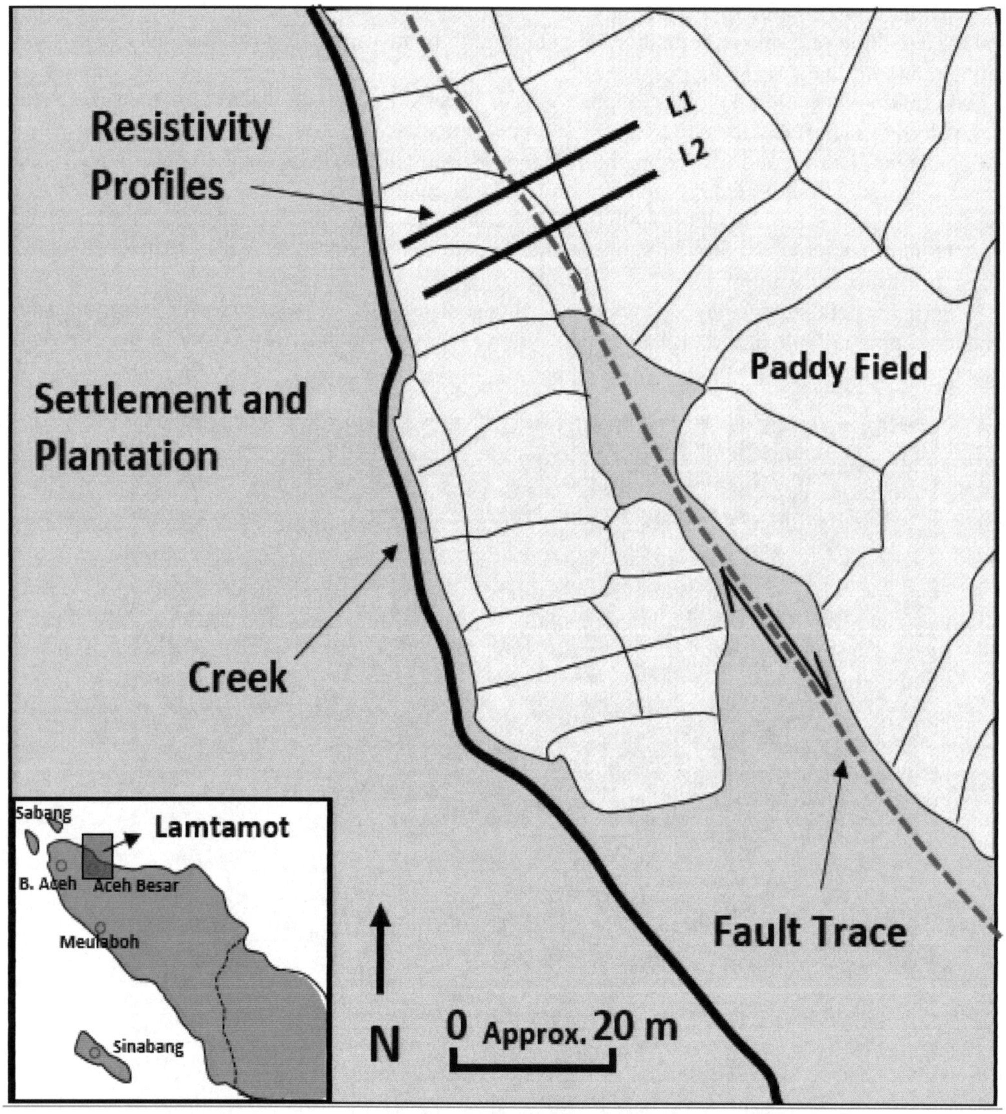

Figure 2. Field measurements

3. Results and Discussion

Apparent resistivity data collected from the field provides a simple image of electrical resistivity pseudo section. However it is not representing the true distribution of subsurface resistivity (the figures are not shown in this paper). True resistivity models were calculated using RES2DINV developed by Loke [8].

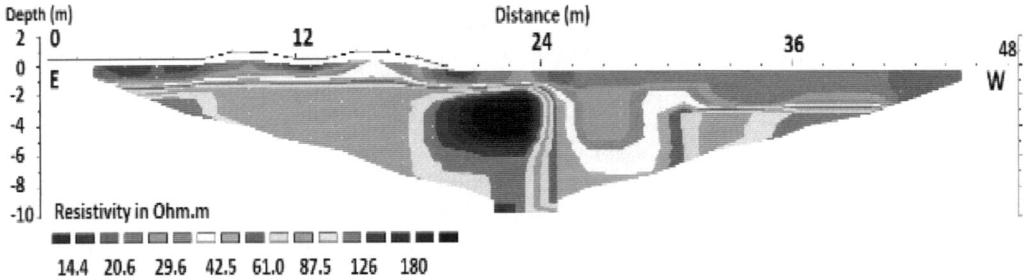

Figure 3. Predicted electrical resistivity model of Profile 1

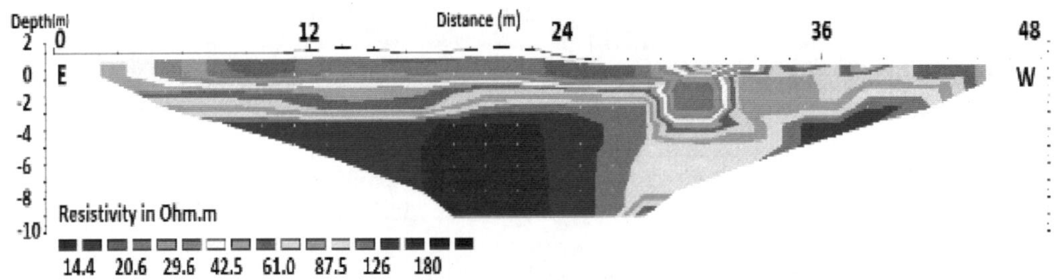

Figure 4. Predicted electrical resistivity model of Profile 2

Figures 3 and 4 show predicted electrical resistivity model derived from Profile 1 and Profile 2, respectively. For Profile 1, the inversion process stops after 6th iteration with absolute error model of 2.0 percent. While for Profile 2, the absolute error model of 1.5 percent was reached after 6th iterations.

The inverted models obtained from both profiles show a good agreement of electrical resistivity values distribution. The models are generally made up by three resistivity layers. The uppermost layer shows a relatively thin (i.e. < 0.5 m) and low resistivity values (i.e. < 40 Ohm-m). The middle layer dominated by moderate resistivity values (i.e. 40 – 100 Ohm-m) overlain by the layer with relatively high resistivity values (i.e. above 100 Ohm-m).

The first layer with relatively high resistivity values represent as fluvial sediment made up by clay and sand as found at the surface. This sediments are considered as disturbed layer by agriculture activities. The medium resistivity layer is interpreted as sand and gravel deposits which is proved from coring information using hand Auger at the site.

Both profiles at this site shows a relatively high resistivity contrast formed vertically at distances 24 and 30 meters for Profile 1 and Profile 2, respectively. This anomaly roughly coincides with the topographic scarp of fault deduced from the southern part of the site as shown in Figure 1. However, the model does not resolve well evidences from minor faults that can be related to the presence of surface ruptures.

On the right side of the profiles, a gently dipping electric contrast could correspond to a stratigraphical discontinuity such as the erosion surface overlain by sands and gravels observed in the neighboring outcrops. The 'U' shape of low resistivity values may correspond to old channel situated beside the fault scarp. The channel is directed in north-south following stepping of the surface elevation from the high land in the northern part where the Seulawah Agam Volcano located (see Figure 1).

Based on the estimated resistivity models obtained from both profiles, we believe that further excavation is possible to be dug at this site. The predicted buried scarp found at the middle part along the profiles may record sequences of historical ruptures that occurred in the past.

4. Conclusion

Application of geoelectrical resistivity method for locating near surface structures of Seulimuem segment has been proven its advantages at Lamtamot area. The 2D electrical resistivity model provided evidence of some resistivity anomalies by high lateral contrast. This anomaly almost coincides with the topographic scarp which is modified by agriculture on the surface at the northern part of Lamtamot. The steep dipping electrical contrast may correspond to a fault. However, the model does not resolve well evidences from minor faults that can be related to the presence of surface ruptures. A near fault paleoseismic investigation requires trenching across the fault in order to detect and analyze the geological record of the past large earthquakes along the Seulimuem segment.

Acknowledgments

This study is part of research project of Penelitian Unggulan Unsyiah (P2U) funded by DIPA Satker of Syiah Kuala University in 2015. The authors would like to thank Prof. Hiroyuki Tsutsumi and Yoshio Soeda for their support. The authors are also grateful to all those who helped during the field work in Lamtamot, Aceh Besar.

References

[1] K. Sieh dan D. Natawidjaja, "Neotectonics of the Sumatran fault, Indonesia," *Journal of Geophysical Research,* vol. 105, no. B12, pp. 295 - 326, 2000.
[2] O. Bellier, M. Sebrier, S. Pramumijoyo, T. Beaudouin, H. Harjono, I. Bahar dan O. Forni, "Paleoseismicity and seismic hazard along the Great Sumatran Fault (indonesia)," *Journal of Geodynamics,* vol. 24, pp. 169 - 183, 1997.
[3] D. Natawidjaja dan K. Sieh, "Slip rates along the sumatran trans current fault and its tectonics significance," dalam *Conference on Tectonic Evolution of Southeast Asia*, London, 1994.
[4] K. Sieh, J. Zachariasen, Y. Bock, L. Edwards, F. Taylor dan P. Gans, "Active tectonics of Sumatra," dalam *Geological Society of America 1994 annual meeting.*, 1994
[5] T. Ito, E. Gunawan, F. Kimata, T. Tabei, M. Simons, I. Meilano, A. Agustan, Y. Ohta, I. Nurdin dan D. Sugiyanto, "Isolating along-strike variations in the depth extent of shallow creep and fault locking on the northern Great Sumatran Fault," *Journal Geophysical Research,* vol. 117, 2012.
[6] H. M. Kelsey, S. E. Engelhart, J. E. Pilarczyk, B. P. Horton, C. M. Rubin, M. R. Daryono, N. Ismail, H. D. Hawkes, C. E. Bernhardt dan N. Cahill, "Accommodation space, relative sea level, and the archiving of paleo-earthquakes along subduction zones," *Geology,* vol. 43, no. 8, pp. 675 - 678 , 2015.
[7] K. Sieh, D. Patrick, E. E. McKinnon, J. E. Pilarczyk, C. Hong-Wei, B. Horton, C. M. Rubin, S. Chuan-Chou, N. Ismail, C. . H. Vane dan M. R. Feener, "Penultimate predecessors of the 2004 Indian Ocean tsunami in Aceh, Sumatra: Stratigraphic, archeological, and historical evidence," *J. Geophys. Res. Solid Earth,* vol. 120, 2015.
[8] M. H. Loke, Electrical imaging surveys for environmental and engineering studies– A practical guide to 2D and 3D surveys, 2002.

Deformation Pattern of Nickel Slag Bonding on the Development of Concrete Construction

E. H. Sujiono[1,*], **H. Husain**[1], **M. Mulyadi**[1], **S. Samnur**[2], **M. Arsyad**[1]

[1]Laboratory of Material Physics, Department of Physics, Universitas Negeri Makassar, Makassar Indonesia

[2]Department of Mechanical Engineering, Universitas Negeri Makassar, Makassar Indonesia

*E-mail: e.h.sujiono@unm.ac.id

Abstract. This paper presents an experimental work to study the deformation and compressive strength on Portland cement concrete with nickel slag aggregate. The amount of nickel slag varied were towards the total mass of coarse aggregate are 0%, 20%, 40%, 60%, 80%, and 100%, respectively. Each variation of the samples was made with a dimension of 15 cm x 15 cm x 15 cm, and then through the curing process. After 28 days, the sample was checked using mechanical testing conducted to investigate the compressive strength. The surface of the concrete fracture after mechanical testing process shows that the bonding between the matrix of Portland cement and nickel slag is a very strong. The bonding has connected very well. Therefore, when the force was given, then the fractions of nickel slag aggregate will hold the connectivity until to the maximum of the pressure force value before the materials are a damaged. The maximum of pressure force caused by the cracks will follow the fracture pattern of the concrete materials. This indicates that the bonding between matrix Portland cement and nickel slag has become the key factor in construction high-quality concrete.

1. Introduction

The development of science in the field of concrete technology allows the use of solid waste as the filling material of concrete forming [1]. Concrete is the building material which is often used in the construction world. Besides being cheap and easy to obtain, the mechanical strength of concrete is one of the reasons why concrete becomes the first choice in the construction industry. The type of aggregate is one of the factors that affect the concrete's mechanical strength [2]. The compressive strength of concrete can also be influenced by the shape, and the size of the sample tested. Many studies reveal that the bigger the cross-section size of the specimen tested, the stronger the compressive strength of that concrete [3]. The mechanical nature of concrete such as tensile, modulus of elasticity, and compressive strength of concrete is very important in designing a building, and the factor which can be optimized to obtain concrete structural strength is the strength of bonding matrix between paste and aggregate [4].

Aggregate is a granular material such as sand, gravel, and crushed stone used as filling materials with cement bonding media. The characteristic of aggregate also affects the planning of concrete design and its economic value. The aggregate used in this study is the gravel of natural stone and added with nickel slag. Nickel slag is the solid waste of nickel industry. Nickel slag contains many elements of silicates so that it can contribute to improving the interface between aggregate and cement paste. Thus nickel slag is very potential to be used as concrete aggregate and as a geopolymer aggregate application [5, 6, 7].

Content from this work may be used under the terms of the Creative Commons Attribution 3.0 licence. Any further distribution of this work must maintain attribution to the author(s) and the title of the work, journal citation and DOI.
Published under licence by IOP Publishing Ltd

In this article, it is discussed in detail the production process, the formation pattern of cracks, and the morphological relation to the surface with the bonding formation between the composite of Portland cement and the nickel slag aggregate and mechanical properties such as the concrete compressive strength with the additional variety of nickel slag aggregate.

2. Material and Method

The material used as the bonding matrix of coarse and fine aggregate is Portland Composite Cement (PCC) type I. The coarse aggregate of nickel slag sized 10-20 mm is obtained through the fraction process of slag nickel chunk. The fine aggregate used in this study is sand of 4.75 mm in sieve size. Based on the production standard of concrete, there are certain limitations on the aggregate gradation, both fine aggregate and coarse aggregate have the gradation and the various level of grain uniformity.

The concrete forming of cement Portland with nickel slag coarse aggregate goes through several stages. First, weighing the required material in accordance with predetermined standard dose. Second, mix all materials together while stirring for ± 3 minutes, adding water as a mixture solvent. Third, the mixture of materials and solvent is put into a molded cube with a dimension of 15x15x15 cm^3. According to reduce pores in the mold, the sample is vibrated for 15 minutes. The mixture composition of material can be seen in table 1.

The 24-hour curing process is done by covering the mold using plaster to maintain the water content of the mixture. Concrete is removed from the mold after 3 days than to maintain the inertia structure; the concrete is soaked for ±7 days. Mechanical testing is done after the sample reaches 28 days.

TABLE 1. Material mixture composition of Portland cement concrete production.

Composition	Portland Cement (kg)	Sand (Kg)	Crushed gravel stone (kg)	Nickel slag aggregate (kg)	Water (Litre)
BSP 0%	1.50	2.20	3.30	0	0.72
BSP 20%	1.50	2.20	2.70	0.67	0.72
BSP 40%	1.50	2.20	2.02	1.34	0.72
BSP 60%	1.50	2.20	1.34	2.02	0.72
BSP 80%	1.50	2.20	0.67	2.70	0.72
BSP 100%	1.50	2.20	0	3.30	0.72

3. Results and Discussion

3.1 The Surface Structure of Portland Cement Concrete Fracture

The bond between Portland cement matrix and nickel slag coarse aggregate is observed through the sample fracture after compressive testing. Compressive strength test is done by giving compressive force, or load with a compressive force on the surface of a sample until a fracture crack has observed.

Figure 1 shows the macroscopic structure of Portland Cement bonded with coarse aggregate for all variations of the addition of nickel slag aggregate and crushed stone. The figure of the morphological surface structure shows that the bond between Portland cement matrix and nickel slag aggregate has been strongly formed.

Figure 1 A is the concrete surface without nickel slag aggregate with material composition BSP 0% as shown in table 1. The fracture pattern on this material surface follows the pattern of crushed stone aggregate which spreads in all part of the sample. It is caused by the bonding pattern between Portland cement composite with crushed stone aggregate is not strong enough to withstand the compressive force was given, so this material has the lowest mechanical properties.

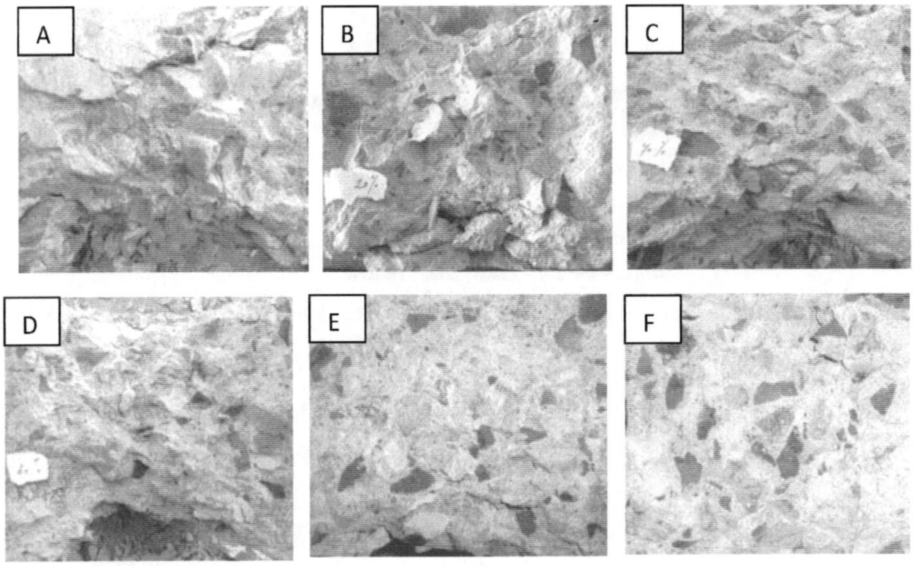

FIGURE 1. The surface structure of Portland Cement Concrete after compressive testing (a) BSP 0%; (b) BSP 20%;
(c) BSP 40%; (d) BSP 60% ; (e) BSP 80% dan (f) BSP 100%, respectively.

Furthermore, in figure 1B, 1C, 1D, 1E, and 1F is a concrete surface with the varied percentage of nickel slag aggregate. The strong bond between nickel slag aggregate and Portland cement matrix makes this material able to maintain fracture until to the maximum limits before nickel slag aggregate crushed. This result will improve compressive strength. The bigger of the percentage of nickel slag aggregate, be affect samples with the biggest value of compressive strength.

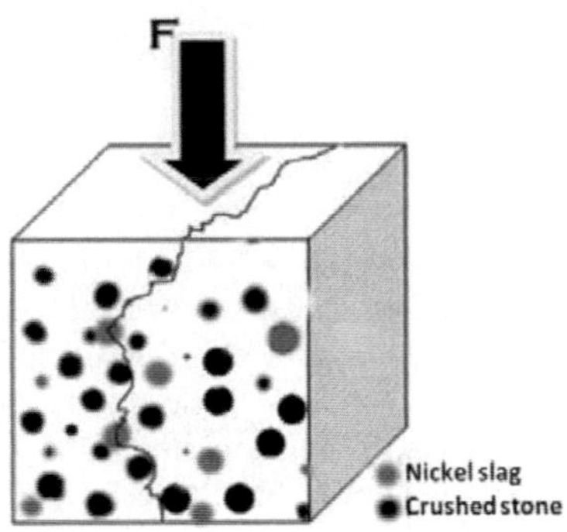

FIGURE 2. Illustration of the fracture on Portland cement concrete sample with the substitution of nickel slag and crushed stone aggregate.

Figure 2 shown the illustration of the occurrence of cracks during the load process of a concrete sample. This illustrates the forming pattern of cracks in load process; the cracks follow the shape of usual stone aggregate. Meanwhile, because the bond between cement matrix and nickel slag is very strong, the fracture occur in the nickel slag aggregate. In contrast to the usual stone aggregate does not bind quite well with Portland cement, therefore if the compressive force is given, the crack will turn to follow the shape of the usual stone aggregate, these results be caused by the lower compressive strength.

FIGURE 3. Portland Cement Concrete Surface with Nickel Slag Aggregate after the compressive test.

Figure 3 shows the crushed nickel slag aggregate due to the withstand of compressive force in the compressive strength test. The strong bond between nickel slag aggregate and Portland cement matrix is formed by the silica contents in the nickel slag surface [5,6,7]. The bigger and more contact between nickel slag surface and cement matrix, make the bond between Portland cement paste and aggregate will also be stronger.

4. Conclusion

The addition of nickel slag aggregate in Portland cement concrete affects the bond between the matrix of aggregate and cement paste matrix, becomes increasingly more strength and compact. These results indicate that the value of concrete compressive strength is better than used of usual stone aggregate. In this study, it is revealed that the mechanical strength of concrete material obtained is higher than the Indonesian nationalize standard for the concrete industry.

Acknowledgments
This research was funded by the Directorate General of Higher Education. Ministry of research, Technology, and Higher Education, through Priority Research Scheme "MP3EI" the fiscal year 2015.

References
1. Bilir, T. "Effects of Non-Ground Slag and Bottom Ash As fine Aggregate on Concrete Permeability Properties." Construction and Building Materials, 26, 2012: 730–734.
2. Maier, P. L., and Durham, S. A. "Beneficial Use of Recycled Materials in Concrete Mixtures." Construction and Building Materials, 29, 2012 : 428–437.
3. Zabihi, N. and Eren, O. "Compressive Strength Conversion Factors of Concrete as Affected by Specimen Shape and Size." Journal of Applied Sciences, Engineering, and Technology, 7, 20, 2014: 4251-4257.
4. Yi, S. T., Yang, E. I. and Choi, J. C. "Effect of specimen sizes, specimen shapes, and placement directions on compressive strength of concrete." Nuclear Engineering and Design, 236, 2006: 115–127.
5. Sugiri, S. "An Application of Nickel Slag as Aggregate and Cement Mixture for High-Quality Concrete." *Jurnal Infrastruktur dan Lingkungan Binaan*, 1, 1, 2005 (*in Indonesia*).
6. E. H. Sujiono, A. Setiawan, H. Husain, A. Irhamsyah, S. Samnur, and S. Subaer. "The influence of nickel slag aggregate concentration to compressive and flexural strength on fly ash-based geopolymer composite." The 3rd International Conference on Advance Materials Science and Technology (ICAMST), Sutikno et.al (Ed.), 2016. ISBN:978-0-7354-1372-6, AIP Conference Proceedings **1725**, 020083 (2016); doi: 10.1063/1.4945537.
7. Yang,T., Yao, X., and Zhang, Z. "Geopolymer prepared with high-magnesium nickel slag: Characterization of properties and microstructure." Construction and Building Materials, 59, 2014, 188–194.

IOP Conf. Series: Journal of Physics: Conf. Series **846** (2017) 012018 doi:10.1088/1742-6596/846/1/012018

Typology of Slum Management in Coastal Settlement as a Reference of Neighborhood Planning in Konawe

Santi [1], Ratna Bachrun[2], Kurniati Ornam[1]

[1]Department of Architecture, Faculty of Engineering, Halu Oleo University, Indonesia,
[2] Environmental Engineering Department, Faculty of Engineering, University of Muhammadiyah Kendari, Indonesia.

* Email: santi_muli@yahoo.co.id

Abstract. Handling slum area actually needs to be done not only in slum areas that are part of a metropolitan city or a large town, but it also need to be conducted in the slum areas in the middle city, small town, and coastal areas. A slum generally includes physical condition, socio-economic conditions of cultural communities who live in the settlements, and the impact of both of these conditions. Environmental coastal settlements in Konawe located in the riverside that lead to loss of quality due to environmental stress of the condition of settlement that are not habitable, can be seen from the problems that exist, namely the existence of housing that goes into river body which causes malfunction of watershed river (DAS), and that is prone to flooding. The purpose of this study is to identify slums in coastal settlements in an effort to improve the quality of structured settlements to obtain a picture of the slum in Konawe, to know the problems regarding the existence of slums and future recommendation of the slum settlement in Konawe. Normative analysis was carried out in this research together with quantitative and qualitative description, as well as methods of spatial analysis (spatial). Identification of slum areas was analyzed based on indicators and parameters issued by the Directorate General of Human Settlements. Problems in some districts in Konawe located in the coastal rivers and coastal areas, consist of drainage, disposal of waste water, waste, and the settlement which is not in accordance with the Spatial Konawe. Among those, there are settlements categorized as slums and some are recommended for refurbishment.

1. Introduction

Handling slum area actually needs to be done not only in slum areas that are part of a metropolitan city or a large town, but it also needs to be done in areas existing slums in medium and small cities. Handling of slum areas in cities large, medium and small cities become a strategic region when it has a direct connection with parts of the metropolitan city such as downtown metropolitan area, metropolitan area growth centers, as well as other regions e.g. industrial zones, trade, warehousing, and offices.

Slum is a problem faced by almost all major cities in Indonesia even big cities in other developing countries. Study of slums generally includes three terms, the first physical condition, and both socio-economic conditions of cultural communities who live in the settlements and the third by the impact of both conditions.

Content from this work may be used under the terms of the Creative Commons Attribution 3.0 licence. Any further distribution of this work must maintain attribution to the author(s) and the title of the work, journal citation and DOI.
Published under licence by IOP Publishing Ltd

The physical condition, among others, appeared on the condition of the building is very dense with low-quality construction, the road network is not patterned and hardened, general sanitation and drainage are not functioning as well as the trash is not managed properly.

Konawe which is one of the districts located in Southeast Sulawesi continues to grow in terms of both development (infrastructure) and the total population. Based on population data Konawe in 2011, the population of Konawe many as 246 798 inhabitants with 679 245 ha area. [2]

The formation of slums, often seen as a potential cause many problems, because it is the source of the emergence of a variety of deviant behavior, such as crime and other social ills sources. There is some identification problems in slum areas Konawe which is located on the outskirts of the river and seafront which resulted in a loss of quality due to environmental stress of the condition of settlement that are not habitable, can be seen from the problems that exist, namely the existence of housing coming into river body that cause the failure of the watershed (DAS), prone to flooding, irregular houses in the coastal ocean.

Another problem is the settlement of the city government's lack of concern about the standardization of housing. The standardization among others for bathing, washing and toilet (MCK), the availability of clean water, availability of air vents, as well as minimum standards for each individual room. The provision of housing for the people by the government less qualified ideal housing and lack of fulfillment of the settlement amount to the community. So that these communities individually or in groups to provide their own settlement.

2. Material and Method

2.1 Basic Theory

Slum settlements are uninhabitable because of the irregularity of the building, high building density levels, and the quality of the buildings and facilities are not eligible. Slum housing is housing that has decreased the quality function as a shelter.

According to Law No. 4 Article 22 of 1992 on housing and settlement, where the slum is a settlement that is unfit for habitation among other things because it is on land that is not in accordance with the designation or zoning, building density is very high in the area are very limited, vulnerable to social ills and diseases environment, general quality of low buildings, underserved adequate environmental infrastructure, endanger the continuity of life and its inhabitants.

In its development, the growth of slums can be caused by several factors. According Constantinos A.Doxiadis (1968), stated that the growth of slums is influenced by several factors:

- Growth of density
  With the population growth that is from birth and for the growing number of families, it will bring new problems. Humanly they want to occupy their own homes. There by increasing the amount of housing in the settlement area this led to the growth of residential housing.
- Urbanization
  Given the attractiveness of the city center, it will cause the flow of rural-urban migration and from outside the city into the city center. The urbanization works in the city center or community who opened the business in the downtown, of course, have to live in settlements around the city center. It will also lead to the growth of residential housing in the downtown area.

Slum characteristics according Silas (1996) is as follows:

a. The state of the house in a slum forced below the standard, an average of 6 m^2 / person. While the urban facilities directly unserved because it is not available. However, because of its location close to existing settlements, the environmental facilities are not difficult to find.
b. The neighborhood was physically provide basic benefits, which is close to workplaces (opportunity value) and home prices are also cheap (the principle of affordability) either buy or rent.
c. The settlement benefits in addition to considerations of employment and low price is an opportunity to get or high accessibility.

According to Law No. 4/1992 Housing and Settlement, improving the quality of housing can be activities, repair or restoration, rejuvenation and management / maintenance of sustainable. The program to improve the quality of housing and settlement which has been the government's attention is the area of housing and settlements category slum areas, which is characterized among other things by the condition of infrastructure and facilities are inadequate both in quality and quantity, socio-economic conditions of society is low, social conditions cultural, and environmental conditions are prone to disasters, diseases and security [3]

Forms of environmental improvements settlement based PU. Cipta Karya, there are several forms of business implementation of slum upgrading, as follows:

a. The restoration of the house, mean reversion physical state as before,
b. Kampung Improvement Program (KIP); KIP is a program that aims to improve the quality of life and the greening of society by improving the physical environment. Its main purpose is to improve the environmental health of the village. Basic components are the improvement of regional infrastructure such as road vehicles, footpaths, drainage channels, toilets and so on,
c. Environmental improvement market area (MIP); environmental improvement market area is the improvement of settlements around the market, made as a result of the additional burden of the community received about the market because it has no means of support such as drainage channels, parking lot, trash, stalls irregular and ineligible / less function , Markets and market society is an integral and interdependent both positive and negative,
d. Housing construction; is or forms of urban renewal by building housing through the arrangement of slum physically in order to accommodate more occupants or others in need. Advantage of this program is relatively fast and immediately visible result,
e. Land consolidation; an integrated activity to restructure the ownership patterns of land in an area that is lacking / irregular,
f. Controlled land development; an arrangement further efforts in the framework of spatial development of the city, especially the area of the city in a more implementable, if necessary through the transfer / development of the suburbs. In general, uncontrolled land development aims to encourage participatory climate in development by involving potential and wishes of the people, especially the private sector, small businesses and consumers,
g. Construction of flats; build residential environment as a whole with the aim to restructure an area of the city, both physically and functional and economic advantages.

2.2 Methodology

Data collection is an important first step, because of the completeness and quality of data and information will determine the sharpness of analysis and ultimately determine the quality of the plan is also to be prepared. The scope of activities includes:

a. Secondary Data Collection
   These activities include the collection of secondary data including document development policy and spatial policy, basic and thematic maps, aerial photos, satellite imagery, a study/research related to themes and research area.
b. Pre-Field Preparation (instruments of design)
   Several survey instruments include a checklist, a questionnaire about the data to be collected, the questionnaire contains a list of topical questions, interviews, unstructured interviews, and a field measurement sheets.
c. Field survey activities include:
   1) A review of the study area (site-visit), to get a directly visual experience above the location of the study;
   2) The on-site introduction of neighbourhood conditions;
   3) Documentation of the an area condition in the form of photographs, slides, and sketches;
   4) Measurements of the field;
   5) Interviews and questionnaires (indicators and parameters issued by the Directorate General of Human Settlements);
   6) Community participation through discussions.

Research conducted in the coastal areas and coastal rivers Konawe. Primary sampling units (Primary Sampling) is 4 district, Sampara District and Bondoala District as the coastal areas of the river, and the Lalonggasumeeto District, Soropia District as marine coastal areas. Sampling has been proportionately and randomly based on the total RT in all RW in the village / village that has been determined to be the survey area. This stage produces an activity profile based verification slum area that has been done through the stages of survey and inventory based on indicators and parameters of the Department of Public Works for Human Settlements. The data obtained will be carried out the analysis stage to get a level of untidiness typology (determination of variables) which will produce a recommendation and priority handling of slums.

IOP Conf. Series: Journal of Physics: Conf. Series **846** (2017) 012018 doi:10.1088/1742-6596/846/1/012018

3. Results and Discussion

3.1 Sampara District

1) Geographical

Sampara sub-district is one of the Districts directly adjacent to the city of Kendari. Geographically villages/ wards in the district Sampara relatively flat terrain with up to mountainous topography. Sampara subdistrict boundaries in the north bordering the District Bondoala, east with the city of Kendari, the southern border with South Konawe. The west is bordered by Besulutu District.

The total area of the Sampara District 60.01 km^2 or 6,001 hectares, or 2.89 percent of the total Konawe. Access from the city of Kendari (Southeast Sulawesi provincial capital) to the villages / wards in the District of Sampara relatively easy, because almost all the villages can be bypassed by four-wheeled vehicles and motorcycles. Except Puusangi village, Tabanggele, Bondoala and villages Puuloro, due to damaged roads.

2) Population

Based on the results of the 2010 Population Census Sampara number of District residents in 2010 was 11 709 inhabitants. In general, the population density of 195.1 inhabitants for each kilometer Sampara District of square which means that every square kilometers inhabited by an average of 195 inhabitants. While the number of households reached 2,432 households with an average of 4.8 people per household.

3) Physical condition

In general, the buildings in Sampara district 35-65% of the buildings has a regularity to the level of building density <80 units / ha. The physical condition of buildings> 60% is semi-permanent environmental conditions of the road surface is> 60% bad and <30% of the area include the channel.

In Sub Sampara> 60% of the area to obtain fresh water from wells. Household waste management is not served by municipal waste management so that people tend to burn household waste produced. At this area not served hydrant firefighters.

Figure 1 Settlement area in Sampara District
Source: Field Survey, 2014

4) Non Physical

In general, ownership and use of residential space / house in District Sampara belongs to the people. Population density in the Sampara District i.e. <400 inhabitants / ha with the livelihoods of> 60% work the informal sector. With the livelihoods of the average household income is below the UMR District. Distribution of settlement areas in the district Sampara largely in accordance with the Spatial generally are slum/illegal located along the river, among others: the settlement area around times Pohara, residential area around the small river. In general, settlements existing in the surrounding area less attention to the aspects of safety, sanitation and river border line rules so that communities living can create the danger of flooding and river pollution. This area generally has a legal as community land.

3.2 Bondoala District

1) Geographical

Bondoala sub-district is one of the Districts directly adjacent to the city of Kendari. Geographically the whole village in Bondoala district relatively flat-lying topography limits the District of Bondoala in the north bordering the District of Sawa, the east with Kapoiala and Kendari District, the southern border with Sampara District, and the west is bordered by Besulutu District. The total area of 133.35 km^2 Bondoala sub district or 2.00 percent of total Konawe area. Access from Kendari city (Southeast Sulawesi provincial capital) to some villages in Bondoala sub district relatively easy, but there are some other villages that are not accessible by four-wheel vehicles, can only are passed with a two-wheeled vehicle or boat / Pincara.

2) Population
Bondoala number of District residents in 2010 reached 8.199 inhabitants or an increase of 2.8 percent from the year 2009 of 223 inhabitants. In general, the population density increased from 59.81 inhabitants for each kilometer square in 2009 to 61.48 inhabitants for each kilometer square in 2010. The number of households in 2010 reached 1,890 households or decreased 6.9 percent when compared with the number of households in 2009 that of 2,020 households with an average of 4 person household members of each household. The decrease in the number of households in addition to the change in the marital status of the population is also caused by the presence of inter-regional migration.

3) Physical condition
In general, the buildings in the district Bondoala 35-65% of the buildings has a regularity to the level of building density <80 units / ha. The physical condition of buildings> 60% is semi-permanent environmental conditions of the road surface is> 60% bad and <30% of the area include the channel.
in sub Bondoala> 60% of the area to obtain fresh water from wells and mountain water. The management of household waste only most underserved so people tend to burn household waste produced. At this area not served hydrant firefighters.

Figure 2 Settlement area in Bondoala District
Source: Field Survey, 2014

Figure 3 Condition of roads in Bondoala District
Source: Field Survey, 2014

4) Non Physical
In general, ownership and use of residential space / house in Bondoala District belongs to the people. Population density in Bondoala District i.e. <400 inhabitants / ha with the livelihoods of> 60% work the informal sector. With the livelihoods of the average household income is below the UMR District. Distribution of settlement areas in Bondoala district largely in accordance with the Spatial generally are slum / illegal located along the river, among others: the settlement area around times Pohara, residential area around the small river. In general, settlements existing in the surrounding area less attention to the aspects of safety, sanitation and river border line rules so that communities living can create the danger of flooding and river pollution. This area generally has a legal as community land.

IOP Conf. Series: Journal of Physics: Conf. Series **846** (2017) 012018 doi:10.1088/1742-6596/846/1/012018

3.3 Lalonggasumeeto District

1) Geographical
   Lalonggasumeeto District geographically a coastal area with a flat and hilly topography so the potential for the development of the fisheries sector. Lalonggasumeeto subdistrict boundaries in the north bordering the Banda Sea, the east with District of Soropia, the south area bordered by the city of Kendari and the west by Kapoiala district and Bondoala District.
   Total area of Lalonggasumeeto District 4,157 hectares or 0.62 percent of Konawe land area. The village with the largest area is the village Lalombonda with an area 686 hectares or 16.50 percent of the area of Lalonggasumeeto District, while Puuwonua village is the village with the smallest area that is 137 hectares or 3.30 percent of the total area. Access from the capital district to all villages in the sub district of Lalonggasumeeto relatively easy, while access from the city of Kendari (Southeast Sulawesi provincial capital) throughout the village in the sub district very close. Because all the villages can be bypassed by four wheel drive and two-wheel throughout the year.
2) Population
   In 2010 based on the results of the Population Census the population of Lalonggasumeeto District many as 4,608 people with an average population density of 110 inhabitants for each square kilometer.
   The village with the densest population density is Puuwonua village of 349 inhabitants per square kilometer while Wawobungi village is the village with the smallest population density is 69 people per square kilometer. The number of households in 2010 reached 1,036 households with an average household of 4.45 members. The increase in the number of households in the side by a change in the marital status of the population is also caused by the additional population through migration between regions.
3) Physical condition
   In general, the buildings in the Lalonggasumeeto district 35-65% of the buildings has a regularity to the level of building density <80 units / ha. The physical condition of buildings> 60% is semi-permanent environmental conditions of the road surface is> 60% good and <30% of the area include the channel in Sub Lalonggasumeeto> 60% of the area to obtain fresh water from wells and mountain water. The management of household waste only most underserved so people tend to burn household waste produced. At this area not served hydrant firefighters.

Figure 4 Settlement area in Lalonggasumeeto District
Source: Field Survey, 2014

Figure 5 Condition of roads in Lalonggasumeeto District
Source: Field Survey, 2014

IOP Conf. Series: Journal of Physics: Conf. Series **846** (2017) 012018 doi:10.1088/1742-6596/846/1/012018

4) Non Physical

In general, ownership and use of residential space / house in Lalonggasumeeto District belongs to the people. Population density in the Lalonggasumeeto District i.e. <400 inhabitants / ha with the livelihoods of> 60% work the informal sector. The average family income is below the UMR District. Distribution of settlement areas in Lalonggasumeeto district largely in accordance with the Spatial generally are slum / illegal located along the river, among others: the residential area around the coast, the residential area around the small river. In general, existing settlements in the surrounding area less attention to the aspects of safety, sanitation and coastal border line rules so that communities living can create the danger of flooding and river pollution. This region has largely legality as state land as being the seashores.

Table 1 Typology, Recommendations and Priorities Management

No.	Regions Slum (Districts)	Compliance with the City Spatial Plan	Value Strategic Location	Economic Activity In the Region	Slums Level	Typology	Handling recommendations slum area	
							Recommendation	Event Handling
1	**Bondoala District**							
	Diolo Village	appropriate	medium	housing	medium slum	IV	**RESTORATION**	Improved regional infrastructure Improved housing conditions
	Tanggobu Village	appropriate	medium	housing	medium slum	IV	**RESTORATION**	Improved regional infrastructure Improved housing conditions
	Lalonggaluku Village	appropriate	medium	housing	weight slum	III	**RENOVATION**	Improved regional infrastructure Improved housing conditions
	Pebunooha Dalam Village	appropriate	medium	housing	weight slum	III	**RENOVATION**	Improved regional infrastructure Improved housing conditions
	Tondowatu Village	appropriate	medium	housing	weight slum	III	**RENOVATION**	Improved regional infrastructure Improved housing conditions
	Laosu Village	appropriate	medium	housing	weight slum	III	**RENOVATION**	Improved regional infrastructure Improved housing conditions
2	**Sampara District**							
	Pohara village	appropriate	High	housing	medium slum	IV	**RESTORATION**	Improved regional infrastructure Improved housing conditions
3	**Lalonggasumeeto District**							
	Wawobungi Village	appropriate	Low	housing	medium slum	VII	**RESTORATION**	The function remains Improved settlement areas with a focus on environmental infrastructure improvement and housing conditions
	Watunggarandu Village	appropriate	Low	housing	medium slum	VII	**RESTORATION**	The function remains Improved settlement areas with a focus on environmental infrastructure improvement and housing conditions
	Nii Tanasa Village	appropriate	Low	housing	medium slum	VII	**RESTORATION**	The function remains Improved settlement areas with a focus on environmental infrastructure improvement and housing conditions
	Lalombonda Village	appropriate	Low	housing	medium slum	VII	**RESTORATION**	The function remains Improved settlement

No.	Regions Slum (Districts)	Compliance with the City Spatial Plan	Value Strategic Location	Economic Activity In the Region	Slums Level	Typology	Handling recommendations slum area	
							Recommendation	Event Handling
								areas with a focus on environmental infrastructure improvement and housing conditions
	Tolo-Toli Village	appropriate	Low	housing	weight slum	VI	**RENOVATION**	Improved regional infrastructure Improved housing conditions
	Rapambinopaka Village	appropriate	Low	housing	weight slum	VI	**RENOVATION**	Improved regional infrastructure Improved housing conditions
	Puuwonua Village	appropriate	Low	housing	weight slum	VI	**RESTORATION**	Improved regional infrastructure Improved housing conditions
					weight slum	I	**RELOCATION**	Returns to the function of non woke region (coastal) Acquisition of land with buildings thereon compensation
					weight slum	I	**RELOCATION**	Returns to the function of non woke region (coastal) Acquisition of land with buildings thereon compensation
					weight slum	I	**RELOCATION**	Returns to the function of non woke region (coastal) Acquisition of land with buildings thereon compensation
					weight slum	I	**RELOCATION**	Returns to the function of non woke region (coastal) Acquisition of land with buildings thereon compensation
					medium slum	VII	**RESTORATION**	Improved regional infrastructure Improved housing conditions
					medium slum	VII	**RESTORATION**	Improved regional infrastructure Improved housing conditions
					medium slum	VII	**RESTORATION**	Improved regional infrastructure Improved housing conditions
					medium slum	VII	**RESTORATION**	Improved regional infrastructure Improved housing conditions

4. Conclusion

The conclusion of this study is in Sampara and Bondoala District as the coastal areas of the river with a designation as residential areas in accordance with the Spatial Plan of the City Konawe with an average value of strategic cities are moderate, levels of untidiness is a rundown medium (typology IV) with a recommendation handling area i.e. slum in the coastal restoration and renovation. For coastal areas sea which is in Lalonggasumeeto District and Soropia District designation as a residential area in Lalonggasumeeto district accordance with the Spatial Plan of the Konawe City, whereas in Soropia District precisely in Bajo Indah village, Mekar Jaya village, Bajoe Village, and Leppe Village not in accordance with the city Spatial Plan Konawe, with an average value of strategic cities is low, the level of untidiness is a rundown weight (typology I) with treatment recommendations in the coastal slum is the relocation.

Acknowledgments

Acknowledgements of this paper goes to the Department of Planning and Regional Development Konawe for the financial support given in this study. As well as the communities in Konawe.

References

1. _______ , Undang-Undang Republik Indonesia Nomor 4 Tahun 1992 Tentang Perumahan Dan Permukiman
2. BPS Kabupaten Konawe, Konawe Dalam Angka 2012,
3. Departemen PU. Dirjen Cipta Karya (1999). *Petunjuk Teknis Pembangunan Perumahan Nelayan.* Dep. PU. Jakarta.
4. Doxiadis, Constantinos, A, (1968), *Ekistic, An Introduction to the Science of Human Settlements*, Hutchinson and Co, Ltd, London
5. Silas, Johan, (1996), *Kampung Surabaya menuju Metropolitan*, Yayasan keluarga Bhakti, Surabaya

IOP Conf. Series: Journal of Physics: Conf. Series **846** (2017) 012019 doi:10.1088/1742-6596/846/1/012019

Tank Model Application for Runoff and Infiltration Analysis on Sub-Watersheds in Lalindu River in South East Sulawesi Indonesia

Sitti Wirdhana Ahmad*

Faculty of Mathematic and Natural Science, Halu Oleo University, Kampus Bumi Tridharma Anduonohu, Kendari 93232, Indonesia

*E-mail : wirdhanaaxtalora@yahoo.com

Abstract. Improper land management often causes flood, this is due to uncontrolled runoff. Runoff is affected by the management of the land cover. The phenomena also occurred in South East Sulawesi, Indonesia. This study aims to analyze the flow rate of water in watershed of Lalindu River in North Konawe, South East Sulawesi by using a Tank Model. The model determined the magnitude of the hydrologic runoff, infiltration capacity and soil water content several land uses were evaluated in the study area. The experimental and calculation results show that the runoff in the forest is 2,639.21 mm /year, in the reed is 2,517.05 mm/year, in the oil palm with a slope more than 45% is 2,715.36 mm/year, and in the oil palm with slopes less than 45% is 2,709.59 mm/year. Infiltration in the forest is 30.70 mm/year, in the reed is 7.51 mm/year, in the palm oil with a slope more than 45% is 24.13 mm/year and in the palm oil with slopes less than 45% is 29.67 mm/year. Runoff contributes to stream flow for water availability.

1. Introduction

Changes in land use in the watersheds (WS) provide a dominant influence on flood discharge [1,2]. The greatest influence changes in land use on the sustainability of water resources is the changes from forest land to other uses such as agriculture, farming, residence or industry. If these activities are not well-managed, it will result in water excess (flood) during the rainy season and drought during the dry season. Unwisely changes in land use and is not accompanied by conservation measures will largely become runoff. Watersheds can be viewed as a hydrological system in which precipitation is the input from streams, and evapotranspiration is the system output. Furthermore, the watershed is the place where the simultaneous processes happen and become a part of the hydrological cycle.

The expansion of oil palm plantations keeps going in Indonesia, including in Southeast Sulawesi, which is spread almost all over North Konawe Regency, especially in Wiwirano District [3]. The changes of land cover grow rapidly and it is one of areas used as a priority area of oil palm development. It can be viewed from the increasingly growing oil palm plantation during the last five years, both operated by the community and by private plantations. The land use ignoring sustainability aspect, such as deforestation will put impact on the reduced water discharge in Lalindu watersheds. The deforestation is a consequence of the land function transfer. Land use transfer has resulted in changes of land structure. During dry season, the soil becomes hard and barren due to absorption strength of the sun, penetrating into the soil. If various trees still exist, the sun heat will be muffled in the foliage and the soil will stay fertile for the stable humidity. In addition, travel time of the water is relatively short, thus the rate of water infiltration by land is slower than the runoff rate. It is influenced by a number factors, among others; by concave topography of Mount Wiwirano (although the nature of the soil supports the absorption of water into the soil) resulting in most rainwater that will be lost through runoff

Content from this work may be used under the terms of the Creative Commons Attribution 3.0 licence. Any further distribution of this work must maintain attribution to the author(s) and the title of the work, journal citation and DOI.
Published under licence by IOP Publishing Ltd

and later run towards the river and eventually into the sea, before the rain water fills the underground water through infiltration and percolation process. A settlement is located surrounding this valley, thus during heavy rains, floods will occur.

To investigate the problem, a tank model is applied to determine watershed parameters in this area. The tank model base on the hypothesis that the runoff flow and infiltration is a function of the water amount presents in the soil [4-6]. The tank model can be constructed in such a way, thus representing the function of sub-watersheds area, or represents the difference in the structure/ type of land in each layer. Besides explaining the lost of initial rainfall and the dependence on prior rain, the tank model can also present some of the components forming the runoff flow, which have specific period and time lag. The tank model structure is the closest model to each watershed [4,5]. A tank with wasting channel on its side represents runoff, lower wasting channel represents infiltration, and saving component represents the runoff processes in one watershed or in watershed partly. Several parallel similar tanks can represent a large watershed [6]. In this study flow rate is analyzed by using a model consisted of runoff parameter, infiltration capacity and ground water content.

2. Material and Method

The research was conducted on sub-watersheds in Lalindu, North Konawe, one of the districts in Southeast Sulawesi, Indonesia. Observations made on hydrological conditions on 4 sub-watersheds which were parts of the Lalindu sub-watershed. The placement of 4 sample plots in each sub-watershed represented the three types of vegetation or land cover in the sub-watershed. The vegetations are the forest, reeds and palm oil. Palm oil itself was divided into two by the steepness, i.e. palm with steepness > 15% and palm oil with steepness <15%. The model was verified by observing the sample plots to obtain parameter value of the first tank (Fig. 1)

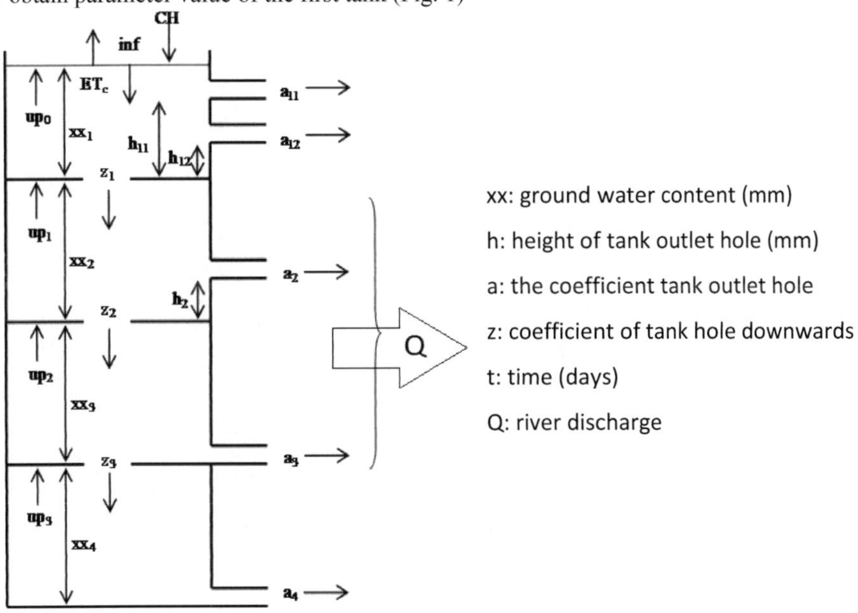

Figure 1. Tank model used in the research

The runoff parameters can be formulated as following equations. The basic equation for the first tank is as follows:

$$xx_1(t) = xx_1(t-1) + CH - ET_c - z_1.xx_1(t-1) \\ -[(xx_1(t) - h_{11})a_{11} + (xx_1(t) - h_{12})a_{12}] \quad (1)$$

The equation for the second tank is as follows:

$$xx_2(t) = xx_2(t-1) - z_2.xx_2(t-1) + z_1.xx_1(t-1) \\ -[(xx_2(t) - h_2)a_2] \quad (2)$$

The equation for the third tank is as follows:

$$xx_3 = xx_3(t-1) - z_3.xx_3(t-1) + z_2.xx_2(t-1) - xx_3(t).a_3 \quad (3)$$

The equation for the fourth tank is as follows:

$$xx_4(t) = xx_4(t-1) + z_3.\,xx_3(t-1) - xx_4(t).\,a_4 \quad (4)$$

The runoff discharge from river (Q) is calculated by the following equation:

$$Q(t) = [(xx_1(t) - h_{11})a_{11} + (xx_1(t) - h_{12})a_{12}] + [(xx_2(t) - h_2)a_2] + xx_3(t).\,a_3 + xx_4(t).\,a_4 \quad (5)$$

where:

- xt : height of ground water content
- h_{ij} : height of stored water (outlet hole level)
- z_i : infiltration coefficient
- a, b : outlet hole coefficient
- CH : rainfall depth
- ETc : actual evapotranspiration
- t : time (days)
- i :1,2,3,4

The data used in this research were primary and secondary data. The primary data included water discharge data on sample plots, sub- watersheds and sub-watershed of one year observation. Secondary data was the climate data, soil-type data and biophysical conditions. The data analysis was performed on the amount of rainfall, infiltration, evapotranspiration and runoff. Furthermore, tank model creation was made to describe runoff processes occurring in the watershed.

Tank model was validated by using daily rainfall and daily actual discharge data from the results of direct measurements in the field. The actual discharge data used in this validation process was the actual discharge data in the sub watersheds of gardens and other lands located in the surrounding garden location of the research.

3. Results and Discussion

Daily rainfall and evapotranspiration (ETc) data from the calculation results of climate data in the garden in 2011 are presented in Fig. 2. The data is used as an input value in the analysis of the tank model.

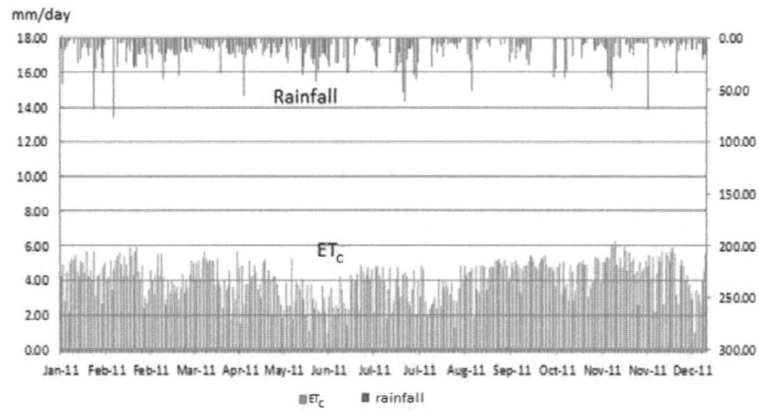

Figure 2. Daily rainfall and evapotranspiration data of ETc value in the garden of 2011

The discharge measurement of sample plots on four types of land cover is forest, reeds, and palm oil land, with a steepness > 15% and <15%. The discharge data for sample plots of the size 16 m^2 is shown in Fig 3.

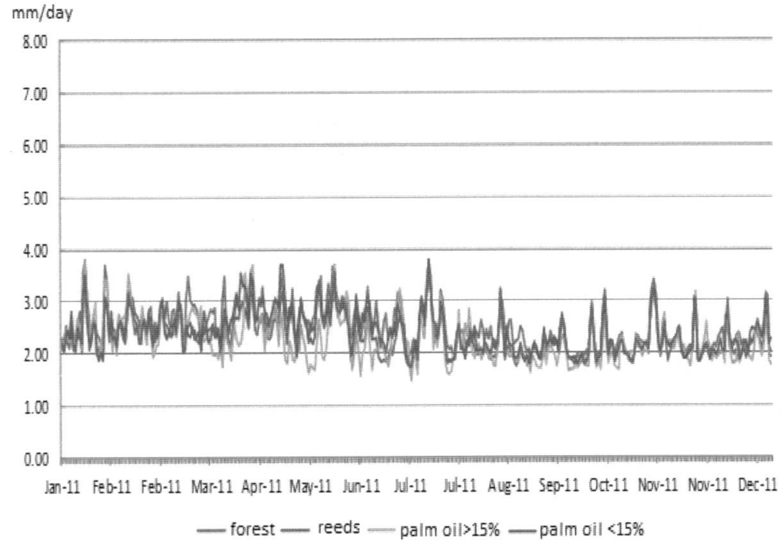

Figure 3. Discharge measurement graphs for each land-use in sample plots

In addition, the discharge measurement is conducted at sub-watersheds in the area of 20.000 ha. The discharge measurement data at the sub-watersheds is influenced also by run-off from the garden area in the surrounding area of the watershed. The discharge data of sub-watersheds from direct measurement is presented in Fig. 4.

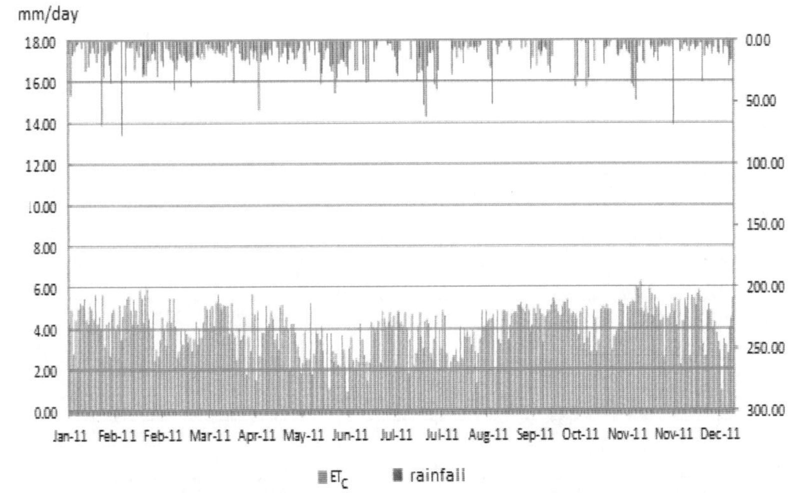

Figure 4. Discharge data of direct measurement on sub-watersheds

3.1 Tank Model Calibration in Sample Plots

The measurement result data in sample plots is used as a basis for calibrating the tank model to determine the magnitude of the tank model coefficient in the first tank. The calibration of each type of land result on sample plots for the first tank on tank models are presented in Table 1. The analysis of tank model of on the sample plots is conducted to determine the land cover characteristic towards the water.

Table 1. Calibration result of each land type on sample plots for the first tank

Parameter	Coefficient Value	Forest	Reeds	Palm oil > 15%	Palm oil < 15%
Ground water content of tank 1	xx1	200	150	180	185
The outlet coefficient of tank 1-1	a11	0.65	0.7	0.6	0.65
The outlet coefficient of tank 1-2	a12	0.55	0.6	0.5	0.45
The outlet height of tank 1-1	h11	75	50	85	80
The outlet height of tank 1-2	h12	60	45	70	65
The coefficient of tank 1 downwards	z1	0.009	0.003	0.004	0.005

3.2 Tank Model Calibration in Sub-Watersheds

In Table 2, the parameter coefficient values of the tank model calibration results in sub-watersheds is determined by using the data of 2011. From the calibration results, the coefficient of determination is obtained to be more than 70% for each land use.

Table 2. Tank model coefficient value of calibration results at the sub-watersheds

Parameter	SW I	SW II	SW III	SW IV
Ground water content of tank 1	500	300	850	900
Ground water content of tank 2	1100	800	1400	1300
Ground water content of tank 3	1500	1300	1800	1800
Ground water content of tank 4	2100	1800	2100	2000
The outlet coefficient of tank 1-2	0.8100	0.8500	0.8100	0.7900
The outlet coefficient of tank 1-2	0.7700	0.8200	0.7800	0.7700
The outlet coefficient of tank 2	0.7400	0.8000	0.7500	0.7600
The outlet coefficient of tank 3	0.0500	0.0500	0.0500	0.0500
The outlet coefficient of tank 4	0.0300	0.0300	0.0300	0.0300
The outlet height of tank 1-1	210	175	205	200
The outlet height of tank 1-2	180	170	190	195
The outlet height of tank 2	170	160	175	180
The coefficient of tank 1 downwards	0.0009	0.0003	0.0007	0.0008
The coefficient of tank 2 downwards	0.00005	0.00002	0.00005	0.00006
The coefficient of tank 3 downwards	0.00003	0.00001	0.00004	0.00005

Based on coefficient values of tank model from the calibration result, it is shown that the coefficient of the tank outlet is smaller to the lower part. It is due to the deeper soil layers, the ability of the soil to have water run-off is smaller. Similarly, the coefficient value of the tank downward (z) is smaller downward. That is due to the deeper layers of the soil, the soil capacity to carry water into the deeper layers (percolation) is getting smaller. From the results of the calibration tank model, it is known that ground water content (xx) is mostly found in the fourth tank. The water content in the soil at each level of the tank depth is greatly influenced by the type of plants that living above, as each model of discharge model outcome approaches the actual discharge model. In the process of model validation, the coefficient value of tank models that have been obtained from the calibration results in 2011 from the previous stage is used. The calibration results of tank in model of sub-watersheds are later validated in the downstream region which is a part of the Lalindu Watershed. The discharge validation data results in gardens sub-watershed and the surrounding other land of 2011 are presented in Fig. 5. The Figure shows the relationship among rainfall, discharge models, and actual discharge of the model validation process.

This validation process is obtained from the model discharge in the sub-watersheds of gardens and other surrounding land, describing the response towards the rainfall. The validation result has determination coefficient of 74.52%, and a tank model is then used to perform scenario of land conversion into oil palm plantation using conservation methods.

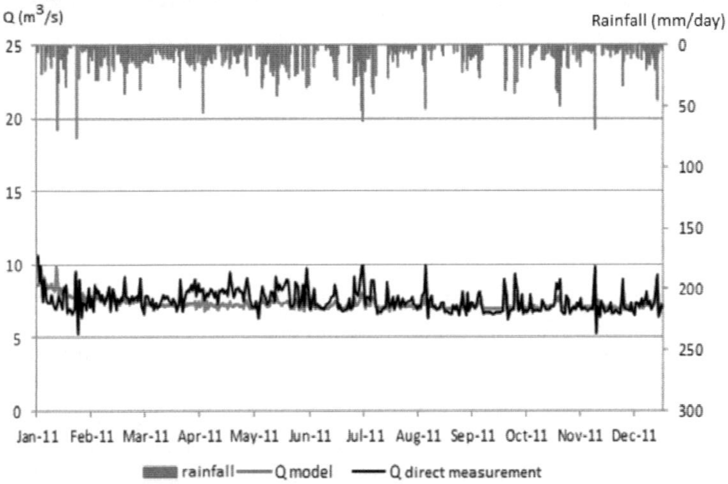

Figure 5. Discharge validation result data in sub-watersheds of gardens and other surrounding land

3.3 The Analysis of Land Cover Changes on Water Supply

The next stage is to analyze the model in order to determine the total runoff and the total infiltration value of each different land cover. The total runoff and total infiltration value from the model analysis results in sub-watersheds at each different land cover is presented in Figure 6 and 7. The amount of runoff and infiltration rate is used to determine the state of water in Lalindu sub-watersheds for the amount comparison of the runoff value, and the infiltration is being simulated for area of 500 hectares from each land cover type.

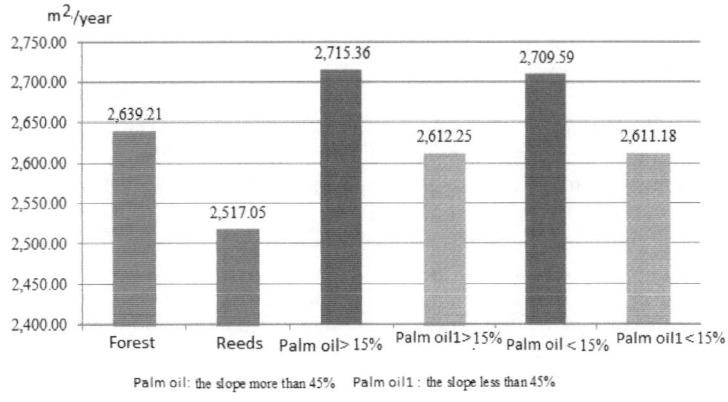

Figure 6. Total discharge runoff of each land cover at sub- watersheds

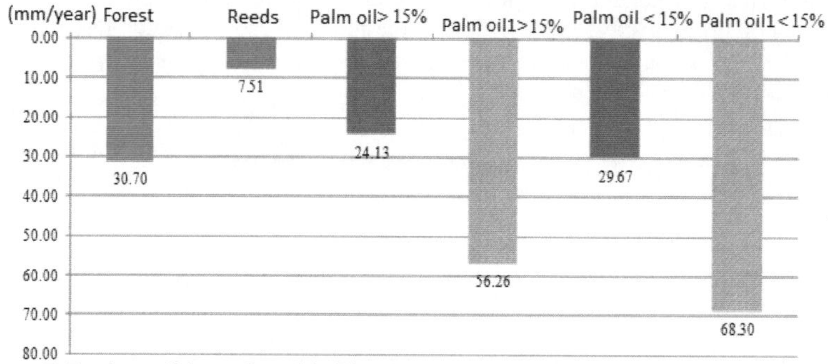

Figure 7. Real condition of total infiltration of each land cover at sub-watersheds

3.4 Analysis of the relationship among soil components, water and vegetation

The runoff total and infiltration on model analysis are then used to examine the relationship among land of soil components, water and vegetation of each unit. There is a close relationship among the soil components, water and vegetations. Land is a medium for the vegetation growth. Different soil types will have distinctive characteristics in terms of the soil physical, biological, and chemical properties. Soil properties can determine the type of nutrients in the soil, amount of water that can be stored in the soil, and root systems reflecting the circulation of water movement in the soil. The soil ability in absorbing water is reflected in the vegetation types at the ground level. The vegetation function can effectively reflect the ability of the soil to absorb rainfall, maintain or increase the infiltration rate, and demonstrate the ability in restraining water or water retention capacity [7,8].

4. Conclusion

Tank Model was successfully applied in Lalindu watersheds to determine magnitude of the hydrologic runoff, infiltration capacity, and soil water content.

1. The hydrological condition in Lalindu sub-watersheds can be presented by Tank Model with coefficient calibration (R^2) of 74.52%.
2. The analysis of runoff the Tank Model at each land cover showed that the runoff discharge in the forest is 2,639.21 mm/ year, in the reeds is 2,517.05 mm /year, in the oil palm with a steepness of more than 45% is 2,715.36 mm/ year and in the oil palm with steepness less than 45% is 2,709.59 mm/ year.
3. The analysis of infiltration in the tank model at each land cover resulted that the infiltration in the forest is 30.70 mm/ year, in the reeds is 7.51 mm/ year, in the palm oil with steepness more than 45% is 24.13 mm/ year and in the oil with steepness of less than 45% is 29.67 mm/ year.

Acknowledgments
This work has been partially supported by the Directorate of Higher Education, at the Ministry of Research Technology and Higher Education, Republic of Indonesia through Research Grant "Hibah Bersaing".

References

[1] R. Jayadi, Application of Fuzzy Relationship Theory to Analyze the Similarity of Two Catchment Areas, J. the Faculty of Engineering, Universitas Gadjah Mada, Vol. 23, pp. 187-196 (2008).
[2] A. Hayakawa, M. Shimizu, K. Woli, K. Kuramochi, R. Hatano, Evaluating stream water quality through land use analysis in two grassland catchments: Impact of wetlands on stream nitrogen concentration, J. Environmental Quality, 35, 617–627 (2006).
[3] S. Wirdhana Ahmad, Water resource conservation model on sustainable palm oil (Case study: Sub-watershed Lalindu, North Konawe, South East Sulawesi province), Doctoral Thesis, Bogor Agricultural University (2013).
[4] M. Sugawara, Tank model and its application to Bird Creek, Wolombi Brook, bikin River, Kitsu River, Sanaga River and Nam Mune. Research Note of national Research Center for Disaster

Prevention, No. 11, pp. 1-64 (1974)

[5] T. Kenji, O. Yuzo, X. Jun, K. Tomofumi, Tank Model And Its Application To Predicting Groundwater Table In Slope, Chinese Journal of Rock Mechanics and Engineering, 27[12],pp.2501-2508 (2008).

[6] A.N. Mulyana, C. Kusuma, K. Abdullah, L.B. Prasetyo, B.I. Setiawan, Aplication of Tank Model for Predicting Water Balance and Flow Discharge Component of Cisadane Upper Catchment. J. Manajemen Hutan Tropika. 17(2), 63-70[2011].

[7] G. O. Schwab, D.D. Fangmeir, W.J. Elliot, R.K. Frevert, Soil and Water Conservation Engineering. Four Edition, John Wiley & Sons. Inc, NewYork, 1992.

[8] T. F. Qiang, H. H. Ping, L. Z. Dong, Thermodynamic watershed hydrological model: Constitutive relationship, Science in China Series E: Technological Sciences, 51[9], pp. 1353-1369 (2008).

[9] M. Hernandez, D.J. Semmens, S.N.Miller, D.C. Goodrich, and W.G. Kepner, Development and application of the automated geospatial watershed assessment tool. Modeling and Remote Sensing Applied to Agriculture, USDA-INIFAP, U.S. and Mexico, pp. 127–158 (2008).

IOP Conf. Series: Journal of Physics: Conf. Series **846** (2017) 012020 doi:10.1088/1742-6596/846/1/012020

Study on Emission Spectral Lines of Iron, Fe in Laser-Induced Breakdown Spectroscopy (LIBS) on Soil Samples

Nasrullah Idris[1]*, Kurnia Lahna[1], Fadhli[1], and Muliadi Ramli[2]

[1]Department of Physics, Faculty of Mathematics and Natural Sciences, Syiah Kuala University, Jl. Syech Abdurrauf No. 3 Darussalam, 23111 Banda Aceh, Aceh, Indonesia
[2]Department of Chemistry, Faculty of Mathematics and Natural Sciences, Syiah Kuala University, Jl. Syech Abdurrauf No. 3 Darussalam, 23111 Banda Aceh, Aceh, Indonesia

[a)]Corresponding author: nasrullah.idris@unsyiah.ac.id

Abstract. . In this work, LIBS technique has been used for detection of heavy metal especially iron, Fe in soil sample. As there are a large number of emission spectral lines due to Fe and other constituents in soil, this study is intended to identify emission spectral lines of Fe and finally to find best fit emission spectral lines for carrying out a qualitative and quantitative analysis. LIBS apparatus used in this work consists of a laser system (Neodymium Yttrium Aluminum Garnet, Nd-YAG: Quanta Ray; LAB SERIES; 1,064 nm; 500 mJ; 8 ns) and an optical multichannel analyzer (OMA) system consisting of a spectrograph (McPherson model 2061; 1,000 mm focal length; f/8.6 Czerny-Turner) and an intensified charge coupled device (ICCD) 1024x256 pixels (Andor I*Star). The soil sample was collected from Banda Aceh city, Aceh, Indonesia. For spectral data acquisition, the soil sample has been prepared by a pressing machine in the form of pellet. The laser beam was focused using a high density lens (f=+150 mm) and irradiated on the surface of the pellet for generating luminous plasma under 1 atmosphere of air surrounding. The plasma emission was collected by an optical fiber and then sent to the optical multichannel analyzer (OMA) system for acquisition of the emission spectra. It was found that there are many Fe emission lines both atomic lines (Fe I) and ionic lines (Fe II) appeared in all detection windows in the wavelength regions, ranging from 200 nm to 1000 nm. The emission lines of Fe with strong intensities occurs together with emission lines due to other atoms such as Mg, Ca, and Si. Thus, the identification of emission lines from Fe is complicated by presence of many other lines due to other major and minor elements in soil. Considering the features of the detected emission lines, several emission spectral lines of Fe I (atomic emission line), especially Fe I 404.58 nm occurring at visible range are potential to be good candidate of analytical lines in relation to detection heavy metal pollution, Fe, in soil sample.

1. Introduction

Iron, Fe is actually essential in soil since it is generally considered as micronutrient for plant growth [1]. Iron deficiency causes chlorosis and a reduced rate of the growth plants [2]. However, it turns into toxicity in excess. The excessive Fe in soil would be exposed to the grown plant, accumulating Fe in the plant tissues at high concentration causing many kinds of metabolic disorders of plants [3-6]. In general, iron concentrations in soils range from 0.2% to 55% (20,000 to 550,000 mg/kg) [7] and the concentrations can vary differently by regions, even within localized areas, due to soil types and the presence of other natural and anthropogenic sources. On average, common concentration of iron in soil including agriculture soil is about 4%. Therefore, periodical analysis of Fe content in soil is very important for both assuring iron concentration meets the essentiality and hinders the toxicity.

Content from this work may be used under the terms of the Creative Commons Attribution 3.0 licence. Any further distribution of this work must maintain attribution to the author(s) and the title of the work, journal citation and DOI.
Published under licence by IOP Publishing Ltd

Tsunami is one source altering significantly chemical composition the impacted soil including heavy metal elements such as Fe as it brings very large volume of water and various materials including sediment from the sea into the impacted land. The giant Indian Ocean tsunami, Aceh 26 December 2004 have not only brought about deadliest, catastrophic effect to human being and living things, but it have also caused severe environmental problems, including erosion and pollution of the impacted land, water source and surface water due to the sea water and the contaminated sediment and other material depositions [8-10].

Analytical studies conducted on sediment samples taken from the tsunami impacted coastal region of Thailand and Banda Aceh, Indonesia after the giant tsunami found that concentration of salts and heavy metals in the tsunami-impacted sand and soil was highly elevated [8-10]. This condition is persistent even 1.5 years and 3.5 years after the tsunami [10]. Therefore, detection of heavy metal including Fe in the tsunami-impacted soil is very important. Nowadays various laboratory tools such as Fourier transform infrared (FTIR) spectroscopy, X-Ray Difraction (XRD) spectroscopy and atomic absorption spectroscopy (AAS) has been used for soil analysis. However, these conventional techniques are impractical considering number of sample and disability for in situ analysis. Therefore, it is required to develop a fast, efficient and effective analytical tool. A new technique analytical tool, namely laser induced breakdown spectroscopy (LIBS) has emerged, a future super star analytical tool, promising unequivocally advantages such as very efficient and effective nondestructive analysis [11]. This technique based on using laser to generate luminous plasma as optical emission source for detection and analysis spectrochemically the constituent of the questioned samples. In previous study, a transversely excited atmospheric pressure (TEA) carbon dioxide (CO2) laser, LIBS technique has succeed to detect heavy metal in soil [12]. This result promises a similar succeed analysis for Fe in soil. In this work, LIBS technique has been used to detect Fe in soil sample. However, referring to atomic spectra database, there are many emission spectral lines of Fe appear in the wavelength range of 200-1000 nm [13]. Therefore, prior to application LIBS technique to detect and analyze iron, Fe in the tsunami-impacted soil, at first the LIBS study will be made on agriculture soil samples from the region which is not affected by the tsunami, the tsunami-unimpacted soil, to detect and identify Fe emission spectral lines and finally conclude the best analytical lines for carrying out a qualitative and quantitative analysis.

2. Experimental Setup

Basic LIBS experimental setup used in this work is similar to that used in previous study elsewhere [14]. The LIBS system consists a Nd-YAG laser (Quanta Ray; LAB SERIES; λ=1,064 nm; E=500 mJ, Δt=8 ns, f=10 Hz) as plasma excitation source, a lens of +250 mm for focusing laser beam, an optical fiber for collecting and delivering plasma emission, and an optical multichannel analyzer (OMA) system as optical detector of the plasma emission. The OMA system composes of a spectrograph (McPherson model 2061; 1,000 mm focal length; f/8.6 Czerny-Turner) and ICCD having 1024x256 pixels (Andor I*Star), 26-mm square (960 x 256 active pixels). The laser beam energy was measured by a joulemeter (Duo watt-joule meter). The energy can be varied in the range from 1 mJ to 500mJ/pulse. In the whole experiment, the laser beam energy was made constant at 68 mJ.

The soil samples used in this work was taken in an agriculture region. The region is the tsunami un-impacted zone in Aceh Besar Regency, Aceh, Indonesia. The collected samples has been prepared in such that the samples are free from leafs, root and other organic matters. The soil sample was pressed into the form of pellet using a hydraulic pressing apparatus with a force of 10 ton. The dimension of pellet is 20 mm in diameter with thickness of 3mm. The soil pellet was fixed on a sample holder placed in a chamber. The ambient gas in the chamber actually can be vacuumed and then exchanged with different kind of gas. The pressure of the ambient gas can be adjusted. However, this experiment was carried out under air surrounding gas at 1 atmosphere. The sample chamber was attached on a movable x-y platform controlled with a stepping motor.

For producing plasma, the laser beam was focused using the lens (+250 mm) onto the pellet surface. The spot size of the focused laser beam is about 300µm, resulting in a power density of around 1 GW/cm2. The sample was rotated during data acquisition to assure new position for each acquisition. The emission from the plasma was collected and delivered by using an optical fiber bundle to the entrance slit of the monochromator (spectrograph). Optical signal has been detected from the exit slit of the monochromator using intensified charge coupled device (ICCD) camera of the OMA system for detecting and acquiring emission spectral lines. During data acquisition, the OMA system was run under 1800 groove/mm grating mode having a useful wavelength window of about 12 nm. Gate delay and gate width of the OMA system were set at 3 µs and 50 µs, respectively during data acquisition.

3. Results and Discussion

In order to carry out a qualitative and quantitative analysis using LIBS, the first step is identification elements through assignment of the emission spectral lines of the elements. Identification and confirmation of an emission spectral line requires knowledge, experience and skill [15]. Thus, in order to detect and make a qualitative and quantitative analysis of Fe in soil sample, we should have strategy or procedures, namely knowing about (1) major and minor elements in the soil sample, (2) relative intensities of the emission spectral lines from reference/database wavelength tables, (3) the ionization stage of the emission spectral lines of the element, (4) the specific experimental conditions, and (5) observation of multiple strong emission spectral lines [15]. Thus, it is really crucial step to find ionization stage and relative intensities of the emission lines from reference/database to know strong emission lines before starting detection and assignment of any emission lines.

It well known that main constituents of soil are Si, Fe, Ca, Al, Ti, Mg, Na, Sr, etc. Inspection on atomic spectra database [13] revealed that there are 14310 atomic (Fe I) and ionic (Fe II) emission spectral lines in the wavelength range from 200 nm-1000 nm with relative intensities vary extremely from 1 (the lowest) till 2510000 (the highest). Although our final goal is to find best fit Fe emission line for carrying out qualitative and quantitative analysis, however we face a problem of dealing with such number of emission spectral lines. Therefore, considering the procedure of identifying element mentioned above, we then decided to narrow the interest emission lines, namely at first to work simply on the emission lines having very strong relative \intensities (≥ 1000000). In this case, there are 27 emission lines consisting of 20 atomic emission lines (Fe I) and 7 ionic emission lines (Fe II), as displayed in Table 1. It can be seen that the strongest emission line is Fe atomic line, Fe I 374.55 nm, having relative intensity of 2510000. In general, based on the tabulated relative emission intensities of the emission lines, which are extremely strong as compared to the lowest relative emission intensity, namely 1, it is expected all the tabulated emission lines could be detected. Excitation energy of the emission lines are also shown in Table 1.

Table 1. List of strong emission lines of Fe (relative intensity ≥1000000) extracted from reference database [13].

No	Wavelength (nm)	Relative Intensity (a. u.)	Excitation Energy (eV)	No	Wavelength (nm)	Relative Intensity (a. u.)	Excitation Energy (eV)
1	Fe II 234.34	1000000	5.28	15	Fe I 373.33	1000000	3.32
2	Fe II 238.20	1800000	5.20	16	Fe I 374.55	2510000	3.31
3	Fe II 239.56	1500000	5.18	17	Fe I 374.58	1100000	3.31
4	Fe II 240.48	1100000	5.15	18	Fe I 374.82	1910000	3.30
5	Fe II 259.93	1000000	4.76	19	Fe I 374.94	1150000	3.31
6	Fe II 273.95	1000000	4.53	20	Fe I 382.44	1000000	3.24
7	Fe II 275.57	1100000	4.50	21	Fe I 385.63	1100000	3.21
8	Fe I 344.06	1350000	3.60	22	Fe I 387.85	1290000	3.20
9	Fe I 344.09	1290000	3.60	23	Fe I 389.97	1070000	3.18
10	Fe I 347.54	1200000	3.57	24	Fe I 392.29	1000000	3.16
11	Fe I 349.05	1320000	3.55	25	Fe I 393.02	1150000	3.16
12	Fe I 363.14	1150000	3.42	26	Fe I 404.58	1000000	3.06
13	Fe I 370.55	1290000	3.34	27	Fe I 526.95	1020000	2.36
14	Fe I 372.25	1290000	3.33				

Figure 1 exhibits emission spectra taken from plasma produced on the soil sample. The measurement window was centered at the wavelength of 373 nm. The emission spectra contain high density of emission lines. These lines are due to Fe and other constituents of soil. Using the above mentioned procedures [15], emission lines due to Fe in this wavelength region were able to be identified. It can be seen clearly that there are many Fe emission lines can be found in the spectra with strong intensities, namely Fe I 370.55 nm, Fe I 372.25 nm, Fe I 373.33 nm, Fe I 374.55 nm, Fe I 374.58 nm, Fe I 374.82 nm, Fe 374.94 nm. The emission lines of Fe I 374.55 nm and Fe I 374.58 nm are not resolved since their separation very narrow, only 0.03 nm, thus the two lines are heavily overlap. The measured intensity of

Fe I 374.55 nm line is very strong in comparison with the other lines (Fe I 370.55 nm, Fe I 372.25 nm, Fe I 373.33 nm) just as expected from the database. While Fe I 374.94 nm shows significantly higher emission intensity as compared to the other Fe lines even though, based on the reference database, its relative intensity is actually lower than that of Fe I 374.55 nm. This is probably due to interference from other constituent emission line. However, in general the detected wavelengths and the measured emission intensities of the emission lines are in good coincidence with that of the database.

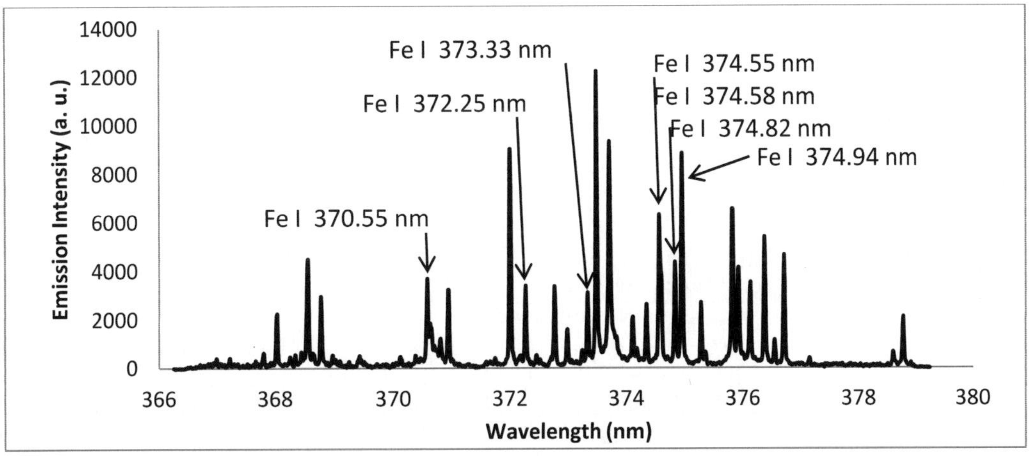

Figure 1. Emission spectra taken from plasma induced by irradiation of the focused Nd-YAG laser beam on the surface of the soil pellet sample under air surrounding gas at atmospheric pressure. The measurement window of the OMA system is centered at the wavelength of 373 nm. The soil sample was collected from agriculture region in Banda Aceh city, Aceh, Indonesia.

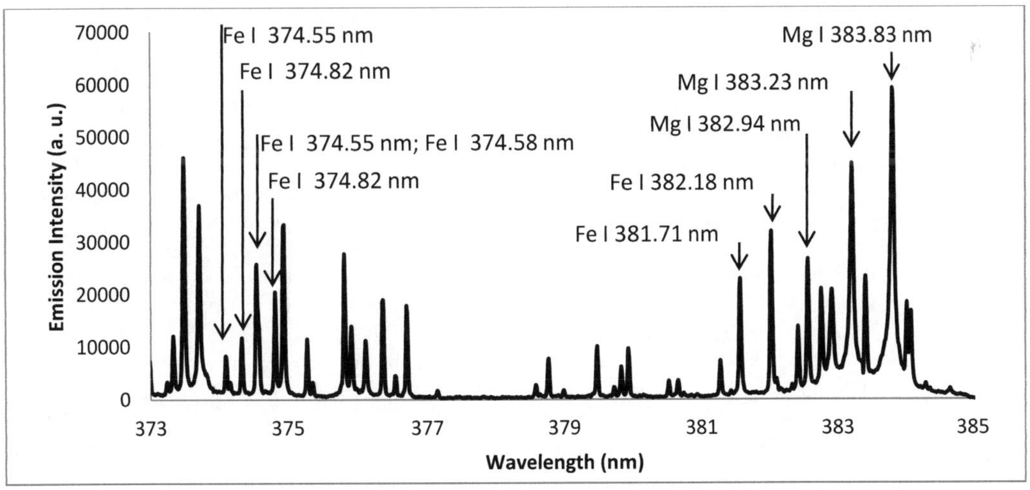

Figure 2. Emission spectra detected from the same soil sample under similar experimental conditions as for the Figure 1. The measurement window of the OMA system was shifted to a wavelength region centered at 379 nm.

Figure 2 displays emission spectra taken from the same soil sample under similar experimental condition as as for the Figure 1. The center of measurement window was shifted to 379 nm. At this wavelength region, the spectrum also contains high density of emission lines, especially on both sides of the spectrum. On the left part of the spectrum, the spectral lines are due to Fe atomic emission, the same as shown in Figure 1 since the detection window was shift only 6 nm from that of Figure 1. The other Fe I emission lines, namely Fe I 381.71 nm and Fe I 382.18 nm, can be found on the right part of the spectrum together with very strong Mg atomic emission lines (Mg I 382.94 nm, Mg I 383.23 nm, and Mg I 383.83 nm). The emission lines of Fe I 381.71 nm and Fe I 382.18 nm are not enlisted in Table 1 because their relative intensities are lower than those listed lines. Nevertheless after detailed examination on the wavelength database [13] the emission lines are confirmed due to Fe. Although their relative intensities lower compared to the strong emission line listed in Table 1, these emission lines appeared clearly with very strong intensities in the detected spectrum.

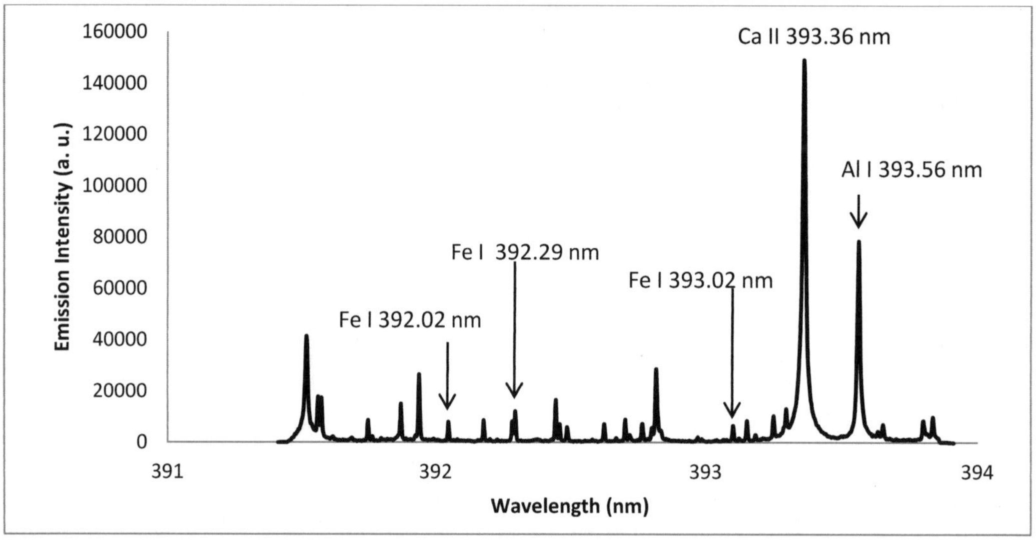

Figure 3. The emission spectra taken from the same plasma using the detection window of the OMA system set at center wavelength of 390 nm.

Next measurement was made on the same sample and experimental conditions by shifting spectral measurement to a detection window centered at the wavelength of 390 nm. Figure 3 presents the detected emission spectra in the wavelength region. It can be seen spectral lines due to atomic emission (Fe I), namely Fe I 392.02 nm, Fe I 392.29 nm, and Fe I 393.02 nm. In the measured spectra, intensities of the emission lines seem superficially weak. However, a careful inspection on the spectra figures out that their intensities are actually very strong, similar to that of Fe emission lines appeared in Figures 1 and 2. This is caused by the presence of the very strong emission line of Ca ionic emision (Ca II 393.36 nm), having about 10 times higher emission intensity, dominating the emission spectra at the region. In the spectrum, along with Fe and Ca lines, emission lines due to Al (Al I 393.56 nm) also occurs with strong emission intensity as soil contains Ca and Al as minor constituents. Figure 4 demonstrates emission spectra measured from the plasma when the spectral window was shifted further to the wavelength region centered at 405 nm. It can be found clearly several Fe atomic lines namely Fe I 400.52 nm, Fe I 404.58 nm, Fe I 406.35 nm and Fe I 407.17 nm. As shown in Table 1, the strongest Fe emission line in this wavelength region is Fe I 404.58 nm and it just coincides with the detected spectra displaying Fe I 404.58 nm as the line with highest emission intensity. The other Fe lines (Fe I 400.52 nm, Fe I 406.35 nm and Fe I 407.17 nm), even though not listed in the wavelength table above, Table 1, since their relative intensities are lower than 1000000, were confirmed as the correct assignment after detailed inspection on the wavelength database [13].

As observed above, the spectra detected at different wavelength regions from soil sample consist of high density of emission lines, especially Figures 1 and 2. After applying the above mentioned identifying procedures, we have able to assign the emission lines of Fe as shown in Figures 1, 2, 3, dan 4. For quantitative analysis, intensity of interest lines and set of lines are useful. The interested line should be free from interference of other lines (overlap), well isolated from other emission lines. The emission lines should also have high signal to background (S/B) ratio[15]. A careful examination on the emission lines of Fe displayed in Figures 1, 2, 3, and 4 found that the wavelength measurement region centered at 405 nm (Figure 4) is best spectral window for detecting Fe emission lines. This is because the detected emission spectra, in general the intensities of the emission lines are very strong in this spectral window. This strong emission intensity is especially observed for emission lines of Fe I 400.52 nm, Fe I 404.58 nm, Fe I 406.35 nm and Fe I 407.17 nm. The detected background emission intensity is relatively low, resulting in good signal to background (S/B) ratio in the spectral region. Moreover, in this wavelength region, the emission lines are clean, well isolated from other emission lines. The strongest Fe emission line in this wavelength region is Fe I 404.58 nm, leading this line to be best candidate as analytical line for detecting and making a qualitative and quantitative analysis of Fe. Along with Fe atomic lines, strong and well isolated lines due to Ti (Ti 398.98 nm and Ti I 399.86 nm) can be observed. In addition, Mn atomic triplet lines also appear in the detected spectrum. These lines (Ti and Mn) especially the triplet Mn could be used to confirm and calibrate the assignment of the emission lines of Fe in the wavelength region.

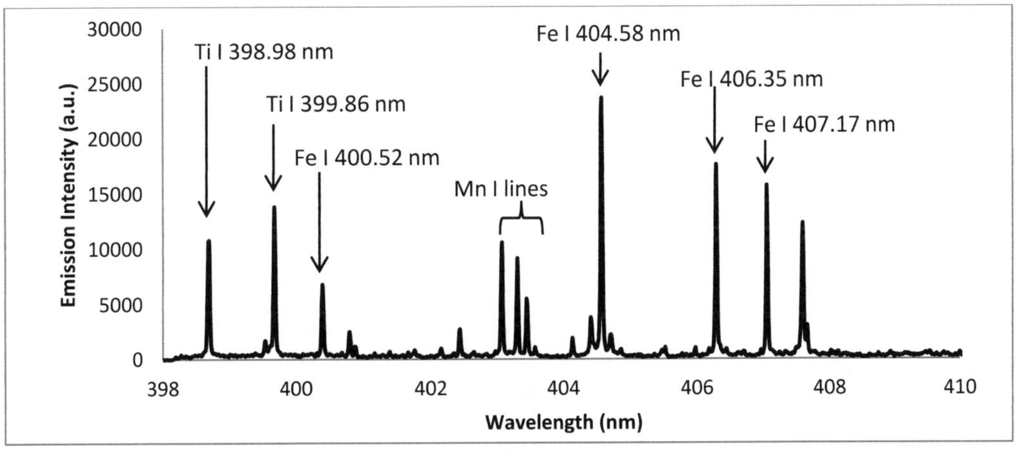

Figure 4. The spectra of the plasma emission detected using the spectral window of the OMA system centered at the wavelength of 405 nm.

4. Conclusion

The Fe emission lines with strong intensities can clearly be detected at various wavelength ranges especially the theoretically strongest atomic line of Fe I 374.55 nm appeared in the spectral windows centered at 373 nm or 376 nm. However, the strongest line Fe I 374.55 nm is not resolved well from Fe I 374.58 nm emission line since their separation is only about 0.03 nm. Profiles of the emission intensities of the detected Fe lines are in a good agreement with that of reference/database where the lines with high relative intensities in the database were found to display high intensities in the measured spectra and vice versa. A number of emission lines such as Fe I 392.02 nm and Fe I 392.29 nm are not well isolated. Several emission lines such as Fe I 374.94 nm probably overlap with other lines (interference), since its measured intensity profile is different from that of the database. Considering the spectral windows examined in this work, the spectral window centered at 405 nm is the best spectral window for detecting Fe emission lines as this region displays only several Fe emission lines featuring high emission intensities, well isolated, and high S/B ratio. The atomic emission line of Fe I 404.58 nm is the best line for detection and making a qualitative and quantitative analysis of Fe in soil sample as will be conducted in our near future work.

Acknowledgments

The authors (N. I., K. L., F., and M. R.) thank for the fundamental research grants from Research and Society Service Directorate (DP2M), Director of Higher Education, Ministry of Research, Technology and Higher Education, Government of Republic Indonesia, through Research Institute of Syiah Kuala University, Aceh, Indonesia with contract no.: 082/UN11.2/LT/SP3/2014 and contract no.: 061/UN 11.2/LT/SP3/2015.

References

1. Thompson, L. M. and F. R. Troeh. 1973. Soils and Soil Fertility, third ed. McGraw-Hill Book Company.
2. Wallihan, E. F. 1966. Iron. In: Chapman, H. D. Diagnostic Criteria for Plants and Soils. University of California, Div. Agric. Sci., Riverside, CA, pp. 203-212.
3. Brown, J. C., and W. E. Jones. 1977. Commun. Soil Sci. Plant Anal. 8: 1-15.

4. Chin, T. F. 1966. Iron and manganese absorption by rice plants. Soils Fert. Taiwan. 1-6.
5. Clements, H. F., Putnam, E. W., Suehisa, R. G., Yee, G. L. N., and M. L. Wehling. 1974. Hawaii Agric. Exp. Stn. Tech. Bull. 88. 52 pp.
6. De Dorlodot, S.; Lutts, S.; Bertin, P. J. Plant Nutr. 2005, 28, 1.
7. Bodek, I., Lyman, W. J., Reehl, W. F., and D.H. Rosenblatt. 1988. Environmental Inorganic Chemistry: Properties, Processes, and Estimation Methods. SETAC Special Publication Series, B.T. Walton and R.A. Conway, editors. Pergamon Press. New York.
8. Szczuciński, W., N. Chaimanee, P. Niedzielski, G. Rachlewicz, D. Saisuttichai, T. Tepsuwan, S. Lorenc and J. Siepak, 2006, Polish Journal of Environmental Study 15(5), 793-810.
9. Szczucinski, W., P. Niedzielski, G. Rachlewicz, T. Sobczynski, A. Zioła, A. Kowalski, S. Lorenc and J. Siepak, 2005, Environmental Geology 49, 321–33.
10. Chaerun, S. K., William B. Whitman, Stephan J. Wirth, and Ruth H. Ellerbrock, the 2009 National Meeting of the American Society of Mining and Reclamation, Billings, MT, Revitalizing the Environment: Proven Solutions and Innovative Approaches May 30 – June 5, 2009. R.I. Barnhisel (Ed.) Published by ASMR, 3134 Montavesta Rd., Lexington, KY 40502.
11. Winefordner, J. D., I. B. Gornushkin, T. Correll, E. Gibb, B. W. Smith, and N. Omenetto, 2004, Journal of Analytical Atomic. Spectrometry 19, 106-108.
12. Idris, N., K Kagawa, F. Sakan, K. Tsuyuki dan S. Miura, 2007, Applied Spectroscopy 61 (12) 1344-1351.
13. Kramida, A., Yu. Ralchenko, J. Reader, and NIST ASD Team (2014). NIST Atomic Spectra Database (ver. 5.2), [Online]. Available: http://physics.nist.gov/asd [2015, November 18]. National Institute of Standards and Technology, Gaithersburg, MD.
14. Idris, N., M. Ramli, R. Hedwig, Z. S. Lie, K. H. Kurniawan and K. Kagawa, American Institute of Physics (AIP) Proceeding, AIP Conference Proceedings 1719, 030051 (2016); doi: 10.1063/1.4943746.
15. Cremers, D. A. and L. J. Radziemski, 2006, Handbook of Laser-Induced Breakdown Spectroscopy, John Wiley and Sons, Ltd, England.

IOP Conf. Series: Journal of Physics: Conf. Series **846** (2017) 012021 doi:10.1088/1742-6596/846/1/012021

Study of Hydrogen and Oxygen and Its Reaction With Host Elements in Sandstone by Laser-Induced Breakdown Spectroscopy (LIBS)

Hery Suyanto*

Physics department, Faculty of Mathematics and Nature Sciences, Udayana University. Kampus Bukit Jimbaran, Jl. Kampus Bukit Jimbaran-Badung, Bali, Indonesia,80361

*email : hery6@yahoo.com

Abstract. A study of hydrogen and oxygen and its reaction with host elements in a sandstone has been done by laser-induced breakdown spectroscopy (LIBS). The sandstone was irradiated by Nd-YAG laser (1064 nm, 7 ns) with varied energy of 60 mJ till 140 mJ in surrounding air gas pressure of 1 atm and produced plasma. The emission intensities of hydrogen H I 656.2 nm and oxygen O I 777.2 nm in the plasma were captured by HR 2500^{+} spectrometer and displayed in intensity as a function of wavelength. The data show that the emission intensities of hydrogen and oxygen increase with increasing laser energy at a gradient of 5.4 and 11.8 respectively every increasing laser energy of 20 mJ. To characterize the reaction process between hydrogen and oxygen with the host elements of the sandstone, a 0.2 ml demineralized water was dropped on the sandstone surface and was analyzed as a function of delay time reaction and temperature. The data show that the oxidation reaction between host elements and oxygen occurred after 25 minutes that the oxygen emission intensity increases and the hydrogen emission intensity decreases. Another data also show that the increasing temperature of sandstone until 80 C increased intermolecular bond between oxygen and host element and dehydrogenation took place after reaching this temperature

1. Introduction

Laser induced breakdown spectroscopy (LIBS) has become an analytical tool in detection of major, minor elements, as well as, light element such as hydrogen presents in solid, liquid and gases. But rare report was found in case of studying reaction hydrogen and oxygen with the host elements. This mainly due to hydrogen is very light and fast-moving atom, as well as, easily reacts to other atoms. As a result, it is very difficult to know the hydrogen as host element or as impurity, and also a weak intensity observed in the emission

Many researchers had done different methods to improve the detection of Hydrogen element. D.A. Cremers et al[1], using the delay time detection method, the background intensity reduce and atomic intensity increase. Kurniawan et al[2-5], using low-pressure (5 torr) ambient helium had demonstrated the detection of the sharp H I 656.2 nm and D I 656.1 nm emission lines from the zircaloy sample. Z.S. Lie et al[6,7] study excitation mechanism using TEA Co_2 laser. Hery et al[8], using an orthogonal double-pulse laser-induced breakdown spectroscopy (LIBS) technique. It was

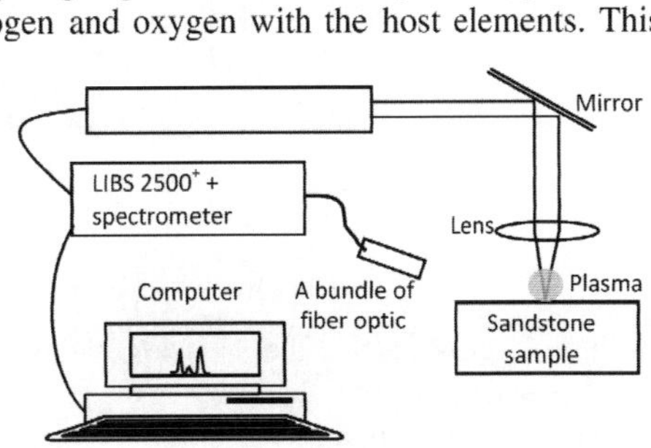

Figure 1. Experimental setup

Content from this work may be used under the terms of the Creative Commons Attribution 3.0 licence. Any further distribution of this work must maintain attribution to the author(s) and the title of the work, journal citation and DOI.
Published under licence by IOP Publishing Ltd

demonstrated, specifically, that a strong and sharp D emission line was detected with a time difference (τ = -1) μs and free from the interfering H I 656.2 nm spectral line. Based on evidences above, this research focused on the characteristics of hydrogen and oxygen in a sandstone sample and its reaction to other host elements called oxidation and reduction processes.

2. Material and Method

2.1 Sample Preparation

In order to understand the characteristics or the reaction process of hydrogen and oxygen in the sandstone, three treatments were conducted in this experiment: first, without added demineralized water to sandstone. This sample used to characterize hydrogen and oxygen as host elements in the sandstone. Second without heating the sandstone but varying the delay time reaction before laser irradiation. In this treatment 0.2 ml demineralized water was dropped onto the surface of sandstone and time lag of 5, 15, 25, 35 and 45 minutes before laser irradiation ; third, by varying the heating temperature (45, 65, 75, 80 and 90^0C) after certain time of optimum delay time reaction process in the second treatment.

2.2 Experimental Setup

The experimental set up used in this study is illustrated in Figure 1. The commercial LIBS system consists of Ocean Optics HR 2500$^+$ spectrometer, a sample chamber, Nd:YAG laser (model CRF 200, 1,064 nm, 7 ns) and OOILIBS software. For this experiment, the Nd-YAG laser was operated at 10 Hz repetition rate with energy of 60 - 140 mJ. Using 150 magnifier included commercial LIBS, the laser irradiation was guided and focused through 10 cm focal length biconvex lens onto a sand on sandstone sample surface and the resulted emission radiation was analyzed by optical multichannel analyzer containing 14,336 CCD pixels in the wavelength region of 200 – 980 nm with resolution of 0.1 nm. All the experiments were conducted in surrounding gas of air at 760 Torr (1 atm) pressure.

3. Results and Discussion

When a pulse laser is focused on surface sample of a sand on sandstone, the leading edge of the pulse rapidly heats, melts and vaporizes a small amount of material sample into a layer just above the surface. This material, then moves fast and compresses a surrounding gas to produce shockwave. The energy of shockwave is used to excite electrons in the atom. These electrons will decay to ground state and emit a photon with a specific wavelength. The emissions are captured by spectrometer and displayed in intensity as a function of wavelength as shown in figure 2.

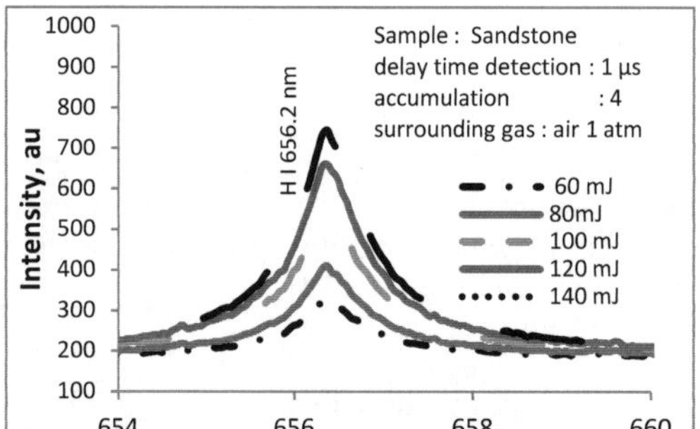

Figure 2. A Spectrum of Hydrogen in a various laser energy.

Figure 2 shows that the emission intensity of Hydrogen increases with the increasing laser energy. The lower laser energy, the lower energy of shockwave deposited in the gas behind it and a consequence a small amount of hydrogen atoms excited.[9] The higher energy laser, the energy of shockwave is strong enough and a result a lot of hydrogen atoms excited. Hence, the emission intensity of Hydrogen increases.

Figure 2 is part of spectrum emission of sandstone in hydrogen region. Actually, the sandstone contains some emission lines of elements such as Si, Na, H, K, Ca, O, N, Fe, Al, Ba, and Mg. To understand the characteristics of Hydrogen and Oxygen elements, first, it is plotted the emission intensity of Hydrogen and Oxygen as host elements in the sandstone as a function of laser energy displayed in Figure 3.

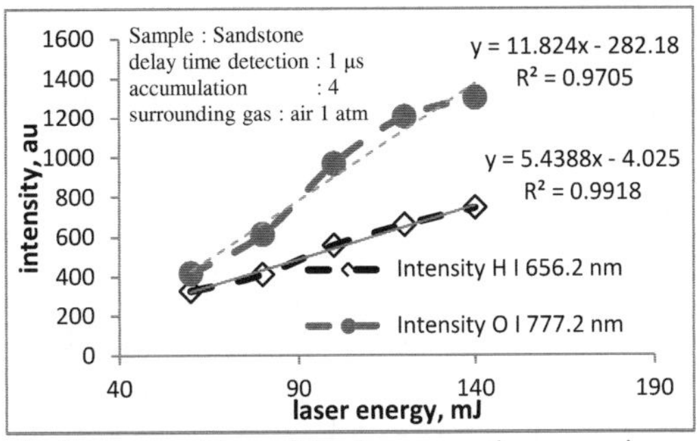

Figure 3. The emission intensities of Hydrogen and oxygen in various laser energy.

Figure 3, shows a linear relationship between emission intensity and laser energy. The data state that the intensity of both hydrogen and oxygen increase with increasing laser energy with a slop of 5.4 and 11.8 respectively, every increasing laser energy of 20 mJ.

To characterize the reaction process between hydrogen and oxygen with other host elements in the sandstone, a 0.2 ml demineralized water was dropped on the sandstone surface and was analyzed as a function of delay-time reaction and temperature as shown in Figure 4 and 5.

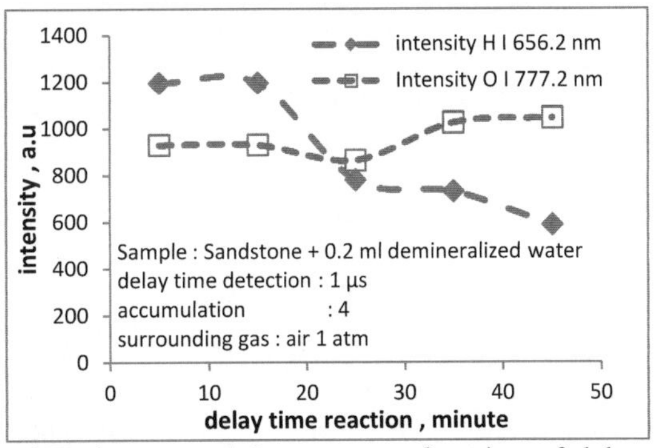

Figure 4. Emission Intensities of Hydrogen and Oxygen as a function of delay-time reaction process radiated with laser energy of 120 mJ.

$$2X + 2H_2O \rightarrow 2XOH + H_{2(gas)} \quad (1a)$$

$$\downarrow$$

$$2XO + H_2 \quad (1b)$$

Data in Figure 4 state that in early time 5 – 15 minutes, the emission intensity of Hydrogen increases about twice compare to hydrogen as host element in figure 3 at the same laser energy of 120 mJ. To explain this phenomenon, it is better to show a redox reaction as shown in equation 1. The reaction in equation 1a is illustration to explain redox reaction between H_20 with host element (X) and produce base (XOH) and H_2 gas. Before leaving the sandstone, the H_2 gas was irradiated by laser and as result the emission intensity of hydrogen rises. However, after 25 minutes, this emission intensity reduces that can be caused by releasing of this gas and after 45 minutes the intensity of hydrogen is the same as in host element.

On the other hand, in the time of 5 till 25 minutes, the emission intensity of Oxygen decreases about three fourth compare to oxygen in host element in figure 3 in the same laser energy. This can be caused by the wet condition of the sandstone that it has weaker intermolecular bond and it also absorbs laser energy. But this intensity increases after 25 minutes and it is caused by increasing intermolecular bond of the base. To understand more detail the Hydrogen and Oxygen characteristics, a 0.2 ml demineralized water was dropped on sandstone surface and after 15 minutes this stone was heated 5 minutes with varied the temperature of 45 ^{0}C till 90 ^{0}C and the result is in figure 5.

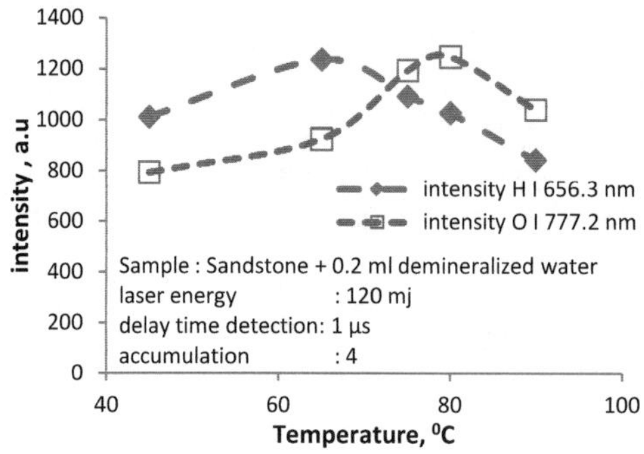

Figure 5.. Emission intensities of Hydrogen and oxygen as a function of temperature, radiated with laser energy of 120 mJ.

Data in figure 5 show that in the early temperature 45 ^{0}C , the emission intensity of both hydrogen and oxygen decrease compare to the condition 15 minutes in data figure 4 but nearly the same value with condition in 25 minutes in the same figure. The reason is some Hydrogen gas was released from the sandstone and also the intermolecular bond in the base was not strong enough to react with laser. At the temperature of 65 ^{0}C, the emission intensity of both hydrogen and oxygen increase but reversed phenomena with the figure 4 after 25 minutes delay time reaction. This phenomena can be explained that emission intensity of hydrogen in figure 4 only comes from H_2 gas resulted from redox reaction (eq. 1a), however, in figure 5, hydrogen emission is also from H_2 gas from dehydrogenation (eq. 1b).
When the stone was heated up until 80 ^{0}C, the emission of hydrogen decreased as hydrogen gas in the stone was released. In other hand, the same reason in the figure 4, the emission intensity of oxygen in the figure 5 continue increasing, but it's value is higher than in the case of delay time reaction. At higher temperature of 80 ^{0}C, the oxygen and hydrogen dissociated and come out from stone before irradiating laser. As a result the emission intensity of these elements decreased.

4. Conclusion and Future Work

Based on data, it can be concluded that the redox reaction process can be studied using LIBS method. Redox phenomenon was analyzed with variables of delay time detection and temperature of sample. The data show that the oxidation reaction between host elements and oxygen occurred after 25 minutes that the oxygen emission intensity increases and the hydrogen emission intensity decreases. Another data also show that the increasing temperature of sandstone until 80 C increased intermolecular bond between oxygen and host element and dehydrogenation took place after reaching this temperature.

Acknowledgments

This work was supported by Lembaga Penelitian dan Pengabdian Kepada Masyarakat Unud under contract no : 246-308/UN14.2/PNL.01.03.00/2015. Special thanks to integrated laboratory FMIPA-Udayana University for LIBS equipment.

References

1. David A. Creamers and Leon J. Radziemski, *Handbook of Laser-Induced Breakdown Spectroscopy*, John Wiley & Sons, Ltd, 2006, p.24
2. Kurniawan, K.H.; Kagawa, K. *Appl. Spectroscopy,* rev. 2006,41,99-130
3. Kurniawan, K.H.; Lie,T.J.; Suliyanti, M.M., Hedwig, R.; Pardede, M.; Kurniawan, D.P.;Kusumoto, Y.; Kagawa, K. *Anal. Chem.* 2006,78,5768-5773
4. Koo H. Kurniawan, T.J. Lie, M.M. Suliyanti, R. Hedwig, M. Pardede, M. Ramli, H. Niki, S.N. Abdulmadjid, N. Idris, K. Lahna, Y. Kusumoto, K. Kagawa, M.O. Tjia, *The Role of He in Enhancing the Intensity and Lifetime of H and D Emissions from Laser-Induced Atmospheric-Pressure Plasma,* J. Appl. Phys. **105** (2009) pp. 103303-1-6.
5. Koo H. Kurniawan, T.J. Lie, M.M. Suliyanti, M. Pardede, S.N. Abdulmadjid, K. Kagawa, M.O. Tjia, *Quenching of He-Induced Intensity Enhancement Effect in H and D Emission Produced by Nd-YAG Laser Irradiation on Solid Targets in Low Pressure Helium Gas*, J. Appl. Phys. **105** (2009) pp. 013301-1-7.

6. Z.S. Lie, A. Khumaeni, H. Niki, K. Kurihara, K.H. Kurniawan, R. Hedwig, K. Fukumoto, K. Kagawa, Y.I. Lee, *A Comprehensive Study of H Emission in TEA CO_2 Laser-Induced Helium Gas Plasma for Highly Sensitive Analysis of Hydrogen in Metal Samples*, J. Korean Phys. Soc. **61**, 1, (2012) pp.49-54.
7. Z.S. Lie, A. Khumaeni, K. Kurihara, K.H. Kurniawan, Y.I. Lee, K. Fukumoto, K. Kagawa, H. Niki, *Excitation Mechanism of H, He, C, and F Atoms in Metal-Assisted Atmospheric Helium Gas Plasma Induced by Transversely Excited Atmospheric - Pressure CO_2 Laser Bombardment*, Jpn. J. Appl. Phys., **50** (2011) pp. 122701 1-7.
8. H. Suyanto; Z.S. Lie; H. Niki; K. Kagawa; K. Fukumoto; R. Hedwig; H.K. Kurniawan, *Analytical Chemistry* 2012,84,2224-2231
9. K. Kagawa and Koo H. Kurniawan: *Laser Induced Shock-Wave Plasma Spectroscopy, Research Trends: Trends in Applied Spectroscopy*, (Research Trends, India, Eds. R.M. Barnes and M.W. Blades, vol. **2**, 1998) pp. 1-36.

Proximate and The Calorific Value Analysis of Brown Coal for High-Calorie Hybrid Briquette Application

Erzam Sahaluddin Hasan[1], Muhammad Jahiding[1], Mashuni[2], WOS Ilmawati[1], Wa Wati[1], I Nyoman Sudiana[1]

[1]Department of Physics, Faculty of Mathematics and Natural Sciences, Halu Oleo University
[2]Department of Chemistry, Faculty of Mathematics and Natural Sciences, Halu Oleo University

E-mail : erzamshasan@yahoo.com

Abstract. A study has been conducted about the quality of young coal (brown coal) briquettes from North Kolaka to determine the effect of varied adhesive on the proximate characteristics and calorific value. The young coal briquettes were made by using adhesives of starch, cassava starch and Castor oil plant starch at a concentration of 5 to 15% of the total mass. The grain size of young coal and the adhesive used were 60 mesh and 100 mesh, respectively. The samples were molded in a cylindrical mold with a diameter of 2.5 cm and a high of 6 cm, and with a pressure of 100 kg/cm^2. After having been compacted, the young coal samples were then analyzed proximately i.e. moisture content, volatile matter, ash content and fixed carbon, as well as their calorific values calculation. The results showed that the increase of the adhesive could tend to increased the water content and volatile matter, but reduced the ash content, and the fixed carbon tend to constant except coal briquettes using starch adhesive it were increased. The calorific value of the young coal briquettes increased for all kinds of adhesives when the adhesive increased. The calorific value per one gram ranged from 3162.7 cal/g to 4678.7 cal/g. The highest calorific value, 4678.7 cal/g, was observed at the adhesive of 15 % of starch. The characteristics of young coal can be used as a raw material for making high-calorie hybrid briquettes.

1. Introduction

The use of non-renewable energy from fossil continues to rise causing the reserves of this fuel energy sources growingly scarce, and will eventually be finished. On the other hand, the fast increase of world's fuel energy price has impacted on the increase of price of fuel energy including that of fuel and gas in Indonesia. Currently, Indonesian government fuel subsidy is about forty-nine trillion rupiahs per year to meet its community's need of about ten million kilo littre per year [1-3]. Therefore, there needs to be innovative research to develop new energy sources with high calorie as alternative sources of fuel for households and small industries, so the energy inventory well sustains.

One of the alternatives energy sources potential for fuel development is brown coal which can be blended with organic wastes (husk of rice, sago waste, and cashewnut husk). As an agricultural country, Indonesia is riches in agricultural wastes less utilized. Statistical data show that there are 110,498 ha of rice farmland in Southeast Sulawesi producing 423.317 tonnes of dried rice in 2007 [4-5]. In addition to rice husk waste, wastes from sago and cashewnut are also commonly found in this province which is one of the biggest producers of sago and cashewnut in Indonesia. The Agricultural wastes, which are biomass, become the huge source of alternative energy containing relatively high energy These biomass have been processed and utilized as a ceramic and other more valuable materials [6-8] .

Content from this work may be used under the terms of the Creative Commons Attribution 3.0 licence. Any further distribution of this work must maintain attribution to the author(s) and the title of the work, journal citation and DOI.
Published under licence by IOP Publishing Ltd

Southeast Sulawesi has coal potential about 9,000,000 tonnes spreads in Wawonii, Konawe dan North Kolaka [9]. Brown coal has high fixed carbon and it is suitable for briquette application, but contains low volatile matter causing it inefficient in its use, while organic wastes from rice hull, for instance, contain high volatile matter, but has low fixed carbon [10]. To improve the quality of briquette, we can combine the materials or carefully control the processing method [10-11]. For example we can control briquette's microstructure by using certain heating rate or by using appropriate technology such as microwave energy. Application microwave for processing materials have been reported by several scholars [12-14]. By using this waves as an alternative energy for heating, the microstructure as well as mechanical property of ceramics can be improved [15-17]. The use microwave for processing of briquette and other biomass based materials has been performing in Halu Oleo University and will be reported in separated papers.

1.1 Coal and Its Classifications

In general, coal is classified into 5 levels (from the highest to the lowest level) based on the relative content of the elements carbon (C) and water (H_2O) contained in coal, namely: anthracite, bituminous, sub-bituminous, lignite and peat (peat). According to Bateman, in anthracite, C content is relatively higher than H_2O, while in bituminous and peat, C content is relatively lower than H_2O. In bituminous, C content is relatively lower than C content in anthracite, whereas in bituminous H_2O content is relatively higher than H_2O content in anthracite [11]. Lignite coal is the softest, and its density is still in the first level. From geological point of view of, lignite coal is the youngest because it is composed of volatile and water content with low levels of fixed carbon. Bituminous coal is also young coal typically used in households and factories because it contains enough volatile matter, but is relatively high in calorific values, so it can produce a higher flame temperature [11]. On the other hand, anthracite coal is the hardest coal and is smoky when burned. One characteristic of the anthracite coal is that it has low levels of hydrocarbons [11]. This type of anthracite coal is the oldest coal from the geological point of view because it is hard coal, comprises of major components of carbon with little volatile matter content and contains almost no water.

Coal is heterogeneous, both in terms of chemical composition and physical features. The physical features of coal include heat value, moisture content, volatile matter and ash. The chemical properties of coal depend on the content of various chemical substances such as carbon, hydrogen, oxygen, and sulfur. The calorific values of coal are different from one mine to another. Anthracite coal's properties are very shiny black and compact, very high in calorific value, in carbon content, and low in water, ash and sulfur content. Bituminous / subbituminous coals are shiny black, but less compact, in their characteristics and high in calorific values with relatively high carbon content, and are little in water, ash and sulfur contents. In contrast, the properties of lignite coal are black and very brittle, low in calorific value with little carbon content, and has a high water content and ash content and sulfur rich [11].

1.2 Coal Briquettes

Coal briquettes are solid fuel made from coal with little mixture of clay, tapioca and cassava starch. The main raw material of coal briquettes are coal sources sufficient for the region of Southeast Sulawesi [1,2]. Briquettes properties are described as follows:

- **Water Content**

The water content is one of the components of solid fuel. The water content of solid fuels is the ratio of the weight of water contained in the solid fuel to the dry weight of the solid fuel. The water content in the solid fuel composed of internal / hygroscopic water and external / mechanical water. The water content will negatively affect the calorific value and combustion characteristics of solid fuel [9].

- **Ash Content**

Ash content is the measure of material contents and a variety of inorganic materials in the material tested. According to Earl [19], ash is the material remaining, for example, in wood heated to a constant weight. Ash content is comparable to the content of inorganic materials in the timber. One of the main elements contained in the ash is silica whose influence is unfavorable on the value of the heat produced. Ash composes of mineral materials such as clay, silica, calcium, magnesium oxide, etc.

- **Volatile Matter**

Volatile matter is material content flammable, and easily forms gas. High levels of volatile matter in agricultural waste indicates that the agricultural waste is easily flamable and burned, even faster burning and difficult to control [20].

Bio-briquette ignition temperature is lower and the burnout time is shorter compared to coal briquettes. When briquettes are heated, the temperature rises, and after reaching a certain temperature, the volatile matter gets out and burns around the briquette. Flame temperature is low if the more volatile matter biomass mixture and biomass flame temperature is lower than that of coal. The addition of biomass on briquette coal can increase the briquette flame.

➢ **Fixed Carbon**

The component which does not form gas when burnt is "fixed carbon" or FC. [21]. FC has an important role in determining the quality of charcoal because it can influence the amount of calorific value produced. The higher the bound carbon in the charcoal, the higher the calorific value produced. Quality charcoal is the one with high calorific value and fixed carbon.

➢ **Calorific Value**

The number of calorie units released by perfect burn of a mass unit or certain fuel volume is defined as burnt calorific value of the fuel. As for solid fuel like coal, the burnt calorific value is measured with calorimeter bomb [22]. Calorimeter bomb is an instrument used to define the heat released by fuel and oxygen at a constant volume. The instrument was invented by Prof. S. W. Parr in 1912, and therefore, it is often called Parr Oxygen Bomb Calorimeter [9]. Burn calorific value of fuel is defined by means of calorimeter bomb, by deheating the burnt down to initial termperature so as the steam produced in the burning of the fuel will condensed and laten heat of the steam will be released, thus, the total burn calorific value contain heat of the steam [22].

2. Material and Method

2.1 Preparation of Brown Coal Briquette

The brown coal was crashed into small pieces, and then put into tins. Further, the coal dried at 150°C for ±2 hours maintaining the oxygen to be at appropriate level, and then removed to cool down, then filtered using 60 mesh sieve.

➢ **Preparation of Adhesive**

Ten grams adhesive (starch, cassava starch, castor oil plant starch) were put into a 250 ml beaker and added with 90 ml of water, then heated in a boiling pot until it forms gel.

➢ **Mixing of Adhesive and Coal Powder**

Sixty-mesh grain coal was mixed with adhesives (starch, cassava starch, castor oil plant starch) each with the composition of 5%, 10% dan 15% consecutively, stirred by means of homogenyzer to obtain homogenous mixtures.

➢ **Coal Compaction**

Adhesive-coal mixture was scaled and then put in the briquette cylinder mold, 2.5 cm in diameter and 6 cm high. The Mold was placed under compaction tool, followed by compaction of samples up to 100 kg/cm^2 pressure.

➢ **Briquette Drying**

The compacted briquettes were then be dried in the open air for four days under the temperature of ± 40°C. After the drying process, analysis of proximate of the briquettes was made.

2.2 Analysis of Brown Coal Briquette

➢ **Water Content**

The water content of brown coal was determined by means of the following equation:

$$\text{Water Content (\%)} = \frac{[\text{BS}-\left(\text{BC+SP }(105^0\text{C})\right)-\text{BCK})]\text{ x }100\ \%}{\text{Samples Weight}} \quad (1)$$

➢ **Ash Content**

The ash content of the brown coal was determined by means of tanur heating with the temperature up tp 600^0 C and coolant desiccator. The percentage of ash content was calculated using formula W1/W2 x (100 %), where W1 was the weight of ash and W2 was the weight of dried sample.

➢ **Volatile Matter**

The content of volatile matter of brown coal can be determined using the following equation:

$$\text{Volatile Matter (\%)} = \text{volatile content }[\text{KZH}(750^0\text{C})] - \text{ water content} \quad (2)$$

$$KZH\ (\%) = \frac{[BS-\left(BC+SP(750^0C)\right)-BCK)]\ x\ 100\ \%}{samples\ weight} \quad (3)$$

➢ **Fixed Carbon**

The content of fixed carbon of brown coal can be determined using formula FC = 100 - (Ka + Vm + Abu) where Ka was water content, Vm volatile matter and Ka was ash content.

➢ **Calorific Value**

The calorific value of brown coal was analyzed using Calorimeter Bomb. The calibration of the burn equipment was done using benzoate acid as the standard to obtain Tare Energy (W). To obtain calorific value of the brown coal briquette the following equation was used:

$$\Delta t = T_2 - T_1 \quad (4)$$

$$W = \frac{6320\ x\ M}{\Delta t}$$

$$E = \frac{W\ x\ \Delta t}{M}\ Cal/gram \quad (5)$$

Where M was benzoat mass weight, Δt benzoat acid temperature, W energy, E calorific value of the burn and 6320 was calorific value /1 gr benzoat acid.

3. Results and Discussion

3.1 The Quality of Brown Coal

The processing of brown coal into briquettes underwent several steps: brown coal carbonization, smoothing and filtering, mixing adhesive, molding, compacting, and briquette drying. Carbonization is a process to converse raw material into charcoal under the heating process. Carbonization process of coal was done burning coal by supplying limited oxygen. Carbonization aimed to improve the carbon and released volatile matter as well as reducing the water content.

Carbonated coal was made briquette using starch adhesive, cassava starch, and castor oil plant starch. Prior to this, the coal charcoal was filtered 60 mesh to obtain uniform granules. The use of adhesive is intended to placard two objects base on the surface bond so that in the process of briquetting, the coal granules can stick well. In the next step, brown coal was mixed with compaction adhesive at 100 kg/cm^2 pressure to obtain compact briquettes, in the form of solids with high granule density. Further, the briquettes were dried to reduce their acidity caused by addition of adhesive.

Then, the water content, ash content, volatile matter, fixed carbon, and calorific value of the coal briquettes, which has been dried, were analyzed. Coal briquette analysis results are shown in Table 1.

❖ **Water Content**

Water content of the brown coal briquettes are influenced by the type of adhesive used. The adhesive of the starch contain lower water compared to the adhesive of cassava starch as well as castor oil plant starch at the composition of 5 % and 10 % adhesive. At the composition of 15 % adhesive, the highest water content was in starch. Figure 1 shows the result of water content of the brown coal briquettes.

Table 1. The Result of Proximate Analysis and Calorific Value of the Brown Coal Briquettes

Test Para-meter	Types of Adhesive	The composition of Adhesive in Briquette (%)		
		5	10	15
Water content (%)	Starch	2.00	2.67	3.33
	Cassava starch	2.33	2.67	3.00
	Castor oil plant starch	2.33	3.00	3.00
Ash content (%)	Starch	8.33	7.33	5.00
	Cassava starch	12.33	8.33	7.67
	Castor oil plant starch	14.00	11.67	9.67
Volatile matter (%)	Starch	16.67	16.67	18.33
	Cassava starch	14.33	18.00	18.33
	Castor oil plant starch	14.33	16.00	18.33
Fixed carbon (%)	Starch	73.00	73.33	73.34
	Cassava starch	71.01	71.00	71.00
	Castor oil plant starch	69.34	69.33	69.00
Calorie (cal/gr)	Starch	4595.9	4645.8	4678.7
	Cassava starch	3256.9	3925.9	4291.7
	Castor oil plant starch	3405.3	3162.7	3445.1

The bigger the adhesive composition of the water content of brown coal briquettes the bigger it is for all types of adhesive. The water content was high because the adhesive used was not dried before used.

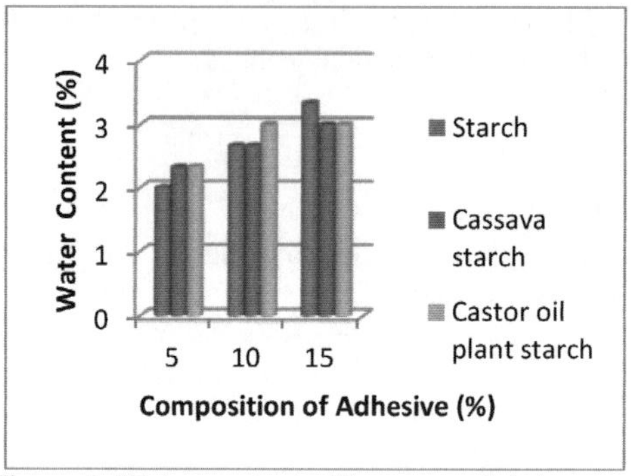

Figure 1. Water content of brown coal briquettes

❖ Ash Content

The ash content of the brown coal briquettes is influenced by adhesive composition, in the sense that the bigger the adhesive composition the higher the ash composition. This is because the adhesive contains high organic matter resulting in a high ash when baking is done. Figure 2 shows brown coal briquettes' ash content for the three types of adhesives.

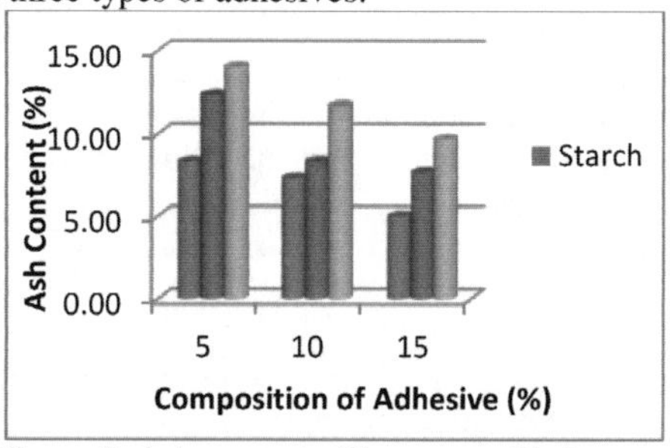

Figure 2. Ash content of brown coal briquettes

❖ Volatile Matter

Volatile matter content in the adhesive contributes positively to brown coal briquettes. The greater the adhesive composition of the greater the volatile matter composition is. The composition of the volatile matter in the brown coal briquettes can improve the quality of the briquette. However, if excessive, the composition can reduce the quality of the briquettes. Of the three types of adhesives. starch is the one that provides most volatile matter, i.e. 18.33%. Figure 3 shows the volatile matter of brown coal briquettes for three adhesive compositions.

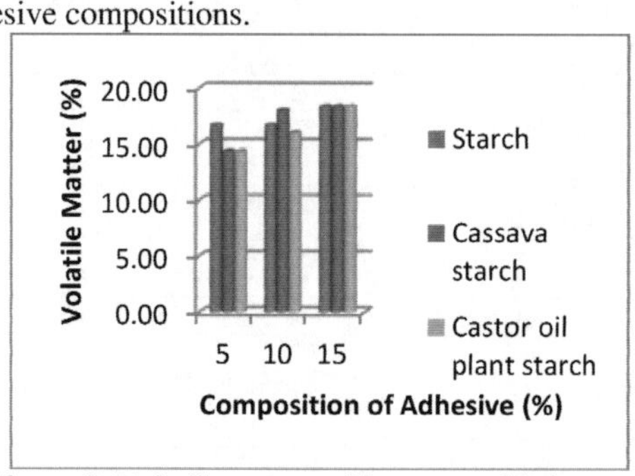

Figure 3. Volatile matter of brown coal briquettes

❖ Fixed Carbon

Fixed carbon content of the briquette determines the calorific value. The addition of adhesive does not directly influence the increase of fixed carbon, but rather due to changes in water content, ash content, and volatile matter of brown coal briquettes. Figure 4 shows the fixed carbon of brown coal for some of the adhesive compositions.

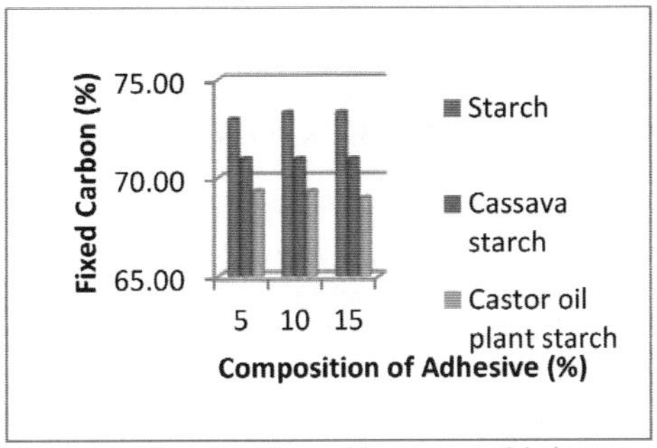

Figure 4. Fixed carbon of brown coal briquettes

3.2 Calorific Value of the Brown Coal Briquettes

The results of the analysis of the coal briquettes using Bomb Calorimeter is presented in Table 1. From the results of this analysis, it is found that the highest calorific value of brown coal briquettes is in the composition of 15% for starch adhesive with a calorific value of 4678.7 cal/gram. The high calorific value in that composition was influenced by the low water content, ash content, and volatile matter values of the briquettes.

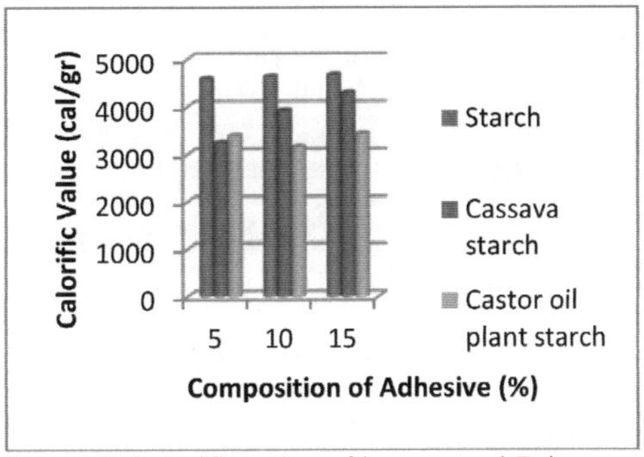

Figure 5. Calorific value of brown coal Briquettes

The relatively small size of the granules of brown coal briquettes (60 mesh) indicated that the pores were so small that the briquettes' moisture (water content) contained in them was difficult to evaporate during the drying process. Consequently, the water content in the briquette was getting higher. High water content reduces the calorific value since briquette's heat is first used to evaporite water that existed prior to radiation emission used as the heat of combustion. Starch and cassava starch adhesives have the highest calorific value, 4678.68 cal/g and 4291.68 cal/gram consecutively, at adhesive composition of 15%. Figure 5 shows the calorific value of brown coal briquettes in various types and composition of adhesive.

Figure 5 shows that the addition of adhesive causes an increase in calorific value. However, the addition of adhesive at higher concentrations reduces calorific value instead [20]. This is because an increase in the volume of adhesive causes the decline of fixed carbon. Thus, it is necessary to determine the optimum composition of the adhesive to obtain the maximum calorific value.

4. Conclusion

The percentage of adhesive affects the calorific value of brown coal briquettes, where higher level of adhesive increases calorific value, that is at the composition of 5% to 15%, the calorific value is 4678.68 cal/gram. Starch adhesives provides the greatest calorific value in all adhesive compositions, of wich they are 4595.91 cal/gram, 4645.77 cal/gram, and 4678.68 cal/gram subsequently, for the composition of 5%, 10% and 15%.

Acknowledgement

The authors acknowledged financial support from Directorate of Higher Education (DIKTI) contract No: 116/SP2H/PL/DIT.LITABMAS/V/2015.

References

1. Arnold. Guy. 1987. Batubara. PT Pradnya Paramita. Jakarta [In Indonesian]
2. Departemen Energi dan Sumber Daya Mineral (DESDM), 2003, Kebijakan Pengembangan Energi Terbarukan dan Konservasi Energi (Energi Hijau).
3. Francis. W. 1996, Solid Fuel, Fuel dan Fuel Technology, 1 st Volume, Pergamon Press, London
4. Dinas Perkebunan dan Hortikultura, Sultra 2013. Statistik Perkebunan Sulawesi Tenggara Th. 2012.
5. Hara. et.al. 1986. Utilization of Agrowastes for Building Materials. J. Ilmu Dasar. Vol. 3 No. 2. 2002: 98-102. International Research and Development Cooperation Division. AIST. MITI. Japan.
6. Dinas Pertambangan dan Energi Propinsi Sulawesi Tenggara. 2007. Laporan Penyelidikan Batubara Kec. Ngapa Kab. Kolaka Utare Provinsi Sulawesi Tenggara. Sulawesi Tenggara. [In Indonesian]
7. H. Aripin, S. Mitsudo, I. N. Sudiana, N. Jumsiah, I. Rahmatia, B. Sunendar, L.Nurdiwijayanto, S. Mitsudo, S.Sabchevski, Preparation of Porous Ceramic with Controllable Additive and Firing Temperature, Advanced Materials Research, Vol. 277 (2011) pp. 151-158
8. I N Sudiana, S. Mitsudo, T. Nishiwaki, P. E. Susilowati, L. Lestari, M. Z. Firihu, H. Aripin, Synthesis and Characterization of Microwave on Sintered Silica Xerogel Produced from Rice Husk Ash, Journal of Physics: Conference Series Vol. 739 No. 1, (2016) 012059
9. Husada. TI. 2008. Arang briquette tongkol jagung sebagai Energi Alternatif. Universitas Negeri Semarang. Semarang. [In Indonesian]
10. M. Jahiding. L.O. Ngkoimani. E.S. Erzam dan S. Maymanah. 2011. Analisis Proksimasi dan Calorific value Bioarang Sekam Padi sebagai Bahan Baku Briquette Hybrid. Jurnal aplikasi Fisika. Vol 7 No. 2. 2011: 77 – 83. Jurusan Fisika FMIPA Universitas Haluoleo. [In Indonesian]
11. Sule. D. dan Sinaga. P.. 1998. Pembuatan Briquette Tanpa Asap dan Tak Berbau dari Batubara Halus dengan Sekam Padi dan Molase. [In Indonesian]
12. I.N. Sudiana, Ryo Ito, S. Inagaki, K. Kuwayama, K. Sako, S. Mitsudo, Densification of Alumina Ceramics Sintered by Using Sub-millimeter Wave Gyrotron, J. Infrared, Millimeter, and Terahertz Waves. 34 (2013), 627-638.
13. I.N. Sudiana, S. Mitsudo, T. Nishiwaki, P. E. Susilowati, L. Lestari, M Zamrun Firihu, H. Aripin, Effect of Microwave Radiation on the Properties of Sintered Oxide Ceramics, Contemporary Engineering Sciences, Vol. 8 No. 34, (2015), 1607-1615
14. I. N. Sudiana, I.N., S. Mitsudo, K. Sako, S. Inagaki, L. O. Ngkoimani, I. Usman, H. Aripin, The microwave effects on the properties of alumina at high frequencies of microwave sintering, American Institute of Physics (AIP) Conference Proceeding, **1719**, 030014 (2016); doi: 10.1063/1.4943709
15. S. Mitsudo, S. Inagaki, I.N. Sudiana, K. Kuwayama, Grain Growth in Millimeter Wave Sintered Alumina Ceramics , Advanced Materials Research, Vol.789 (2013), pp. 279-282.
16. H. Aripin, S. Mitsudo**,** I Nyoman Sudiana, Bambang Sunendar , Svilen Sabchevski, Structural Characterization of a Glass Ceramic Developed from TiO_2 and a Novel Material-Silica Xerogel Converted from Sago Waste Ash, Material Science Forum Vol. 872, 2016, pp.81-86.
17. I. N. Sudiana, S. Mitsudo, M. Z. Firihu, Effect of Initial Green Samples on Mechanical Properties of Alumina Ceramic , Contemporary Engineering Sciences, Vol. 9, 2016, no. 12, 595-602
18. I N Sudiana, S. Mitsudo, M. Z. Firihu, H. Aripin, Effect of High-Frequency Microwave on Micro Hardness of Alumina Ceramic, Material Science Forum , Vo. 872, 2016, pp.114-117.
19. Sukandarrumidi. 2006. Batubara dan Pemanfaatannya. Gajah Mada University Press. Yogyakarta. [In Indonesian]
20. Pebriadi. B. dan Mastur. 2008. Pemanfaatan Batubara dan Sekam Sebagai Energi Alternatif di Rumah Tangga Perdesaan. Balai Pengkajian Taknologi Pertanian Kalimantan Timur. [In Indonesian]
21. Kulshrestha. S.K. 1989. Termodinamika Terpakai. Teknik Uap dan Panas. Universitas Indonesia UI-Press. Jakarta [In Indonesian]
22. Marsudi. Djiteng. 2005. Pembangkit Energi Listrik. Erlangga. Jakarta. [In Indonesian]

IOP Conf. Series: Journal of Physics: Conf. Series **846** (2017) 012023 doi:10.1088/1742-6596/846/1/012023

Fast Drying of Agriculture Commodities by Using Microwave

La Ode Ngkoimani[1*], Megawati[2], Gde Purwana Saputra[3], Edi Cahyono[4], Haji Aripin[5], Komang Gde Suastika[6] , I Nyoman Sudiana[2]

[1]Department of Geophysics, Faculty of Earth Sciences and Technology, Halu Oleo University, Kampus Bumi Tridharma Anduonohu, Kendari 93232, Indonesia
[2]Department of Physics, Faculty of Mathematic and Natural Science, Halu Oleo University, Kampus Bumi Tridharma Anduonohu, Kendari 93232, Indonesia
[3]Faculty of Learning Teacher and Education Science, Sebelas November University
[4]Department of Mathematics, Faculty of Mathematic and Natural Science, Halu Oleo University, Kampus Bumi Tridharma Anduonohu, Kendari 93232, Indonesia
[5]Center for Material Processing and Renewable Energy, Faculty of Learning Teacher and Education Science, Siliwangi University, Jl. Siliwangi 24 Tasikmalaya 46115, West Java, Indonesia
[6] Department of Physics Education, Palangka Raya University, Indonesia, 73112

Email: laodem@yahoo.com

Abstract. Some progress has been made and reported previously due to investigate microwave effects to materials. The microwave applications for material processing by using wide range microwave frequencies such as in sintering, chemical reaction, and drying have been performed. Microwave drying is based on a unique volumetric heating mode with electromagnetic radiation at 2,450 MHz. However, the quest for a what a true microwave effect is still plagued with difficulties .This paper provides a experimental and theoretical analysis of drying materials using microwave. For drying experiments, in this investigation, we were using a domestic microwave oven which operated at three power levels for drying chamber. The samples are agriculture commodity collected from local farmers. The experimental results show that microwave accelerate drying in most materials. The experimental data were analyzed by using an available model constructed from fundamental physics by other scholars. The model has been applied to more understanding the behavior of the microwave drying material.

1. Introduction

Drying of agricultural and food products is one of the most energy intensive processes. These products are dried to inhibit quality decay. The agricultural products are dried after harvest to the moisture content that allows microbial stability during storage [1-2]. For examples, vegetables are blanched before drying to avoid rapid darkening, and drying is not only carried out to inhibit microbial growth, but also to avoid browning during storage. For fruits, the reduction of moisture acts in combination with its acid and sugar contents to provide protection against microbial growth. Other products as crackers are dried beyond the microbial growth threshold to confer a crispy texture, which is liked by consumers. Most farmers use conventional drying method (electric furnace or sun-drying method). In order to improve the performance of these process, new technologies such as microwave heating and a combination of hot air and microwave heating have been used. Their performance and characteristics are evaluated using a mathematical model by scientists. Several theoretical drying models have been developed and most of these models are based on the conventional heat transfer [3].

Content from this work may be used under the terms of the Creative Commons Attribution 3.0 licence. Any further distribution of this work must maintain attribution to the author(s) and the title of the work, journal citation and DOI.
Published under licence by IOP Publishing Ltd

The application of microwave energy for thermal processing of different materials and substances is a rapidly growing trend of modern science and engineering especially for sintering ceramics [4-10] as well as for drying food [11-13]. In a microwave furnace, the material will absorb microwave energy convert the energy into heat. Heat is generated internally within the materials resulting homogenous heating. The benefit of microwave heating is not only rapid processing but also sometimes improving material properties [14-17]. Experiments of drying characteristics of cocoa bean by using a microwave furnace as well its theoretical analysis have been successfully performed and reported previously as shown in Fig. 1 [12]. The faster reduction of moisture content of samples compared to electric furnace drying result was found. In this paper, experiment results of drying various materials by using a microwave furnace will be presented. From the characteristics of microwave drying of these materials may inform us the role of microwave drying mechanisms.

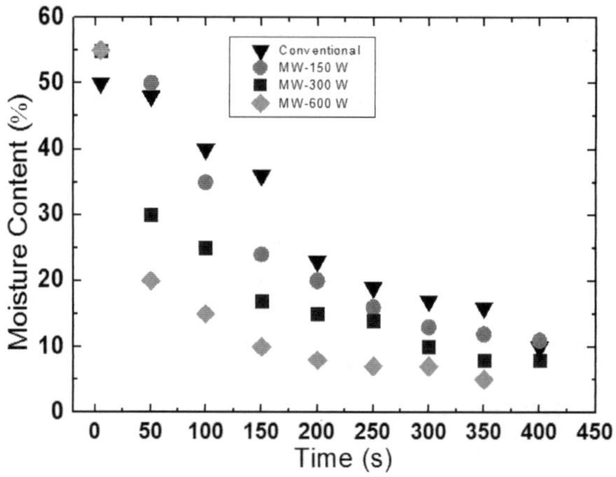

Figure 1. Microwave drying of cocoa bean slabs with various microwave power [12]

Because drying technology is very important in industrial application, an extensive investigations have been performed by researchers. Several theoretical model have been developed by scholars. The model mostly are based on the conventional heat transfer. Review of model of heat and mass transfer in various systems is compiled by Eckert *et al.* [18]. Drying models start from the classical liquid diffusion model based on Fick's law where gradient of liquid content (moisture content) as driving force was reported by Sherwood [19]. King and Harmathy assume that the drying process only based on vapour diffusion [20-21]. Most of the model describe above have not been able to predict drying rates and temperature and moisture content distribution for both hygroscopic and non-hygroscopic materials over a range of boundary conditions. This is because the moisture tansport in materials is caused by a variety of mechanisms, each of which is prevalent under different condition. Temperature, the local moisture content and the material properties influence the contribution of each drying mechanism and affect the drying rate. Izume and Hayakawa [22] develop a unique set of heat and moisture equation transfer which are used to model the drying of cylindrical, elastoplatics food. They are solved by the Galerkin's finite element method and in the time domain by the Crank-Nicholson method. In this paper will be reported experimental results of drying of agriculture commodities by using microwave. Analysis by using available theory was also described.

2. Experimental Method

2.1 Sample preparation

Samples Corn, Coffee, unhulled Rice, and Cacao were collected from local farmers in South East Sulawesi, Indonesia. Prior to drying, samples were selected then washed. All samples then were pellet and cut into slab shape with desired size of 10 mm in diameter and 3 mm in thickness. Silica powder was extracted from rice husk ash where also collected from local farmers. The detail procedure of extraction is published elsewhere [23,24]. Drying experiments were performing by using a microwave oven as well as electric furnace. The electric furnace drying results are used as comparison.

2.2 Drying Experiment

A microwave oven (Panasonic NE-C236, Japan) applied as furnace for drying experiments as shown schematically in Fig. 2. This 2.45 GHz microwave device has maximum output power of 800 W with power consumption about 1.43 kW. The drying experiments were performed by using three power levels: high power of 600 W, medium power of 300 W and low power of 150 W. The microwave oven equipped by a temperature controlling system with shielded thermocouple. Microwave powers were varied for every samples.

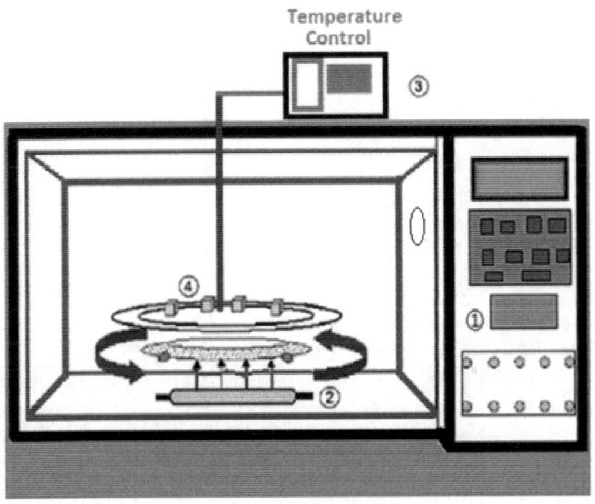

Frequency : 2.45 GHz
Maximum output power : 800 W
(1) control;
(2) lamp;
(3) temperature control;
(4) samples on turntable plat

Figure 2. Schematic picture of a microwave drying system [12]

The moisture content value was determined by using equation:

$$M\ (\%) = (W_t - W_d)/W_d \quad (1)$$

where M is moisture content, W_t is the weight of sample (g) at any time and W_d is the weight of the dried sample.

3. Results and Discussion

3.1 Drying Experiment Results

Drying curves of coffee and corn slabs dried with different methods are presented in Fig. 3 and 4. They shows that the moisture content and drying rate decreased continuously with drying. There are no constant rates drying because most crops as well as coffee exhibit the constant rate drying characteristics at their critical moisture content. Coffee exhibits a constant rate behavior during drying in high moisture content as well as cocoa [1]. However the initial moisture content in this experiment is not up to this range. At the falling rate period the movement of moisture within the coffee to the surface is governed by diffusion since the material is no longer saturated with water. The graph shows that the moisture ratio decreased as the drying time increased. Both graphs also show a faster drying in the microwave drying compared than in the conventional one on all microwave powers.

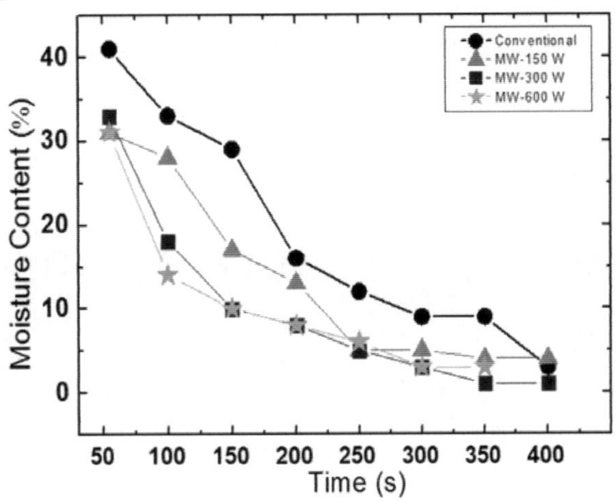

Figure 3. Drying curves of coffee slabs (diameter of 10 mm) in a microwave with different microwave power and in an electric furnace (conventional method)

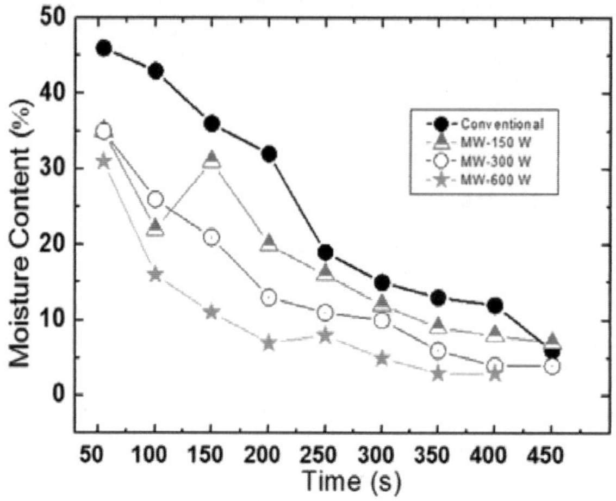

Figure 4. Drying curves of corn slabs (diameter of 10 mm) in a microwave with different microwave power and in an electric furnace (conventional method)

Generally characteristic of microwave processing of different agriculture materials depend on their physical properties such as moisture content, density, thermal conductivity, and chemical substances as well as temperature. Microwave heating is mainly due to polarization and ionic conduction of water molecules in material. Simplify can be described that the ionic conduction losses and due to dipolar rotation towards microwave frequencies with temperature. The absorption of microwave energy and conversion to heat is due to polarization and conduction would result in a rise in temperature, and this is given by the following equation [25]:

$$\frac{\Delta T}{\Delta t} = \frac{2\pi f \varepsilon_0 \varepsilon''_{eff} E_{rms}^2}{\rho c_p} \qquad (2)$$

where, ε''_0 is the permittivity of free space (8.85×10^{-12} V/m^3), ε''_{eff} is the relative effective dielectric loss due to ionic conduction and dipolar reorientation, f is the frequency (Hz), E_{rms} is the root mean square of the electric field within the material (V/m), ρ is the bulk density of dielectric material (kg/m^3) and Cp is the heat capacity of the material at constant pressure(J/kg°C). The water dipole attempts to continuously reorient in microwave's oscillating field. The dipole lags between the dipole and the field leads to an energy loss by heating. The ease of the movement depends on the strength and extent of the hydrogen bonded network.

Understanding and availability the properties of each materials are important to develop drying model of the materials [26-27]. Some dielectric and thermal properties of various materials at microwaves frequencies provides by Komarov in [28]. The dependency of physical parameters on temperature and moisture content are given in form of analytical equations. For example for Corn is shown in Equation 3 and 4 [28].

$\varepsilon'(T) = -0.0005502\,T^2 - 0.03792T + 60.68$ (3)

$\varepsilon''(T) = -0.0005502\,T^2 - 0.03792T + 60.68$ (4)

($5 < T$ °C<120)

Other materials'data mostly available in Komarov and Hippel [28-29]. By using such equations, the drying path of materials can be control and their quality can be improved.

As comparison for microwave drying material of agriculture commodities, there is shown in Figure 5. Silica ceramic was drying by using 2.45 GHz microwave heating system. Effect of microwave was shown as in agriculture commodities materials.

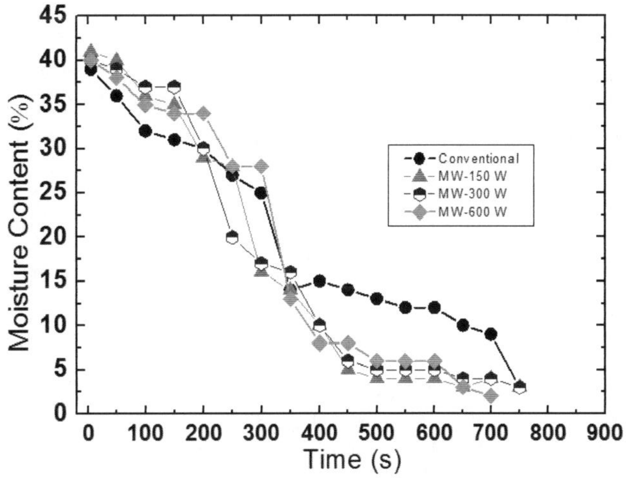

Figure 5. Drying curves of silica slabs (diameter of 10 mm) in a microwave with different microwave power and in an electric furnace (conventional method)

3.2 Theoretical Analysis of Drying Experiment

The microwave-material interaction analysis basically can be evaluated from Maxwell's equations in matter. For microwave drying, the most simple assumption that the absorption rate and the permittivity and permeability of a drying samples are all constant. Thus, Maxwell's equations are reduced to the telegrapher's equation which can be solved without reference to the temperature [30]. The temperature variation is calculated by solving the forced heat equation with the forcing term dependent on the square of the amplitude of the electric field. In microwave drying, most of the theoretical work to date has used the simplifying assumption that the amplitude of the electric field is constant. The assumptions made because for materials properties which are temperature-dependent with variable electric field, the forced heat equation and Maxwell's equation become a highly nonlinear coupled system. The assumptions generally can be applied only for thin materials. One of methods is applied by Ayappa, *et al.* which predict the temperature profiles in multilayer slabs by simultaneously solving Maxwell's equation with the heat transfer equation, using the Galerkin finite element method [31].

Another method for modeling the microwave drying material developed by Izumi and Hayakawa [32]. They solve three equations i.e. : heat transfer equation, mass transfer equation, and electromagnetic field equation. The modeling of the heat and mass transfer in food is based on these equations. The equation include all the thermodynamically interactive heat and mass fluxes and mass flux corrections as described follows.

1. **Mass/Moisture Transfer Equation:**

$$\frac{\partial C}{\partial t} = \nabla.(D_W \nabla C + D_T \nabla T + D_P \nabla P) \quad (5)$$

Where : C = Volumetric moisture concentration (kg m^{-3}), t = time (s), D_W = mass diffusivity (m^2 s^{-1}), D_T = soret mass diffusivity (kg m^{-1} s^{-1} K^{-1}), T = temperature (K), D_p = pressure mass diffusivity (kg m^{-1} s^{-1} Pa^{-1}), and P = water vapour pressure (Pa).

2. **Heat Transfer Equation:**

$$c_p\,\rho_b \frac{\partial T}{\partial t} = \nabla.(k_T \nabla T + k_c \nabla C + k_P \nabla P) + \boldsymbol{E_{MV}} \quad (6)$$

Where: c_p = specific heat (J kg^{-1} K^{-1}), ρ_b=density of dry solid (kg m^3), k_T = thermal conductivity (W m^{-1} K^{-1}), k_c =dufour thermal conductivity (W m^2 kg^{-1}), k_p = pressure thermal conductivity (W m^{-1} Pa^{-1}), and **E_{MV} = internal heat source due to microwave heating.**

The ∇P in the above equation is correlated as follow :

$$\nabla P = \gamma_c \nabla C - \gamma_T \nabla T - \gamma_s \nabla S_v \tag{7}$$

Where: S_v is volumetric shrinkage coefficient and $\gamma_c, \gamma_T, \gamma_s$ are empirical parameters for relating a static pressure change with vaporised or condense mass, temperature change and volumetric shrinkage.

For calculating the moisture concentration (C), it must be converted to moisture potential to state the moisture transport. This conversion can be using the Gibbs free energy of moisture [23].

$$\emptyset = \mu_{Gibbs} + RT \ln a_W \qquad for\ W \leq W_{100\%}$$
$$\emptyset = \left[\frac{W - W_{100\%}}{W_{100\%} - W_{0\%}} + 1\right]\mu_{Gibbs} \quad for\ W > W_{100\%} \tag{8}$$

Where:

$\emptyset$ = moisture potential, a_w = water activity, μ_{Gibbs} = Gibbs free energy of saturated free water (J mol^{-1}) R = universal gas contant (J mol^{-1} K^{-1}), W = moisture content (kg water (kg solid)$^{-1}$), $W_{100\%}$ = food moisture content equilibrate to 100% humidity and $W_{0\%}$ = food moisture content equilibrate to 0% humidity. Water activity is function of temperature and moisture so that difficult to estimated. Some people estimated by using the Gugenheim-Anderson-de Bore equation [33]. This method can be applied to the drying experiments in this research. The details will be published in separated paper.

3.3 Electromagnetic Field Equation

The electromagnetic field equations in matter are also based on Maxwell's electromagnetic theory. However because of complexity of the Maxwell equations in real matter, the assumptions for simplify are always made. For example, the electromagnetic field equation are calculated assuming that a plane electromagnetic wave is propagated in slabs. The slab is divided into layers and each layer has a constant temperature, moisture concentration and constant dielectric properties. This is true may be for thin slab or special materials only. By using these assumptions, the problem become simpler. For example the detail of the calculations and numerical simulation are presented by Tran and Piotrowski based on the following equations [34].

$$N_M = \sqrt{\frac{\varepsilon_M \varepsilon_0}{\mu_0}} \qquad N_0 = \sqrt{\frac{\varepsilon_0}{\mu_0}} \qquad Y_M = N_M \tag{9}$$

$$N_i = \sqrt{\frac{\varepsilon_i \varepsilon_0}{\mu_0}} \qquad Y_i = N_i \frac{Y_{i+1} - N_i \tanh(2\gamma_i(h_i - h_{i-1}))}{N Y_{i+1} - Y_{i+1} \tanh(2\gamma_i(h_i - h_{i-1}))} \tag{10}$$

Where :

$$\gamma_i = -\mu_0 \varepsilon_0 \varepsilon_i (2\pi f)^2$$
$$\varepsilon_i = \varepsilon_{dry}(1 - S_i) + \varepsilon_{water} S_i$$

f = frequency of electromagnetic field (Hz), h_i = position of i-th numerical mode from the surface of the sample (m), S_i = relative water content in i-th numerical node, ε_0 = permittivity of vacum (F m^{-1}), ε_{dry} = relative permittivity of dry material, ε_{water} = relative permittivity of water, μ_0= permeability of vacuum.

By using these theories the experiment results can be predicted and controlled so that microwave drying can be more effective. Generally, by using microwave drying agriculture commodities is faster.

4. Conclusion

Experiments of application microwave energy for drying several agriculture commodities have been performed. The microwave showed a faster drying than that of conventional one. Theoretical analysis of the experimental data was performed based on available theories. The application of the microwave energy forces the moisture content gradient and temperature gradient to act together to speed up the drying process. In general, the results in this experiment suggested that microwave technology is appropriate technology for drying agriculture commodities.

Acknowledgements

Thank to all members of Material Physics Laboratory and Theoretical and computational Physics Laboratory, Department of Physics, Faculty of Mathematics and Natural Sciences, Halu Oleo University for all facility supports.

References

1. D.O. Oke and K.F. Omotayo, . 2012. Effect of forced air artificial intermittent drying on cocoa beans in South Western Nigeria, Journal of Cereals and Oil seeds Vol. 3 (1), pp. 1-5.
2. R.J. Duncan, Godfrey E.G., Yap,T.N, Pettipher, G.L,Tharu-marajah, T. 1989. Improvement of Malaysian cocoa bean flavour by modification of harvesting, fermentation and drying method. The Planter (Malaysia), pp. 57-65.
3. E.R.G. Eckert, R.J. Goldstein, W.E. Ibele, S.V. Patankar, T.W. Simon, P.J. Strykowski, K.K. Tammaj.V.R. Heberlein, D.L. Hofeldt, T.H Kuehn and K.A. Stelson, Heat Transfer- A Review of 1992 Literature, Int. J. Heat Mass transfer, 37,9,1285-1375 (1994).
4. H. Aripin, S. Mitsudo, I.N. Sudiana, T. Saito, S. Sabchevski, 2015. Structure Formation of a Double Sintered Nanocrystalline Silica Xerogel Converted From Sago Waste Ash, Transactions of The Indian Ceramic Society, DOI: 10.1080/0371750X.2014.980850
5. M. Zamrun Firihu, I. N. Sudiana, S. Mitsudo, Microwaves Enhanced Sintering Mechanisms in Alumina Ceramic Sintering Experiments, Contemporary Engineering Sciences, Vol. 9, 2016, **5**, 237 – 247
6. I.N. Sudiana, S. Mitsudo, T. Nishiwaki, P. E. Susilowati, L. Lestari, M. Z. Firihu, H. Aripin, 2015. Effect of Microwave Radiation on the Properties of Sintered Oxide Ceramics, Contemporary Engineering Sciences, Vol. 8 No. 34, 1607-1615
7. S. Mitsudo, S. Inagaki, I.N. Sudiana, K. Kuwayama, Grain Growth in Millimeter Wave Sintered Alumina Ceramics , Advanced Materials Research, Vol.789 (2013), pp. 279-282.
8. I. N. Sudiana, R. Ito, S. Inagaki, K. Kuwayama, K. Sako, S. Mitsudo, 2013. Densification of Alumina Ceramics Sintered by Using Sub-millimeter Wave Gyrotron, Int. J. of Infrared, Millimeter, and Terahertz Waves. 34 627-638.
9. I.N. Sudiana, S. Mitsudo, K. Sako, S. Inagaki, L. O. Ngkoimani, I. Usman, H. Aripin, The microwave effects on the properties of alumina at high frequencies of microwave sintering, American Institute of Physics (AIP) Conference Proceeding, **1719**, 030014 (2016); doi: 10.1063/1.4943709
10. I. N. Sudiana, S. Mitsudo, M. Z. Firihu, H. Aripin, 2016. Effect of High-Frequency Microwave on Micro Hardness of Alumina Ceramic, Material Science Forum , Vo. 872, pp.114-117.
11. Hu, Q.G., M. Zhang, A.S. Mujumdar, G.N. Xiao, and J.G. Sun. 2006. Drying of Edamames by Hot Air and Vacuum Microwave Combination. J. of Food Engineering 77:977-982.
12. M. Z. Firihu, I.N. Sudiana, 2016. 2.45 GHz microwave drying of cocoa bean , ARPN Journal of Engineering and Applied Sciences Vol. 12 No. 19, pp. 11595-11598
13. T.K. Sherwood, The drying of Solid Industrial Engineering. Chemistry 21, p.12 (1929).
14. M. Z. Firihu, I.N. Sudiana, 2016. Properties of 2.45 GHz microwave sintered S_iO_2 from rice husk ash and Al_2O_3, ARPN Journal of Engineering and Applied Sciences Vol. 12 No. 19, pp. 11595-11598
15. H. Aripin, S. Mitsudo, I. N. Sudiana, N. Jumsiah, I. Rahmatia, B. Sunendar, L.Nurdiwijayanto, S. Mitsudo, S.Sabchevski, 2011. Preparation of Porous Ceramic with Controllable Additive and Firing Temperature, Advanced Materials Research. Vol. 277 pp. 151-158
16. I. N. Sudiana, S. Mitsudo, M. Z. Firihu, 2016. Effect of initial green samples on mechanical properties of alumina ceramic, Contemporary Engineering Sciences, Vol. 9, No. 12, 595-602
17. H. Aripin, S. Mitsudo**,** I Nyoman Sudiana, Bambang Sunendar , Svilen Sabchevski, 2016. Structural Characterization of a Glass Ceramic Developed from TiO_2 and a Novel Material-Silica Xerogel Converted from Sago Waste Ash, Material Science Forum Vol. 872, pp.81-86.
18. E.R.G. Eckert, R.J. Goldstein, 1994. Heat transfer- a review of 1992 literature, Int. J. Heat Mass transfer, 37,9,1285-1375 (1994)
19. T.K. Sherwood, Application of the Theoretical Diffusion Equation to The Drying of Solid, Trans, AIChE,27,190-202.
20. C.J. King, 1971. Freeze Drying of Foods, Butterworth, London.
21. T.Z. Harmathy, 1969. Simultaneous Moisture and Heat Transfer in Porous System with Particular References to Drying, Ind. Engng. Chem., 8, pp.92-103.
22. M. Izumi and K.I. Hayakawa, 1995. Heat and Moisture Transfer and Hygrostress Crack Formation and Propagation in Cylindrical, Elastoplastic Food, Int. J. Heat mass transfer, 38,6,1033-1041.
23. I N Sudiana, S. Mitsudo, 2016. T. Nishiwaki, P. E. Susilowati, L. Lestari, M. Z. Firihu, H. Aripin, Synthesis and Characterization of Microwave on Sintered Silica Xerogel Produced from Rice Husk Ash, Journal of Physics: Conference Series Vol. 739 No. 1, 012059
24. I. N. Sudiana, S. Mitsudo, M.Z. Firihu, L. Aba, L. O. Ngkoimani, M.W. Arsana, H. Aripin, Investigation of silica from rice husk ash wastes as an alternative material for microwave absorbers,

American Institute of Physics (AIP) Conference Proceeding, 1801(1), 040003 (2017); DOI : 10.1063/1.4973092

25. W. H. Sutton, Microwave Processing of Ceramic Materials, Microwave Solutions for Ceramic Engineers, Am. Cer. Soc., Ed.by D.E. Clark, D.C. Folz, C.E. Folgar, M.M. Mahmoud , pp.35-65 (2005).
26. McMinn W. A. M., Khraisheh M. A. M., Magee T., R. A., Modeling the mass transfer during convective, microwave and combined microwave-convective drying of solid slabs and cylinders, Food Research International, 36, pp. 977-983 (2003).
27. A.K. Hakhi, and A. Amanifard., Heat and Mass Transfer during drying of food Products, Brazilian Journal of Chemical Engineering, 25, pp. pp. 977-983 (2003).
28. V.V. Komarov, 2012. Handbook of Dielectric and Thermal Properties Materials at Microwaves Frequencies, Artech House, London.
29. A. R. V. Hippel, 1995. Dielectric Materials and Applications, Artech House.
30. https://mysite.du.edu/~jcalvert/tech/cable.htm
31. K.G. Ayappa, H.T. Davis, E.A. Devis, J. Gordon , analysis of microwave heating of materials with temperature-dependent properties, AIChE J., 37,3,313-322 (1991)
32. N. Marcuvitz, 1985. Waveguide handbook, Peter Peregrinus Ltd. New York.
33. F. Prothon, L. M. Ahrné, 2003. Application of the Guggenheim, Anderson and De Boer model to correlate water activity and moisture content during osmotic dehydration of apples, Journal of Food Engineering Vol. 61, Issue 3, February 2004, Pages 467–470
34. N. Tran and A. Piotrowski, 1997. Advances in the Modeling of Microwave RF and Hot Air Drying of Materials, Microwaves: Theory and Application in Materials Processing IV. The American Ceramic Society.

Characterization of Central Kalimantan's Amethysts by Using X-Ray Diffraction

Komang Gde Suastika[1], Lila Yuwana[2], Luqman Hakim[1], Darmaji[3*] and Didik Khusnul[3]

[1] Program Studi Pendidikan Fisika, Universitas Palangka Raya, Indonesia, 73112
[2] Jurusan Fisika FMIPA, ITS Surabaya
[3] Pusat Studi Sains institut Teknologi Sepuluh November (ITS) Surabaya, Indonesia

*E-mail: darmaji@matematika.its.ac.id

Abstract. X-Ray Diffraction (XRD) is one of spectroscopy method that utilize X-Ray as an energy source in the phases identification. This method is used to characterize the type of amethyst of Sukamara, Central Kalimantan. In this paper, the crystallography of three types of Sukamara amethyst is reported. Samples was collected directly from the amethyst mine in the Ajang village in the District of Permata Kecubung. The samples consists of three types i.e. purple amethyst, black amethyst, and violet amethyst. The results show that quartz or silica (SiO_2) is the main phase of the gemstone amethyst. Amethyst cell parameters are $a = b \neq c$ and $\alpha = \beta \neq \gamma$, the crystal system is hexagonal. Amethyst average density is 2.65 g / cm^3 and 112.84 pm^3 cell volume. Peak FWHM value indicates that the degree of crystallinity of the black one is greater than that of the gasoline, while the degree of crystallinity of violet amethyst higher than purple amethyst.

1. Introduction

Amethyst is a gemstone that has a main phase of SiO_2. Its refractive index is between 1544 - 1553, with hardness about 7 mosh. They are available in four main purple colors i.e. lilac, mauve, black and white. This gemstone is often given a symbol of sincerity, a sense of comfort and peace [1]. Although amethyst is well known gemstone and has been used for centuries, but the scientific identification of amethyst is still not reported. Therefore, for maintaining the commercial value of natural gemstones, it is necessary to identify the gemstone. By identifying gemstones we can differ a hardness level of various amethyst. This requires the analytical instruments to measure a hardness level. The instruments are often used in identifying gemstones are refractometer, polariscope, microscopy, and mini spectroscopy. In addition, there are advanced equipment also used for the identification of gemstones such as infrared spectroscopy, ultraviolet, and X-ray fluorescence (XRF) [2-4]. This paper shows scientific report of microstructural properties of amethyst gemstones. In particular, an X-Ray Diffraction (XRD) analysis was applied to determine the crystallographic characteristics of three types of amethyst, i.e. purple, violet and coffee amethyst. XRD method have been applied to scientist to characterized material worldwide such as for microstructural evolution and phase change of ceramics during processing [5-7]. Furthermore, this paper also reveals experimental and analytical methods to examine a hardness level of various amethyst based on XRD measurement results. Another method to determine material's hardness is by using Vicker's Hardness testing as reported previously to alumina ceramics [8-9]. Gemstones' hardness is one of their important property in addition to their beautiful colors. By detail scientific identification of these gemstones, we may be not only can maintenance their valuable properties but also can improve them.

Content from this work may be used under the terms of the Creative Commons Attribution 3.0 licence. Any further distribution of this work must maintain attribution to the author(s) and the title of the work, journal citation and DOI.
Published under licence by IOP Publishing Ltd

2. Material and Method

XRD was utilized to characterize the microstructure of amethysts. The study was conducted through the following steps. Firstly, the collection and retrieval of stones containing three types of amethysts in Ajang village, Sukamara. Secondly, cutting and polishing to obtain the desired amethyst. Thirdly, take small part of each gemstone to be ground into. Lastly, performing XRD testing and analysis for samples. The XRD was setup at voltage of 40 kV with 35 mA in current. Angle shots start from 15-90 deg, a step size of 0.02 deg, time / step 0.5 s, and a scan speed of 0.04 deg/s.

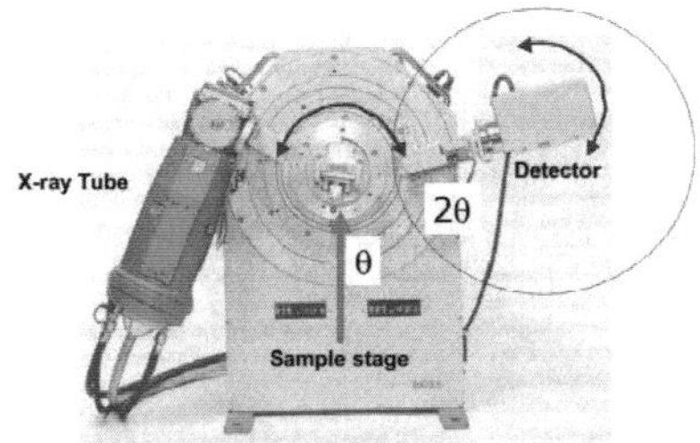

FIGURE 1. Main parts of X-Ray Diffraction (XRD)

3. Results and Discussion

XRD testing conducted on purple, black and violet amethysts to obtain the characteristic of these gemstones. XRD data analysis was performed automatically by using software X'Pert Highscore Plus at the Laboratory of Central Materials and Advanced Minerals in State University of Malang followed by manual calculation. The results are described in detail in following parts.

3.1 XRD analysis of purple amethyst

Purple amethyst stone is a gemstone which has a purple color and white fibers inside (Figure 2). This stone is often found in Kalimantan, Papua and West Java. For jewelry, this stone has been used since ancient Greece era [10].

FIGURE 2. Purple Amethyst

XRD result of purple amethyst can be seen in Figure 3. The pattern identified in purple amethyst is Quartz with a percentage of 100%, has the chemical formula SiO_2. Purple amethyst has a hexagonal crystal system with lattices parameters a = b = 4.9108 Å and c = 5.4028 Å where the angle between a, b and c of $\alpha = \beta = \gamma = 90^o$ and 120°. Volume amounted to 112.84 PM3 cell and density of 2.65 g/cm^3. There are 23 peaks detected in black amethyst, with the distance between atoms in cell units (d) is 1.22814-4.25524 Å. The total value of FWHM peak multiplied by the relative intensity is 26.0056, this value indicates the degree of crystallography. The size of the atomic crystal amethyst is between 0.82 x 10^{-9} m and 2.74 x 10^{-9}m.

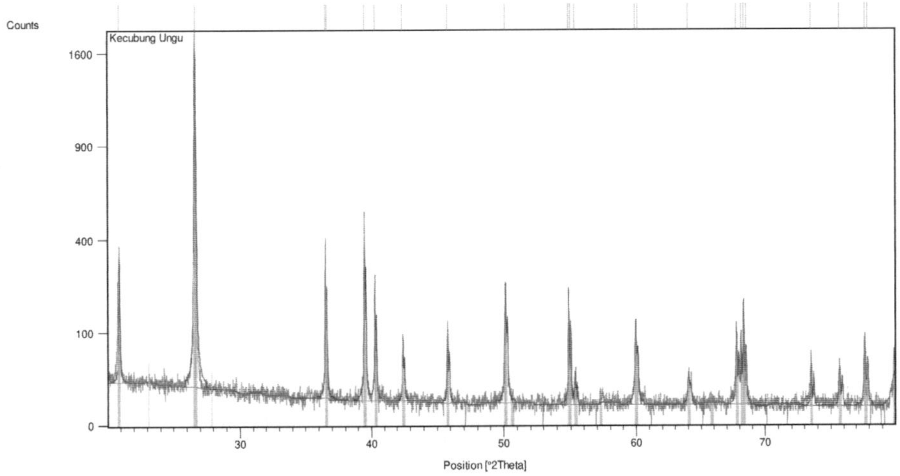

FIGURE 3. XRD result of purple amethyst

3.2 XRD analysis of black amethyst

Black amethyst (Figure 4) is seen to be orange to reddish color if it is exposed by the light. It also has unique fibers therein. Black Amethyst is also often called as Wulung amethyst. In mineral quartz subject, black amethyst has similarities with Black Obsidian [11].

FIGURE 4. Black amethyst

Figure 5 shows XRD result of black amethyst. Crystallography parameter of black amethyst and purple amethyst are the same. The pattern identified in purple amethyst is Quartz with a percentage of 100%, has the chemical formula SiO_2. Coffee amethyst has a hexagonal crystal system with parameters of lattices a = b = 4.9108 Å and c = 5.4028 Å where the angle between a, b and c of $\alpha = \beta = \gamma = 90^o$ and 120°. Volume amounted to 112.84 PM3 cell and density of 2.65 g / cm3. There are 24 peaks detected in black amethyst, with the distance between atoms in cell units (d) is 1.22817-4.24901 Å. The total value of FWHM peak multiplied by the relative intensity is 29.87297, this value indicates the degree of crystallography. The size of the atomic crystal amethyst is between 4.2966 x 10^{-11} m - 1.3886 x 10^{-8} m.

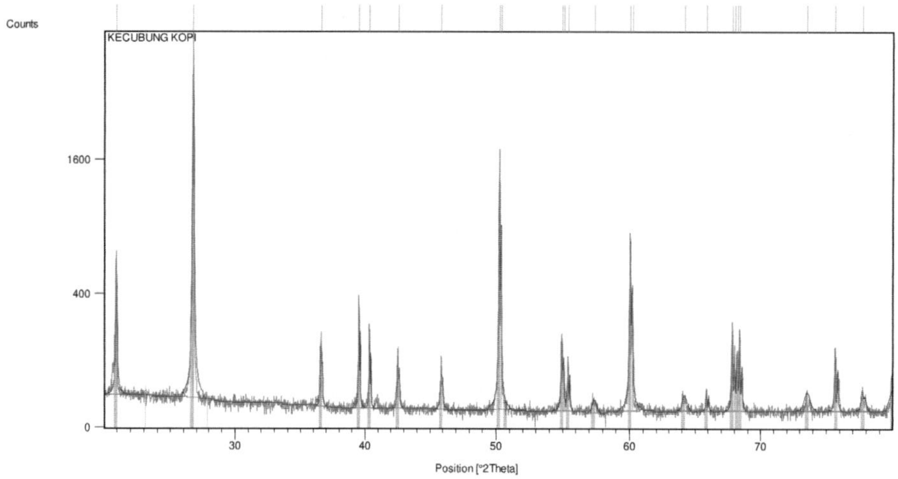

FIGURE 5. XRD result of black amethyst

3.3 XRD analysis violet amethyst

One amethyst with a unique and very spectacular color is violet amethyst (Figure 6). This stone shines when exposed to the light and has ice-like fibers. However, when amethyst have crystallized, fibers do not appear.

FIGURE 6. Violet amethyst

Figure 7 depicts XRD result of violet amethyst. Crystallography parameter of black amethyst, violet amethyst and purple amethyst are the same. The pattern identified in purple amethyst is Quartz with a percentage of 100%, has the chemical formula SiO_2. Violet amethyst has a hexagonal crystal system with parameters of lattices a = b = 4.9108 Å and c = 5.4028 Å where the angle between a, b and c of $\alpha = \beta = \gamma = 90^o$ and 120°. Volume amounted to 112.84 PM3 cell and density of 2.65 g / cm3. There are 26 peaks detected in violet amethyst, with the distance between atoms in cell units (d) is 1.22846-4.25143 Å. The total value of FWHM peak multiplied by the relative intensity is 28.28162, this value indicates the degree of crystallography. The size of the atomic crystal amethyst is between 3.8808×10^{-11} m – 1.386×10^{-8}m. According to XRD analysis, three sample of amethysts has the same crystallographic characteristics, but vary in a number of peaks, the distance between the unit cells, the total value of FWHM and atomic crystals sizes. The greatest number of peaks is detected at violet amethyst followed by black and purple amethyst. The total value of FWHM peak of black amethyst multiplied by relative intensity greater than gasoline and purple amethyst. This value indicates hardness level of a gemstone [12]. Therefore, black amethyst has the highest hardness, followed by violet amethyst and purple amethyst. This conclusion is the consequence of their value of FWHM peak multiplied by the relative intensity of black amethyst, violet amethyst, and purple amethyst. The values are 29.87297, 28.28162, and 26.0056, respectively.

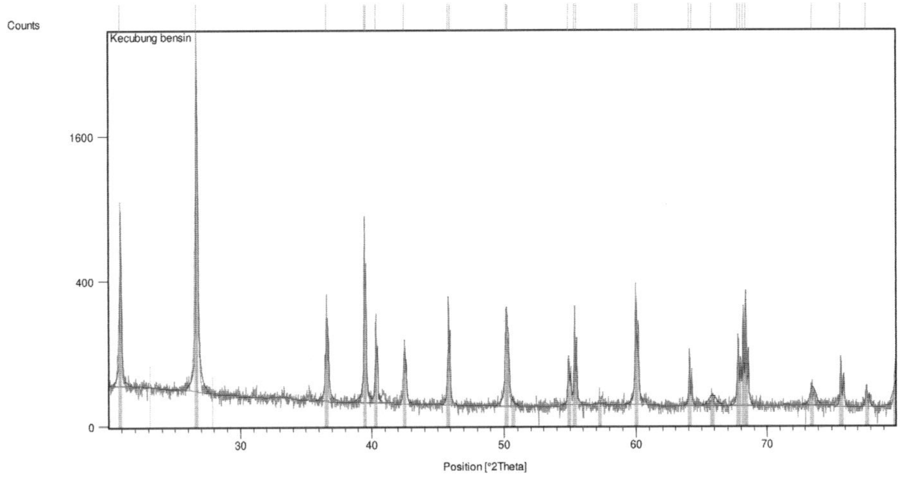

FIGURE 7. XRD result of violet amethyst

4. Conclusion

From the analysis data from experimental results of XRD on coffee, gasoline, and purple amethysts, we can describe as follows:

1. Stone purple, coffee and petrol has a pattern of Quartz with a percentage of 100%.
2. The total value of FWHM peak of black amethyst multiplied by relative intensity greater than gasoline and purple amethyst, i.e. 29.87297, 28.28162, and 26.0056, respectively.
3. Black amethyst has the highest hardness, followed by violet amethyst and purple amethyst.

References

1. Top Lintas. Purple Amethyst Gemstones. http://toplintas.com/batu-kecubung-ungu/. 13 Agustus 2016. [in Indonesian]
2. Ho, Sut kam, Applied Spectroscopy. Vol. 69, No. 2, pp. 268-276, 2015.
3. Karyasa, I Wayan, dkk., Jurnal Sains dan Teknologi. Vol. 4, No. 2, pp. 604-610, 2015 [in Indonesian]
4. Moyo, Stanley, et.al. 2015. Blombos cave: Middle stone age ochre differentiation trough FTIR, ICP OES, ED XRF and XRD. Quanternary International *xxx*. pp. 1-10.
5. Fajrin, R., Analysis of the formation of a nanocrystal material of a zirconium-based amorphous material at a temperature of 390°-410°C, Thesis, Institut Teknologi Sepuluh November. 2008. [in Indonesian]
6. H. Aripin, S. Mitsudo, I. N. Sudiana, N. Jumsiah, I. Rahmatia, B. Sunendar, L.Nurdiwijayanto, S. Mitsudo, S.Sabchevski, Preparation of Porous Ceramic with Controllable Additive and Firing Temperature, Advanced Materials Research. Vol. 277 (2011) pp. 151-158
7. I. N. Sudiana, S. Mitsudo, T. Nishiwaki, P. E. Susilowati, L. Lestari, M. Z. Firihu, H. Aripin, Synthesis and Characterization of Microwave on Sintered Silica Xerogel Produced from Rice Husk Ash, Journal of Physics: Conference Series Vol. 739 No. 1, (2016) 012059
8. I. N. Sudiana, S. Mitsudo, M. Z. Firihu, Effect of initial green samples on mechanical properties of alumina ceramic, Contemporary Engineering Sciences, Vol. 9, 2016, no. 12, 595-602
9. I. N. Sudiana, S. Mitsudo, M. Z. Firihu, H. Aripin, Effect of High-Frequency Microwave on Micro Hardness of Alumina Ceramic, Material Science Forum , Vo. 872, 2016, pp.114-117.
10. _. Precious Lite. Amethyst Gemstones. http://batuakikkecubung.net/. 13 Agustus 2016.
11. Siopung. Black Amethyst Gemstones. http://siopung.com/batu-akik-kecubung-kopi-harga-khasiat. 13 Agustus 2016. [in Indonesian]
12. Rao, R. Venkateswara, *et.al.*, Applied Radiation and Isotopes. Vol. 82, pp. 211-222, 2015.

IOP Conf. Series: Journal of Physics: Conf. Series **846** (2017) 012025 doi:10.1088/1742-6596/846/1/012025

Characterization of Coconut Shell Liquid Volatile Matter (CS-LVM) by Using Gas Chomatroghaphy

Muhammad Jahiding[2)], Mashuni[1)], WOS Ilmawati[2)], Ermawati[2)], Rahmat[2)], Jumiati Arsyad[2)], S S Riskayanti[2)]

[1]Department of Chemistry , Faculty of Mathematics and Natural Sciences, Halu Oleo University (UHO), Kendari, South East Sulawesi, 93231
[2]Department of Physics, Faculty of Mathematics and Natural Sciences, Halu Oleo University (UHO), Kendari, South East Sulawesi, 93231

E-mail: muhjahiding68@yahoo.com

Abstract. Generally, the coconut shell is only used for the fuel of furnace or is just burnt in which this will just create pollution. One way of solving this problem is by re-processing the coconut shell as raw materials for making liquid volatile matter (LVM) by pyrolysis method. Coconut shell is part of the coconut fruit at which having biological function to protect fruit core and is located on the inner side of the fiber with a thickness ranging from 3-6 mm. Coconut shell is classified as hardwood, mainly composed of lignin, cellulose, and hemicellulose, with water content of approximately 6-9 %. Coconut shell is more suitable for pyrolysis process, since they contain less amount of ash, more amount of volatile matter and available with lower cost in rural areas during all the sessions of the year. This research was aimed at determining the influence of pyrolysis temperature towards the LVM volume of coconut shell. LVM was made of condensing the smoke of pyrolysis result from the coconut shell while the analysis of compound composition of LVMcoconut shell used *Gas Chromatography*. Based on the result of the research, it was known that the pyrolysis, at the temperatures of 400°C, 500°C, 600°C and 700°C can create LVM volume as many as 204.167 mL kg^{-1}, 208,33 mL kg^{-1} and 216.67 mL kg^{-1}. The LVM created from the pyrolysis at 400°C was made of ammonia (12. 41%), acetic acid (37.27%), phenol (31.66%), furfural (4.16%), and alcohol (5.01%).The LVM created from the pyrolysis at 500°C was made ofammonia (12.22%), hydrazine (5.61%), acetic acid (40.96%), phenol (32.82%), and alcohol (3.10%), furfural (5.30%).The LVM created from the pyrolysis at 600°C was made ofammonia (15.49%), acetic acid (36,01%), phenol (32.85%), alcohol (6.75%), and furfural (4.62%).The LVM created from the pyrolysis at 700°C was made ofammonia (15,.), acetic acid (35.20%), phenol (22.60%), alcohol (5.07%), and furfural (4,90%). From this result, it can be seen that LVM has big advantages for food flavor and other specific flavor as well as for preservative because of its antimicrobe and anti-oxidant characteristics.

1. Introduction

Biomass, defined as any hydrocarbon material mainly consisting of carbon, hydrogen, oxygen, nitrogen and some other components in small proportions [1], includes wood and its derived wastes, different organic wastes (including biodegradable MSW), agricultural and crop wastes, animal wastes, energy plantations, among others. Cashew nut (CNS) and coconut shells (CcNS) are part of this family [2]. Coconut shell is part of the coconut fruit at which having biological function to protect fruit core and is located on the inner side of the fiber with a thickness ranging from 3-6 mm [3]. Coconut shell is classified as hardwood, mainly composed of lignin, cellulose, and hemicellulose, with water content of approximately 6-9 % [3].

Coconut shell is more suitable for pyrolysis process, since they contain less amount of ash, more amount of volatile matter and available with lower cost in rural areas during all sessions of the year [4].

Content from this work may be used under the terms of the Creative Commons Attribution 3.0 licence. Any further distribution of this work must maintain attribution to the author(s) and the title of the work, journal citation and DOI.
Published under licence by IOP Publishing Ltd

IOP Conf. Series: Journal of Physics: Conf. Series **846** (2017) 012025 doi:10.1088/1742-6596/846/1/012025

Pyrolysis is a thermo-chemical process in which organic material is converted into a carbon rich solid (char) and volatile matter (liquids and gases) by heating in the absence of oxygen [5]. Lignocellulosic biomass basic components are hemicellulose, cellulose and lignin. Researchers have already confirmed that lignin starts decomposing at low temperatures (160–170°C) and continues to decompose at low rate until approx. 900 °C. Hemicellulose is the second component to start 524 A. Tsamba reported that decomposing, followed by cellulose, in a narrow temperature interval from about 200 to 400°C. This is the interval in which the main decomposition takes place and accounts for the greatest decomposition in the biomass pyrolysis process consisting of degradation reactions. Beyond 400 °C, the most important reaction leads to the aromatisation process, at low mass loss rate [6]. Up to now, pyrolisis is commonly equipped by conventional heater (electric heater or furnace). Recently, the use of microwave radiation as a substitute for the conventional heating for synthesizing organic [7-10] ceramic processing [11-13], and accelerate a chemical reaction and atomic diffusion [14-17]. Microwave is not only shortening processing time but also improving material properties such as microstructure and hardness [18-19]. The heating with a microwave depends more on the molecular characteristics and the condition of reactions compared to a conventional heating. The use of this microwave energy for pyrolysis and processing biomass based materials has been performing in Halu Oleo University and will be reported in separated papers.

Smoke is defined as a suspension of solid and liquid particles in a gas medium. The smoke is produced by the pyrolysis process by means of incomplete combustion at which involving reaction of constituent polymers decomposition into organic compounds with low molecular weight due to the effects of heat include oxidation reaction, decomposition, polymerization, and condensation [20]. Liquid smoke is a solution of the result of condensation of vapor of wood smoke which is burned with limited air at high temperatures [21]. Liquid smoke contains many compounds that are formed due to the pyrolysis of three components of wood, namely cellulose, hemicellulose and lignin[22].

Coconut shell pyrolysis process produces three fractions, namely fraction of solid in the form of charcoal, the weight fraction of tar, and the light fraction of gas [23]. More than 400 chemical compounds in the smoke have been successfully identified. These components are found in varying amounts depending on the type of wood, age of the plant of wood sources, and wood growing conditions such as climate and soil. These components include acid, carbonyl, and phenol. Acids affect the taste, pH, and storage duration of smoked products. Carbonyl reacts with proteins to form brown staining. Phenol is a major framer of aroma and showed antioxidant activity [24].

Chromatography is a powerful and versatile tool for separating closely related chemical species. In addition, it can be used for the qualitative identification and quantitative determination of separated species. [24].Some detectors are universal; that is they are sensitive to practically every compound that elutes from the column. On the other hand, there are discriminating (selective) detectors that are sensitive only to specific compounds, yielding a very uncomplicated chromatogram. The ideal situation to quantify an analytic, would be to have a detector which sees only this analytic. They can be also categorized as destructive or non-destructive of the analytic. Detectors are classified into two groups depending on whether they lead only to a single information such as the retention time and those which yield, besides retention time, structural information of the analytic concerned. For this reason, some gas chromatographs are equipped with two or three detectors linked in series. Nonetheless, the response of all detectors is dependent on the molar concentration or on the mass of analytic in the carrier gas [25].

2. Material and Method

2.1 Materials and Tools

Materials used for the manufacture of liquid smoke is a coconut shell waste . The main equipment used to obtain the coconut shell liquid smoke is pyrolysis , and liquid smoke produced in identification using Gas Chromatography Agilent brands.

2.2 Sample Preparation

Preparation of the material in this study is the material taken from the wild form of coconut shell cleaned from coir and cut into small pieces (size 2 - 3 cm) and then dried in the sun for 7-10 days.

2.3 Pyrolysis process

Pyrolysis process is done by using a pyrolysis reactor equipment with the following procedures: 1). Coconut shell sample is weighed as much as 1200 grams, 2) . The sample is introduced into a pyrolysis reactor tube, 3). Pyrolysis appliance connected to the electric current, 4). The process temperature is set by means Termocouple until it reaches a temperature variations 400°C, 500°C, and 600°C .for 1 hour, 5).

Smoke coming out has been condensed and collected in a bottle, 6). After selesasi process , the reactor is cooled, 7). The resulting liquid volatile matter volume was analyzed using GC .

2.4 Identification of Chemical Components by GC

LVM constituent chemical components obtained from the results of pyrolysis GC identified in the laboratory of Forensic Makassar . The equipment uses 2 colum, (1). J & W 122-5731 : 1721.65322 with colum temperature of 400°C with a diameter of 30 m x 250 µm x 0.1 µm, (2). Frontier Alloy 5Ultra UA - 5 with a temperature of 450°C with a diameter of 30m x 0.25 µm 250µmx operating conditions at an oven temperature of 160°C/min , injection 250°C , helium carrier gas flow rate of 150 ml/min .

3. Results and Discussion

3.1 Pyrolysis results

Results coconut shell waste pyrolysis produces liquid distillate in the form of smoke and charcoal residue. It also obtained the gases which can not be condensed by cooling, so it could not be accommodated in the liquid reservoir. Most of these gases are trapped in the container while others apart from the reservoir through the conduit of smoke and escape into the atmosphere .

The volume of liquid smoke produced is an important parameter to determine the outcome of a process. Liquid smoke in this study were generated through a process of condensing the smoke released pyrolysis reactor. During the pyrolysis process of evaporation of various kinds of compounds. Volume smoke liquid smoke produced from each temperature are presented in Table 1 .

Table 1. Volume of liquid smoke cocoa shell waste pyrolysis results with electric reactor at a temperature of 400°C, 500°C and 600°C

Temperature (°C)	Material Weights (kg)	Charcoal weights (kg)	LVM (ml)
400	1.2	0.3	245
500	1.2	0.35	250
600	1.2	0.33	260

Based on Table 1 the average volume of liquid smoke produced from each successive temperature are 204.167 $ml.kg^{-1}$, 208.33 $ml.kg^{-1}$ and 216.67 $ml.kg^{-1}$. In this case the amount of liquid smoke generated in the pyrolysis process is very dependent on temperature pyrolysis and condensation systems used. That is appropriate condition for the formation of liquid smoke used water as the cooling medium so that the heat exchange process can occur relatively quickly. Pyrolysis at temperatures high and too long will lead to the formation of liquid smoke is reduced as the temperature of the cooling water is increasing so that the smoke generated is not completely condensed.

The quality of liquid smoke is very dependent on the composition of the chemical compounds they contain. Criteria for a good quality liquid smoke flavor and aroma as a characteristic possessed the smoke is determined by the class of chemical compounds it contains. Chemical compounds contained in liquid smoke is very dependent on the pyrolysis conditions and raw materials used . In addition, the process of pyrolysis of a material that does not last perfect can cause chemical components generated in the liquid smoke less complete. Chemicals component that have been identified in the liquid smoke include compounds phenol, carbonyl, carboxylic acid, furan, hydrocarbon, alcohol and lactone.

3.2 Chemical components of LVM

LVM generated in the pyrolysis process of waste coconut shall beforehand dissolved in methanol for chemical content are further identified. Determination of the compounds contained in liquid smoke is done by using GC equipment. Depictions of the chromatogram pattern shaped curve as a function of time. Chromatograms in Figures 1 to 3 show that the liquid smoke is generated in the pyrolysis process of waste coconut shell shows the separation of the chemical components through chromatogram peaks appearing in gas chromatography. Results chromatogram liquid smoke at a temperature of 400°C peak chromatogram start appearing at a retention time from 1.86 to 19.27 minutes (Figure 1) and identified as many as 20 compounds. Results chromatogram liquid smoke at a temperature of 500°C chromatogram peaks begin to appear on the retention time from 1.82 to 19.27 minutes (Figure 2) and identified as many as 14 compounds. While the results of the chromatogram liquid smoke at a temperature of 600°C peak cromtogram began to appear on the retention time of 1.85 to 19.27 minutes (Figure 3) and identified as many as 14 compounds.

IOP Conf. Series: Journal of Physics: Conf. Series **846** (2017) 012025 doi:10.1088/1742-6596/846/1/012025

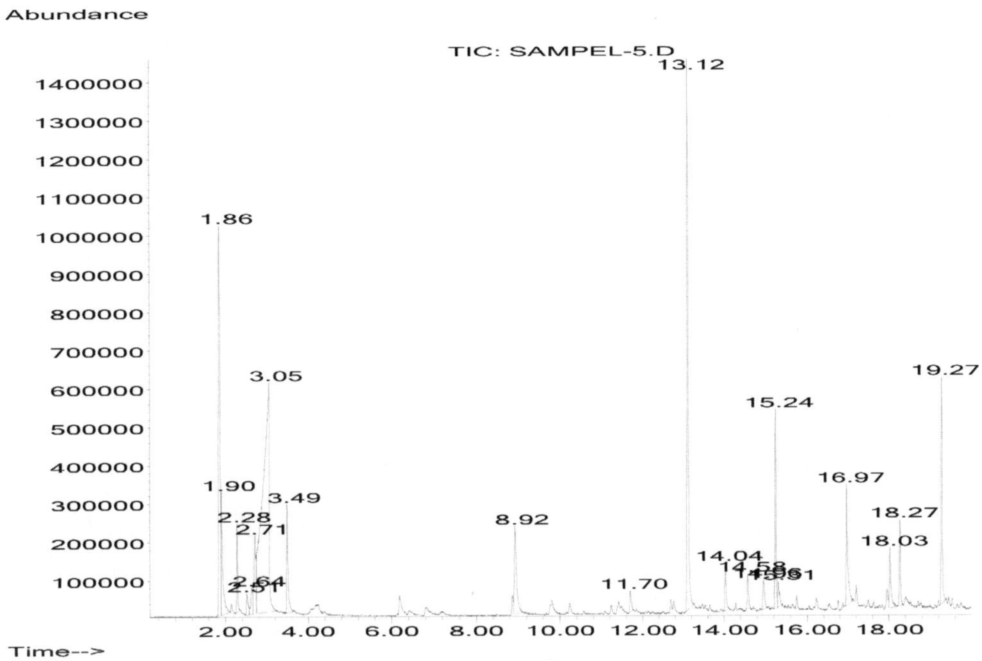

Figure 1. Results chromatogram liquid smoke coconut shell pyrolysis temperature of 400° C

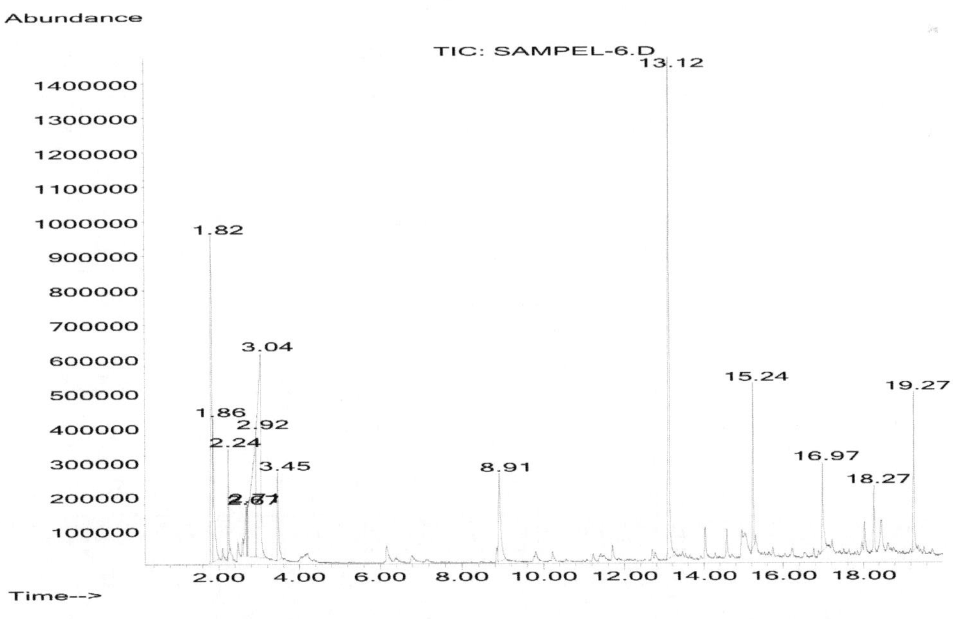

Figure 2. Results chromatogram liquid smoke coconut shell pyrolysis temperature of 500°C

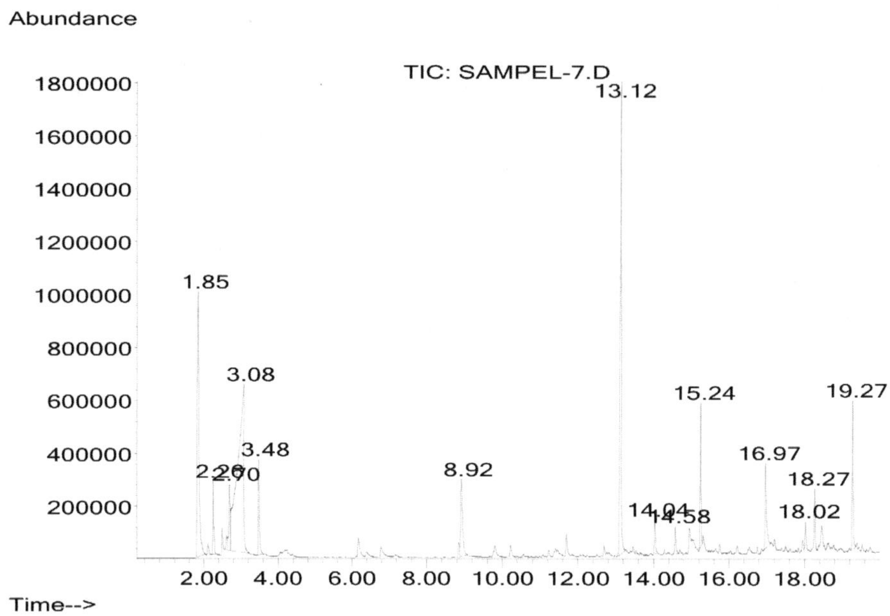

Figure 3. Results chromatogram liquid smoke coconut shell pyrolysis temperature of 600°C

The compounds are believed to have the name and structure as listed in Table 2 .

Table 2. Chemical Ingredients liquid smoke coconut shall

No. peak	Temperature (°C)	Retention time (min)	Name of compound	Concentration (%)
1	400	1.86	Amonia	12.41
2		1.9	Hydrazine	3.87
3		2.28	Acetid acid	1.75
4		2.51	Formic acid, Metil Ester	0.75
5		2.63	Acetid acid	0.95
6		2.71	Acetid acid	3.33
7		3.05	Acetid acid	31.24
8		3.49	2-Propanone	3.04
9		8.92	Furfural, 2-Furancaboxaldehyde	4.16
10		11.7	1.3 Cyclopentanedione	0.64
11		13.12	Phenol	15.41
12		14.04	2-Cyclopenten	1.39
13		14.58	Phenol	1.04
14		14.96	Phenol	1.06
15		15.24	Phenol	4.59
16		15.31	1-Propanol	0.86
17		16.97	Phenol	4.1
18		18.03	1.2-Benzenediol	1.97
19		18.27	Benzeneethanol	1.97
20		19.27	Phenol	5.46

Table 2. (Continued)

No. peak	Temperature (°C)	Retention time (min)	Name of compound	Concentration (%)
1	500	1.82	Amonia	12.22
2		1.86	Hydrazine	5.61
3		2.24	Acetid acid	1.97
4		2.67	Acetid acid	2.15
5		2.71	Acetid acid	1.33
6		2.92	Acetid acid	15.39
7		3.04	Acetid acid	20.12
8		3.45	2-Propanone	3.1
9		8.91	2-Furancarboxaldehyde	5.3
10		13.12	Phenol	18.49
11		15.24	Phenol	4.4
12		16.97	Phenol	3.3
13		18.27	Guaicol, 4- Ethil Phenol	1.98
14		19.27	Phenol	4.65
1	600	1.85	Amonia	15.49
2		2.26	Acetid acid	2.27
3		2.7	Hexane	3.13
4		3.08	Acetid acid	33.74
5		3.48	2-Propanone	3.37
6		8.92	Furfural, 2-Furancarboxaldehyde	4.62
7		13.13	Phenol	19.12
8		14.64	2-Cyclopenten	1.23
9		14.58	Phenol	1.07
10		15.24	2-Methoxyphenol	4.23
11		16.47	4-Methybenzen	3.68
12		1.03	1.2-Benzeneediol	1.15
13		18.27	Benzeneethanol	2.15
14		19.27	Phenol	4.75

From the results of the above identification is found phenol. Where the phenolic compounds can be applied as a food preservative.

4. Conclusion

Results pyrolysis cocoa shell with temperature variations 400°C, 500°C and 600°C gained an average volume of liquid smoke respectively 204.167 ml kg^{-1}, 208.33 ml kg^{-1}, and 216.67 ml kg^{-1}. Results of the Gas Chromatography identification of unknown chemical components of liquid smoke cocoa shell has ammonia, acetic acid, propanone, furfural, and phenol. Phenol compounds can be used as a preservative applications

Acknowledgement
We would like to thank Directorate P2M-Ditjen DIKTI for helping in funding this research under the scheme of Penelitian Unggulan Perguruan Tinggi (PUPT) in year 2015.

References

1. S. Yaman, Energy Conversion and Management 45 (2004) 651–671.
2. P. Das, A. Ganesh, Biomass and Energy 25 (2003) 113–117.
3. Tilman D. 1981. Wood Combustion: Principles, Processes and Economics. Academics Press Inc., New York, USA.
4. Ganapathy E. Sundaram and E. Nataraja. 2009 Pyrolysis of Coconut Shell: An Experimental Investigation *The Journal of Engineering Research Vol. 6, No. 2 (2009) 33-39*
5. Demirbas, A., Arin, G. (2002). An Overview of Biomass Pyrolysis. *Energy Sources.* **24**, 471-482.
6. T. Fisher, M. Hajaligol, B. Waymack, D. Kellog, Journal of Analytical and Applied Pyrolysis 62 (2002) 331–349.
7. Wu, Y., Zhang, C., Liu, Y., Fu,Z., Dai, B. dan Yin, D. 2012. Biomass Char Sulfonic Acids (BC-SO3H)-Catalyzed Hydrolysis of Bamboo under Microwave Irradiation. Bioresources. 7(4), 5950-5959.
8. H. Aripin, S. Mitsudo, I. N. Sudiana, N. Jumsiah, I. Rahmatia, B. Sunendar, L.Nurdiwijayanto, S. Mitsudo, S.Sabchevski, Preparation of Porous Ceramic with Controllable Additive and Firing Temperature, Advanced Materials Research. Vol. 277 (2011) pp. 151-158
9. Stuerga, D. 2006. Microwaves in Organic Synthesis, Second Edition. Wiley-VCH.
10. Dehani, F. R., Argo, B. D. dan Yuliningsih, R. 2013. Utilization of Microwave Irradiation to Maximize Pretreatment Process Degradation of Lignin Jeremi Paddy. Jurnal Biopress Komoditas Tropis Vol.1 No. 1, 175-182 [In Indonesia].
11. I.N. Sudiana, Ryo Ito, S. Inagaki, K. Kuwayama, K. Sako, S. Mitsudo, Densification of Alumina Ceramics Sintered by Using Sub-millimeter Wave Gyrotron, J. Infrared, Millimeter, and Terahertz Waves. 34 (2013), 627-638.
12. Thostenson, E.T. dan Chou, T.W. 1999. Microwave processing : Fundamental and Applications, Elsevier Composites Applied Science and Manufacturing: Part A 30, 1055-1071.
13. S. Mitsudo, S. Inagaki, I.N. Sudiana, K. Kuwayama, Grain Growth in Millimeter Wave Sintered Alumina Ceramics , Advanced Materials Research, Vol.789 (2013), pp. 279-282.
14. H.M.S. Kingston, H.J. Haswell, 1997. Microwave-Enhanced Chemistry: Fundamental, Sample Preparation, and Applications, American Chemical Society.
15. M. Z. Firihu., I. N. Sudiana, S. Mitsudo, Microwaves Enhanced Sintering Mechanisms in Alumina Ceramic Sintering Experiments, Contemporary Engineering Sciences, Vol. 9, 2016, 5, 237 – 247
16. I.N. Sudiana, S. Mitsudo, K. Sako, S. Inagaki, L. O. Ngkoimani, I. Usman, H. Aripin, The microwave effects on the properties of alumina at high frequencies of microwave sintering, American Institute of Physics (AIP) Conference Proceeding, **1719**, 030014 (2016); doi: 10.1063/1.4943709
17. I. N. Sudiana, S. Mitsudo, T. Nishiwaki, P. E. Susilowati, L. Lestari, M. Z. Firihu, H. Aripin, Synthesis and Characterization of Microwave on Sintered Silica Xerogel Produced from Rice Husk Ash, Journal of Physics: Conference Series, Vol. 739 No. 1, (2016) 012059
18. I. N. Sudiana, S. Mitsudo, M. Z. Firihu, Effect of Initial Green Samples on Mechanical Properties of Alumina Ceramic , Contemporary Engineering Sciences, Vol. 9, 2016, no. 12, 595-602.
19. I. N. Sudiana, S. Mitsudo, M. Z. Firihu, H. Aripin, Effect of High-Frequency Microwave on Micro Hardness of Alumina Ceramic, Material Science Forum , Vo. 872, 2016, pp.114-117.
20. Girard J P. 1992. Technology of Meat and Meat Products. Ellis Horwood, New York, USA.
21. Yulistiani. 1997. The inhibition ability of liquid smoke against growth of bacterial pathogens and blight on cow tongue. Master Thesis. Study Program of Food Science and Technology. University of Gadjah Mada.
22. Lawrie R.A. 1983. Meat Science. Pargamon Press, London.
23. Astuti. 2000. Utilization of fiber and shell of coconut and oil palm shell for producing liquid smoke as natural food preservatives.http://alcoconut.multiply.com/journal/item/6.
24. Douglas A. Skoog, Donald M. west, F. James Holler, Stanley r. Crouch. 2004. Fundamentals of analytical chemistry ninth edition. Mary finch. USA
25. Francis Rouessac and Annick Rouessac. 2007. Chemical Analysis. University of Le Mans. France.

Characterization of Liquid Volatile Matter (LVM) of Cashew Nut Shell using Pyrolysis and Gas Chomatroghaphy

Mashuni[1], Muhammad Jahiding[2], Waode Sitti Ilmawati[2], Ita Kurniasih[2], Wa Wati[2], Muzirah[2], Muniati Burhan[2]

[1]Department of Chemistry, Faculty of Mathematics and Natural Sciences, Halu Oleo University (UHO), Kendari, South East Sulawesi, 93231
[2]Department of Physics, Faculty of Mathematics and Natural Sciences, Halu Oleo University (UHO), Kendari, South East Sulawesi, 93231

E-mail: mashuni2696@yahoo.com

Abstract. Cashew nut areexcellent products in Southeast Sulawesi. Cashew nut is one part of the cashew plant untapped waste.Cashew nut shell potential as a producer of liquidvolatile matter (LVM)and charcoal because it contains lignocellulos. LVM is the smoke condensation products obtained from the pyrolysis reactor can used foradhesive of briquettes hybridapplication. The aim of this reseach is to produce LVM of cashew nut shellby pyrolysisand analyze the content byGas Chromatography(GC).The research procedure begin with drying the cashew nut, the sample inserting into the pyrolysis reactor then heating with three variations of temperature respectively is 400°C, 500°C and 600°C.Cashew nutshell have been heating by pyrolisis processwith high temperatures resulting chorcoal and LVM separately. Volume LVM measured,then identify is components using GC. LVM obtained respectively200ml kg^{-1}, 340 ml kg^{-1}, and 340 ml kg^{-1}.Chromatogramof the GCfrom LVM of cashew nut shell has ammonia, hexane, acetic acid, propanone and phenol.Phenol compounds can be used as a adhesive for hybrid briquettes applications.

1. Introduction

Indonesia as an agricultural country, especially a lot of waste southeast Sulawesi plantations underutilized. Cashew nut are excellent products in Southeast Sulawesi. Cashew nut is one part of the cashew plant untapped waste. One way of handling this is to do are-processing to make cashew nut as a raw material for making liquid volatile matter(LVM) [1]. LVM from Cashew nut shell is a versatile material with many applications in the manufacture of friction dust for brake linings and clutch facing, different types of coatings, adhesives, and pesticides among many others [2]. LVM is the result of condensation or condensing steam pyrolysis results of materials containing lignocellulose and other carbon compounds [3]. In the other side, lignocellulosic biomass basic components are hemicellulose, cellulose and lignin so has the potential to be used as raw material for making LVM [4].

Pyrolysis is a process that converts biomass directly into solid, liquid, and gaseous products by thermal decomposition in the absence of oxygen [5]. Moreover, pyrolysis has been the source of many basic organic chemicals such as phenol, acetic acid, and acetone [6]. In this study, the compounds of LVM identified using Gas Chromatography (GC). GC is an instrumental chromatography configuration, meaning we have electronic detection and peaks on a computer screen [7]. Pyrolisis is commonly applied heater (electric heater or furnace) during heating. Recently. progress in electromagnetic radiation source especially in microwave frequency range. open opportunity to heat material by using this waves as n alternative energy for heating. This method was applied and reported for synthesizing organic material [8-10], ceramics processing [12-16], enhancing mechanical properties of ceramics [17-18]. and chemical

Content from this work may be used under the terms of the Creative Commons Attribution 3.0 licence. Any further distribution of this work must maintain attribution to the author(s) and the title of the work, journal citation and DOI.
Published under licence by IOP Publishing Ltd

reaction [19-20]. The application microwave for pyrolysis has been performing in Halu Oleo University and will be reported in separated papers.

LVM that produced from the pyrolysis process cashew nut shell can be made in raw materials for the application of adhesive for hybrid briquettes, because the compound like phenol content than can be used as an adhesive [2].

2. Material and Method

2.1 Materials and Tools

Materials used for the manufacture of LVM is a cashew nut shell waste. The main equipment used to obtain the cashew nut shell LVM is pyrolysis, and LVM produced in identification using Gas Chromatography Agilent brands. Preparation of the material in this study is the material dried in the sun for 1-2 weeks.

2.2 Sample Preparation

Preparation of the material in this study is the material taken from the wild form of coconut shell cleaned from coir and cut into small pieces (size 2 - 3 cm) and then dried in the sun for 7-10 days.

2.3 Pyrolysis process

Pyrolysis process is done by using a pyrolysis reactor equipment with the following procedures: 1). Coconut shell sample is weighed as much as 1200 grams, 2) . The sample is introduced into a pyrolysis reactor tube, 3). Pyrolysis appliance connected to the electric current, 4). The process temperature is set by means Termocouple until it reaches a temperature variations 400°C, 500°C, and 600°C .for 1 hour, 5). Smoke coming out has been condensed and collected in a bottle, 6). After selesasi process , the reactor is cooled, 7). The resulting liquid volatile matter volume was analyzed using GC .

2.4 Identification of Chemical Components by GC

LVM constituent chemical components obtained from the results of pyrolysis GC identified in the laboratory of Forensic Makassar . The equipment uses 2 colum, (1). J & W 122-5731 : 1721.65322 with colum temperature of 400°C with a diameter of 30 m x 250 μm x 0.1 μm, (2). Frontier Alloy 5Ultra UA - 5 with a temperature of 450°C with a diameter of 30m x 0.25 μm 250μmx operating conditions at an oven temperature of 160°C/min , injection 250°C , helium carrier gas flow rate of 150 ml/min .

3. Results and Discussion

3.1 Pyrolysis results

Results cashew nut shell waste pyrolysis produces liquid distillate in the form of smoke and charcoal residue. It also obtained the gases which can not be condensed by cooling, so it could not be accommodated in the liquid reservoir. Most of these gases are trapped in the container while others apart from the reservoir through the conduit of smoke and escape into the atmosphere .

The volume of liquid smoke produced is an important parameter to determine the outcome of a process. Liquid smoke in this study were generated through a process of condensing the smoke released pyrolysis reactor. During the pyrolysis process of evaporation of various kinds of compounds. Volume smoke liquid smoke produced from each temperature are presented in Table 1 .

Table 1. Volume of liquid smoke cashew nut shell waste pyrolysis results with electric reactor at a temperature of 400°C, 500°C and 600°C

Temperatures (°C)	Material Weights (kg)	Charcoal weights (kg)	LVM (ml kg)
500	0.5	0.16	200
600	0.5	0.14	340
700	0.5	0.11	340

Based on Table 1 the average volume of liquid smoke produced from each successive temperature 200 ml kg^{-1}, 340 ml kg^{-1}, and 340 ml kg^{-1}. In this case the amount of LVM generated in the pyrolysis process is very dependent on temperature pyrolysis and condensation systems used. That is appropriate condition for the formation of LVM used water as the cooling medium so that the heat exchange process can occur relatively quickly. Pyrolysis at temperatures high and too long will lead to

IOP Conf. Series: Journal of Physics: Conf. Series **846** (2017) 012026 doi:10.1088/1742-6596/846/1/012026

the formation of LVM is increases. The quality of LVM is very dependent on the composition of the chemical compounds they contain. Criteria for a good quality LVM flavor and aroma as a characteristic possessed the smoke is determined by the class of chemical compounds it contains. Chemical compounds contained in LVM is very dependent on the pyrolysis conditions and raw materials used. In addition, the process of pyrolysis of a material that does not last perfect can cause chemical components generated in the LVM less complete. Chemicals component that have been identified in the LVMinclude compounds ammonia, hexane, phenol, acetid acid, furan and ketones.

3.2 Chemical Components of Liquid Smoke

Liquid smoke generated in the pyrolysis process of waste cashew nut shell first filtered using filter paper for further identified. Determination of the compounds contained in liquid smoke is done by using GC equipment. Depictions of the chromatogram pattern shaped curve as a function of time. Chromatograms in Figures 1 to 3show that the LVM is generated in the pyrolysis process of waste cashew nut shell shows the separation of the chemical components through chromatogram peaks appearing in GC. Results chromatogram LVM at a temperature of 400°C chromatogram peaks start appearing at a retention time from 1.88 to 18.46 minutes (Figure 1)and identified as many as 17 compounds. Results chromatogram LVM at a temperature of 500°C chromatogram peaks begin to appear on the retention time from 1.88 to 16.50 minutes (Figure 2) and identified as many as 10 compounds. Results chromatogram LVM at a temperature of 600°C chromatogram peaks begin to appear on the retention time from 1.88 to 17.17 minutes (Figure 3) and identified as many as 8 compounds

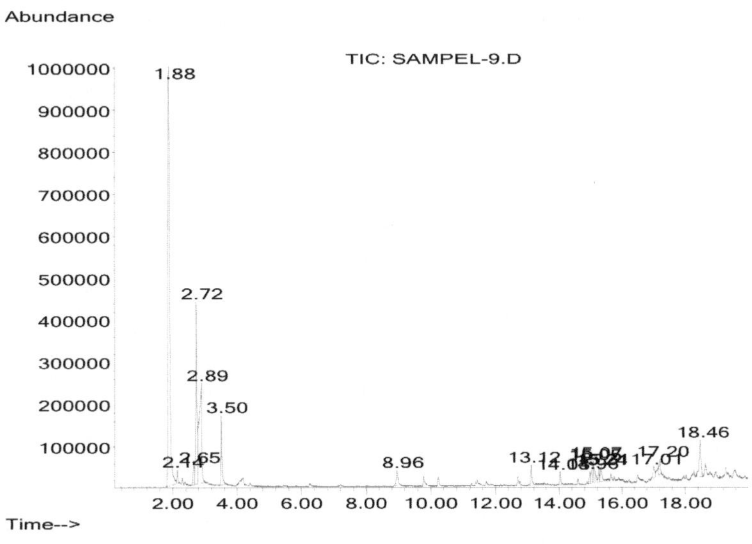

Figure 1. Chromatogram LVM of cashew nut shell at temperature 400°C

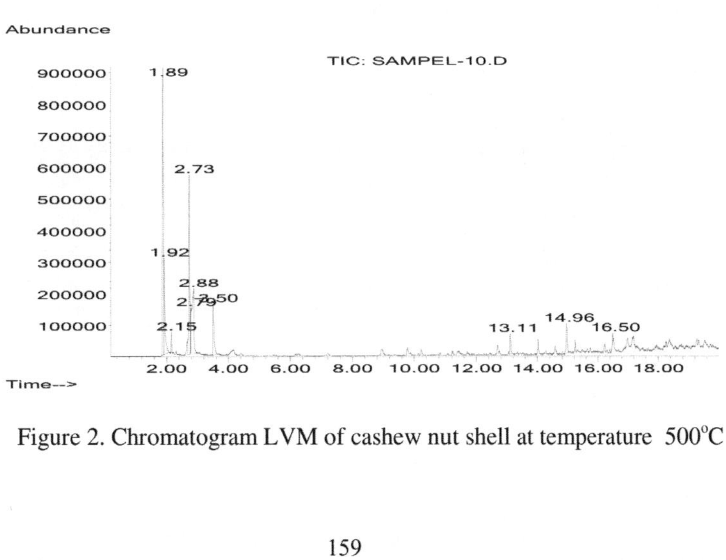

Figure 2. Chromatogram LVM of cashew nut shell at temperature 500°C

IOP Conf. Series: Journal of Physics: Conf. Series **846** (2017) 012026 doi:10.1088/1742-6596/846/1/012026

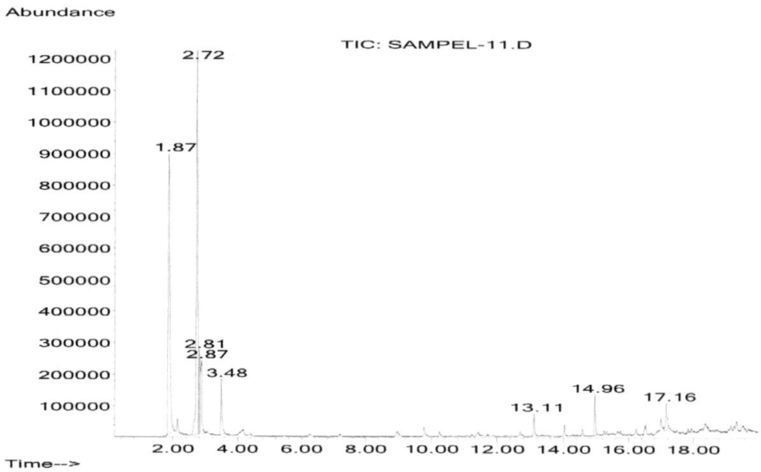

Figure 3. Chromatogram LVM of cashew nutshell at temperature 600°C

The compounds are believed to have the name and structure as listed in Table 2

Table 2. Chemical Ingredients LVM of cashew nut shell

Peak	Temperature (°C)	Retention Time (min)	Name of compound	Concentration (%)
1	400	1.88	Ammonia	45.42
2		2.14	2-Propanone, Aceton (Ketone)	0.79
3		2.65	Acetic Acid	1.32
4		2.72	Hexane	12.92
5		2.89	Acetic Acid	17.5
6		3.5	2-Propanone, Hidroxy Aceton (Ketone)	4.48
7		8.96	Furfural, 2-Furan Karboxaldehide	1.55
8		13.12	Phenol	1.68
9		14.03	1,2-Siklopentanedione	0.83
10		14.96	3-Methyl Phenol	0.85
11		15.05	2-Piran	1.71
12		15.07	3,4-Metoxy benzoik Acid	1.93
13		15.24	Mequinol	0.93
14		15.31	2-Metil Amino	1.45
15		17.01	2-ethoxy Phenol	1.30

Table 2. (continued)

Peak	Temperature (°C)	Retention Time (min)	Name of compound	Concentration (%)
16		17.2	Siringol	1,16
17		18.46	Hidroquinon	4.19
1	500	1.88	Ammonia	35.85
2		1.92	Hydrazine	10.65
3		2.15	2-Propanone, Aceton (Ketone)	1.39
4		2.73	Hexane	18.64
5		2.79	Acetic Acid	4.17
6		2.88	Acetic Acid	15.47
7		3.5	2-Propanone, Hidroxy Aceton (Ketone)	6.81
8		13.11	Phenol	2.64
9		14.96	3-Methyl Phenol	4.03
10		16.5	4-Ethyl Phenol	0.36
1	600	1.88	Ammonia	36.13
2		2.72	Hexane	36.03
3		2.81	Acetic Acid	7.35
4		2.86	Acetic Acid	7.10
5		3.48	2-Propanone, HidroxyAceton (Ketone)	5.29
6		13.11	Phenol	1.89
7		14.96	3-Methyl Phenol	3.21
8		17.,17	2,3-Dihydro-Benzofuran	3.00

From the results of the above identification is found phenol . Where the phenolic compounds can be applied as a adhesive hybridbriquettes.

Table 3. Chemical Phenol LVM of cashew nut shell

Temperature (°C)	Totally of Phenol (%)
400	10.11
500	7.03
600	5.1

4. Conclusion

Results pyrolysis of cashew nutshell with temperature variations 400°C, 500°C and 600°C gained an average volume of LVM respectively200 ml.kg^{-1}, 340 ml.kg^{-1} and 340 ml.kg^{-1}. Pyrolysis process is very dependent on temperature pyrolysis and condensation systems used. That is appropriate condition for the formation of LVM used water as the cooling medium so that the heat exchange process can occur relatively quickly. Pyrolysis at temperatures high and too long will lead to the formation of LVM is increases. Results of the Chromatogram of the GCfrom LVM of cashew nut shell has ammonia, hexane, hydrazine, acetic acid, propanone and phenol. Phenol compounds can be used as a adhesive for hybrid briquettes applications.So, based on the table 3, at temperature pyrolysis 400°C compound phenol very good for adhesive briquettes hybrid application.

Acknowledgement
We would like to thank Directorate P2M-Ditjen DIKTI for helping in funding this research under the scheme of Penelitian Unggulan Perguruan Tinggi (PUPT) 2015.

References

1. Risfaheri, dkk, 2004. Pemisahan Kardanol dari Minyak Kulit Biji Jambu Mete dengan Metode Destilasi Vakum. Jurnal Penelitian Pascapanen Pertanian, 1(1) : 1-11. [In Indonesian]
2. Patrick M. Mwangi, Christopher Aule and George T. Thiong'o Energy Studies Of Some Cashew Nut By-Products In Kenya (Chemistry Department, Jomo Kenyatta University of Agriculture and Technology, P. O. Box 62000-00200, Nairobi, Kenya)
3. HarsonoSoni 2008 Pengembangan Teknologi Produksi Asap Cair (Laporan Hilink II, September 2008 Universitas Jember). [In Indonesian]
4. T. Fisher, M. Hajaligol, B.Waymack, D. Kellog, Journal of Analytical and Applied Pyrolysis 62 (2002) 331–349.
5. Pütün,2007. A.E., Comparision between the Slow and the Fast Pyrolysis of Tabacco Residue. Industrial Crops and Products, 26: p. 307-314.
6. Bhatia, V.K,.1988.Biomass as a non-fossil fuel source - an overview. Res. Ind.,. 33 p. 154-61.
7. John Kenkel, 2013. Analytical Chemistry for Technicians 4th Edition. CRC Press Taylor and Francis Group, London.
8. Wu, Y., Zhang, C., Liu, Y., Fu,Z., Dai, B. dan Yin, D. 2012. Biomass Char Sulfonic Acids (BC-SO3H)-Catalyzed Hydrolysis of Bamboo under Microwave Irradiation. Bioresources. 7(4), 5950-5959.
9. H. Aripin, S. Mitsudo, I. N. Sudiana, N. Jumsiah, I. Rahmatia, B. Sunendar, L. Nurdiwijayanto, S. Mitsudo, S. Sabchevski, Preparation of Porous Ceramic with Controllable Additive and Firing Temperature, Advanced Materials Research. Vol. 277 (2011) pp. 151-158
10. Stuerga, D. 2006. Microwaves in Organic Synthesis, Second Edition. Wiley-VCH.
11. I.N. Sudiana, Ryo Ito, S. Inagaki, K. Kuwayama, K. Sako, S. Mitsudo, Densification of Alumina Ceramics Sintered by Using Sub-millimeter Wave Gyrotron, J. Infrared, Millimeter, and Terahertz Waves. 34 (2013), 627-638.
12. W. H. Sutton, Microwave Processing of Ceramic Materials, Microwave Solutions for Ceramic Engineers, Am. Cer. Soc., Ed.by D.E. Clark, D.C. Folz, C.E. Folgar, M.M. Mahmoud , pp.35-65 (2005).
13. S. Mitsudo, S. Inagaki, I.N. Sudiana, K. Kuwayama, Grain Growth in Millimeter Wave Sintered Alumina Ceramics , Advanced Materials Research, Vol.789 (2013), pp. 279-282.
14. M. Z. Firihu., I. N. Sudiana, S. Mitsudo, Microwaves Enhanced Sintering Mechanisms in Alumina Ceramic Sintering Experiments, Contemporary Engineering Sciences, Vol. 9, 2016, 5, 237 – 247
15. Sudiana, I.N., S. Mitsudo, K. Sako, S. Inagaki, L. O. Ngkoimani, I. Usman, H. Aripin, The microwave effects on the properties of alumina at high frequencies of microwave sintering, American Institute of Physics (AIP) Conference Proceeding, **1719**, 030014 (2016); doi: 10.1063/1.4943709
16. I. N. Sudiana, S. Mitsudo, T. Nishiwaki, P. E. Susilowati, L. Lestari, M. Z. Firihu, H. Aripin, Synthesis and Characterization of Microwave on Sintered Silica Xerogel Produced from Rice Husk Ash, Journal of Physics: Conference Series, Vol. 739 No. 1, (2016) 012059
17. I. N. Sudiana, S. Mitsudo, M. Z. Firihu, Effect of Initial Green Samples on Mechanical Properties of Alumina Ceramic , Contemporary Engineering Sciences, Vol. 9, 2016, no. 12, 595-602.
18. I. N. Sudiana, S. Mitsudo, M. Z. Firihu, H. Aripin, Effect of High-Frequency Microwave on Micro Hardness of Alumina Ceramic, Material Science Forum , Vo. 872, 2016, pp.114-117
19. H.M.S. Kingston, H.J. Haswell, 1997. Microwave-Enhanced Chemistry: Fundamental, Sample Preparation, and Applications, American Chemical Society.
20. Thostenson, E.T. dan Chou, T.W. 1999. Microwave processing : Fundamental and Applications, Elsevier Composites Applied Science and Manufacturing: Part A 30, 1055-1071.

Geometric and Road Environmental Effects against Total Number of Traffic Accidents in Kendari

M. Akbar Kurdin *, **La Welendo, Nur Annisa**

Halu Oleo University, Kampus Bumi Tri Dharma, Anduonohu, Kendari, Indonesia

*E-mail: akbarnanang71@yahoo.com

Abstract. From the large number of traffic accidents that occurred, the carrying of Kendari as the biggest contributor to accidents in the Southeast. The number of accidents in Kendari row since 2011 was recorded at 18 accidents due to the influence of geometric road, in 2012 registered at 13 accident and in 2013 amounted to 6 accidents, with accident data because of the influence Geometric recorded for 3 consecutive years the biggest contributor to accidents because of the influence of geometric is Abeli districts. This study aimed to determine the road which common point of accident-prone (Black spot) in Kecamatan Abeli as accident-prone areas in Kendari, analyze the influence of geometric and road environment against accidents on roads in Kecamatan Abeli, provide alternative treatment based on the causes of accidents on the location of the accident-prone points (blackspot) to reduce the rate of traffic accidents. From the results of a study of 6 curve the accident-prone locations, that the curve I, II, and VI is the "Black Spot" influenced by the amount and condition of traffic accidents, while at the curve II, a traffic accident that occurred also be caused by unsafe geometric where the type of geometric should be changed from Spiral-Spiral type to Spiral-Circle-Spiral type. This indicates geometric effect on the number of accidents.

1. Introduction

Currently there are some indications that land transport services, particularly urban transport decreased or not optimal. Indications that can be perceived by the public is irregularity, congestion, low travel speed, and the accident rate is relatively high, a fairly large number of victims will give economic and social impact. In anticipation of the indications mentioned above, the planning highway becomes very important to get a safe and comfortable roads, where the road geometric aspect becames the main thing in planning roads.

From the large number of traffic accidents that occurred, was recorded of Kendari as the biggest contributor to accidents in the Southeast Sulawesi. Number of accidents in Kendari since 2011 stood at 18 accidents due to the effect of road geometric (sharp turns), in 2012 stood at 13 accidents and in 2013 stood at 6 accidents, with accident data because of the effect of geometric recorded for 3 consecutive years the biggest contributor to accidents because of the effect of geometric is Kecamatan Abeli.

The aim of this study was to answer the above mentioned problems concerning the location of black spots on curve area, as well as analyzing the roads geometric and environmental conditions on the roads black spots under review.

Content from this work may be used under the terms of the Creative Commons Attribution 3.0 licence. Any further distribution of this work must maintain attribution to the author(s) and the title of the work, journal citation and DOI.
Published under licence by IOP Publishing Ltd

IOP Conf. Series: Journal of Physics: Conf. Series **846** (2017) 012027 doi:10.1088/1742-6596/846/1/012027

2. Literature Review

Roads geometric is part road planning emphasis on physical form planning, so that it can fulfill its function to provide optimum services on traffic and access to inter-city [16]. In general, geometric planning involves several aspects such as road width, curve, the flatness of roads, and visibility as well as combinations of these parts, both for the road, as well as for a crossing between two or more road segments [14].

Geometric element in the planning of roads and consists of horizontal alignment and vertical alignment. Planning road pavement and road drainage is not included in the planning of roads geometric although included in the planning of roads completely. For roads in Indonesia, roads geometric planning refers to the rules, namely: (1) Peraturan Perencanaan Geometrik Jalan Raya no. 130/1990; (2) Standar Perencanaan Geometrik untuk Jalan Perkotaan, 1992; (3) Peraturan Perencanaan Geometrik untuk Jalan Antar Kota No. 038/T/BM/1997; and (4) Standar Geometrik Jalan Perkotaan RSNI T-14-2004.

2.1 Horizontal Alignment

Horizontal alignment is mainly focused on the axis roads planning which will see the road is a straight road, swerved to the left or to the right. Horizontal alignment consists of three (3) types of curve are: Full Circle, Spiral-Circle-Spiral and Spiral-Spiral. Full Circle is a form of bow-shaped curve with a characteristic full circle, having a radius of curvature (R) which is large in order to avoid fracture [9], where the super elevation required <3% [15]. While Spiral-Spiral has the characteristics that is composed of 2 (two) curved spiral and has a super elevation (e) is small, and are used if the form Spiral -Circle-Spiral cannot be used. Spiral-Circle-spiral consisting of 1 (one) curved Circle and 2 (two) curved spiral, and is used if the form of Full Circle cannot be used. Geometric forms of each type of the curve can be seen in **FIGURE 1** below:

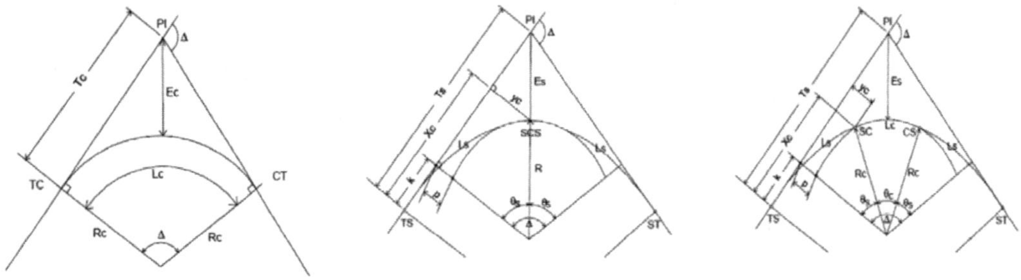

a) Full Circle Curve *b) Spiral – Spiral Curve* *c) Spiral – Circle – Spiral Curve*

FIGURE 1. Forms of Curved, Source: [14]

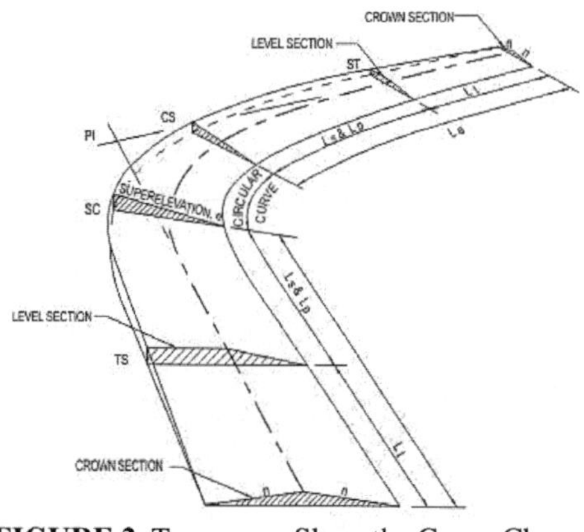

FIGURE 2. Transverse Slope the Curve Changes

2.2 Super Elevation

Super elevation achieved gradually from normal transverse slope on the straight road to full slope (super elevation) in the curved. At curve Spiral-Circle-Spiral and Full Circle super elevation attainment conducted linearly, while at curve Spiral - Spiral super elevation attainment entirely done on the spiral [9].

2.3 Vertical Alignment

Vertical alignment or longitudinal cross section of road which can be a road without slope, incline or decline. Hendarsin (2000) states that vertical alignment is planning elevation axis road at any point in the review, in the form of longitudinal profile. In the planning of vertical alignment is to consider how to put the axis of the according field conditions with regard to the nature of vehicle operations, safety, visibility, and the function of road. Observed from the starting point of planning, the straight sections can be either positive slope (incline), or negative slope (decrease), or zero slope (flat). Vertical curved sections can be concave curved or convex curved [15].

Vertical Curved should be provided at each location experiencing gradient changes, with the aim of reducing the shock due to gradient changes and to provide stopping visibility.

The minimum length of vertical convex curved by stopping visibility, for each design speed (V_R) can be seen in **TABLE 1** below:

TABLE 1. Planning control for Convex Curved Vertical by visibility stopping

Design-speed (km/h)	Visibility of Stopping (m)	Value Curved Vertical (K)
20	20	1
30	35	2
40	50	4
50	65	7
60	85	11
70	105	17
80	130	26
90	160	39
100	185	52

Source: [15]

2.4 Traffic Accident

Traffic accidents are incident on the road traffic where involves one or more vehicles resulting in damage of adverse owner or victim. According to the Law of the Republic of Indonesia Number 22 of 2009 regarding Traffic and Road Transport, traffic accident is an event on the road unexpected and unintentional involve vehicles with or without other road users which resulted in human victims and / or loss of property. Accidents are events that occur in a traffic movement due to an error in the traffic forming system, namely the driver (human), vehicles, road and environment. According Pignataro (1993) that most traffic accidents caused by several factors, the behavior of the driver or the pedestrian, road conditions, poor weather conditions and a bad view. Meanwhile, several other factors are factor condition of the vehicle, other vehicles, signs or traffic control devices, other objects on the road and planning road geometry that is not appropriate, or a combination of the above factors.

There are four factors of environmental conditions that affect human behavior so that potentially cause traffic accidents, namely:

a. Land use and activity, crowded areas, deserted, which instinctively driver reduces speed or vice versa.
b. The weather and the possibilities that looks, for example during fog, heavy smoke, heavy rain in such a way so as to reduce the visibility of the driver.
c. The existing facilities on the road network, the traffic signs, traffic lights and traffic markings.
d. Traffic flow and traffic characteristics, number, type and composition of the vehicle will greatly affect the speed of travel.

The site of a traffic accident can be in various places, at the crossroads, in streets, on bridges, in rural areas, in urban areas, and so on, so it needs identified areas that the frequent occurrence of accidents or accident-prone areas [10]. Areas prone to accidents is an area that has accident rate is high, the risk of accidents is high, the accident can be identified in specific locations on the roads (black spot) and also on certain roads (black site) or on a specific area (black area). The provisions of accident-prone locations (Black spots) based on "Pedoman Penanganan Lokasi Rawan Kecelakaan Lalu Lintas (2004)" can be seen in **TABLE 2** below:

IOP Conf. Series: Journal of Physics: Conf. Series **846** (2017) 012027 doi:10.1088/1742-6596/846/1/012027

TABLE 2 The provisions of Accident Prone Location (Black Spot)

Black Spot	Urban	Suburb
On roads and intersections	At least 2 traffic accidents with consequent death (per year), or At least 5 traffic accident with injuries / loss of material (per year)	At least 3 traffic accidents with consequent death (per year), or At least 5 traffic accident with injuries / loss of material (per year)

3. Research Methodology

The research location is located in the District Abeli Kendari, namely the curve I, II, and III on the main road between Abeli - Nambo which is at the Kelurahan Nambo, curve IV and V on the main road Abeli - Moramo located in the Kelurahan Sambuli, curve VI on the road the main Abeli - Moramo located on the Kelurahan Tondonggeu. Survey data retrieval begins with accident data and identify the accident because of effects of road geometric. The next survey of the condition of geometric and environmental survey on each street curve that has been identified as accident-prone locations. The data obtained from this survey is then analyzed horizontal alignment and vertical alignment. The processing of data using the theories and formulas of the study of literature as well as regulations issued by Bina Marga.

3.1 Geometric Parameters

There are some parameters in geometric analysis namely the design speed (VR), the centrifugal force, visibility, side-curve-free areas.

3.1.1 Design-Speed (VR)

Design-speed (V_R) on a road section is chosen speed as the basic geometric design that allows vehicles to move safely and comfortably in sunny weather conditions, traffic deserted, and road side effects are by no means [14].

TABLE 3. Design-speed (V_R) according to the classification of road in urban areas

Road Functions	V_R (km/h)
Primary Artery	50 – 100
Primary Collector	40 – 80
Secondary Artery	50 – 80
Secondary Collector	30 – 50
Secondary Local	30 – 50

Source: [15]

3.1.2 Centrifugal Force

Centrifugal force is the force exerted on a vehicle moving at a constant speed on the road with a curved-shaped trajectory, pushing the vehicle radially outward of the track road to a direction perpendicular to force the pace. To compensate for the centrifugal force, takes the friction force between the tire and the road surface, as well as the weight of a vehicle component due to the transverse slope of the road surface [14].

3.1.3 Visibility

Saodang (2004) states that visibility is a distance that is required by the driver in response to a dangerous obstacle, to take action to avoid the obstacle, which consists of stopping visibility (J_h) and visibility precedes (J_d).

TABLE 4. Stopping Visibility

V_R (km/h)	100	90	80	70	60	50	40	30
J_h minimum (m)	185	160	130	10 5	85	65	50	35

Source: [15]

3.1.4 Widening Line Traffic in Curve

Widening of the curve is meant to maintain operational service traffic conditions at the curve, so that same with the operational service on the straight road. If it's widening less than 0.60 m can be ignored [15].

4. Results Discussion

4.1 Black Spot Analysis

Based on traffic accident data years 2001 – 2013 were obtained from Kendari Police shows that 6 (six) curve as the research object, only curve I, curve II and curve IV expressed as the Black Spot. Results of the analysis of the determination of the Black Spot can be seen in **TABLE 5**.

TABLE 5. The Results of the Black Spot Analysis

Black Spot Parameters	Curve I	Curve II	Curve III	Curve IV	Curve V	Curve VI
Death	0	1	0	0	1	0
Injuries	5	5	1	3	0	6
Black Spot	YES	YES	NO	NO	NO	YES

4.2 Horizontal Alignment and Vertical Alignment Analysis

From the results of a survey of geometric conditions, it is known that the curve I, curve III, curve IV, and curve VI using Spiral-Circle-Spiral type, whereas the curve II and curve V using Spiral-Spiral type. Road class at a location to be reviewed is a primary arterial road class. Curve I have a hill slope category analyzed using design-speed (V_R) is 60 km/h, curve II have a flat slope category analyzed using design-speed (V_R) is 70 km/h, curve III and curve IV which has slope the flat category is analyzed using design speed (V_R) are 80 km / h, curve V which is a flat area with design-speed (V_R) is 50 km/h, and curve V which is a flat area with design-speed (V_R) is 60 km/h. The value of design speed (V_R) is determined based on the radius of curvature of each curve.

Based on the analysis of vertical alignment, the incline and decline area in the research location is safe. Based on a survey of the road environmental condition, all of the study locations located in the deserted area, without fog and smoke, with the condition of traffic signs, street lights and road markings that are still incomplete, so that the environment does not affect the accident that occurred at research locations. Results of the geometric and the road environment analysis can be seen in **TABLE 6** below:

TABLE 6. Results of the Geometric and the Road Environment Analysis

Geometric Parameter	Curve I	Curve II	Curve III	Curve IV	Curve V	Curve VI
Type of Curve	SCS	SS	SCS	SCS	SS	SCS
Super elevation	Safe	Not Safe	Safe	Safe	Safe	Safe
Widening the bend	0,22 m	0,51 m	0,25 m	No Need	0,58 m	0,56 m
Vertical Alignment	Safe					
Road Environment	No Problem					

From **Table 6** above, it appears that the geometric of curve II is not safe, so the type of curve should be changed from Spiral-Spiral type to Spiral-Circle-Spiral type. After analyzed by Spiral-Circle-Spiral type, the value of super elevation on the Curve II to be safe.

On the other hand, almost all the curve require widening, but its widening less than 0.60 m, so the widening can be ignored.

5. Conclusion

Based on the results of the discussion can be concluded that the curve I, II, and VI is the "Black Spot" influenced by the amount and condition of traffic accidents, while at the curve II, a traffic accident that occurred also be caused by unsafe geometric, where the type of geometric should be changed from Spiral-Spiral type to Spiral-Circle-Spiral type. To maintain the environmental conditions are still safe for traffic, it is advisable to do routine maintenance on the condition of signs, road markings, road shoulders and pavement conditions.For further research, it is advisable to add the factor of pavement conditions, besides factors geometric, to determine Back Spot, Black Site, and Black Area, because the design of geometric and pavement conditions that do not qualify potentially cause of accidents, such as the curve is too sharp, the condition pavement ineligible (slippery and / or damaged).

References

1. Agus. T.M, Audit Keselamatan Infrastruktur Jalan, Yogyakarta: Jurnal Ilmiah Teknik Sipil UGM vol. 16 No.3. 2009
2. Agus, T.M, Kushari, B., Faisol, Kurniawati dan Gunawan, H.E., Modul Pelatihan Inspeksi Keselamatan Jalan (IKJ) dalam Penyelenggaraan Jalan Berkeselamatan, FSTP. 2008
3. Agus. S, Model Kecelakaan Lalu Lintas Di Tikungan Karena Pengaruh Konsistensi Alinyemen Horizontal Dalam Desain Geometrik Jalan Raya, Solo: Jurnal Ilmiah Teknik Sipil Universitas Sebelas Maret Vol. X, 2010.
4. Ditjen Bina Marga, Kajian Kebutuhan Pelaksanaan Keselamatan Jalan di Indonesia, Jakarta: Departemen Pekerjaan Umum, 2006.
5. Ditjen Bina Marga, Penyusunan Sistem Manajemen dan Pedoman Keselamatan Jalan dalam Kegiatan Pembangunan Jalan, Jakarta: Departemen Pekerjaan Umum, 2007.
6. Departemen Pekerjaan Umum, , Manual Kapasitas Jalan Indonesia (MKJI), Jakarta, 1997.
7. Elly. T.P, Pengaruh Geometrik Jalan Terhadap Kecelakaan Lalu Lintas Di Jalan Tol (Studi Kasus Tol Semarang Dan Tol Cikampek, Semarang: Program Pascasarjana, Universitas Diponegoro, 2006.
8. Francesco, S.C, Accidents And Geometric Characteristic Of Road Layout : Some Quantitative Evaluations, journal of Oregon State University, 2000.
9. Hendarsin, S.L, Penuntun Praktir Perencanaan Teknik Jalan Raya, Bandung, Politeknik Negeri Bandung, 2000.
10. NAASRA, Guide to Traffic Engineering Practice, Sydney Australia, NAASRA, 1988
11. Pignataro, Traffic Engineering Category Practice, New Jersey, USA, Ent. in Englood Cliffs, 1993
12. Pedoman Penanganan Lokasi Rawan Kecelakaan Lalu Lintas. Departemen Permukiman Dan Prasarana Wilayah. Pd T-09-2004-13
13. Rasmussen, J, The Definition of Human Error and a Taxonomy for Technical System Design, in New Technology and Human Error, Chicester: John Wiley & Sons, 1987
14. Saodang, H, Konstruksi Jalan Raya, Bandung, Penerbit Nova, 2004
15. Standar Geometri Jalan Perkotaan, RSNI T-14-2004 Departemen Pekerjaan Umum Direktorat Jenderal Bina Marga, 2004.
16. Sukirman, S, Dasar-dasar Perencanaan Geometrik Jalan, Bandung, Penerbit Nova, 1999.
    Tata Cara perencanaan Geometrik Jalan antar Kota DPU Direktorat Jend. Bina Marga, Jakarta, 1997
17. Warpani, P.S, Keselamatan Lalu Lintas, Simpusium ke-4 FSTPT, Bali: Universitas Udayana, 2001.

IOP Conf. Series: Journal of Physics: Conf. Series **846** (2017) 012028 doi:10.1088/1742-6596/846/1/012028

Development of Integrated Natural Science Teaching Materials Webbed Type with Applying Discourse Analysis on Students Grade VIII in Physics Class

Luh Sukariasih *

Physics Education, Haluoleo University, Kendari, South East Sulawesi, 93231

*luhsukariasih76@yahoo.com

Abstract. This study aims to produce teaching materials integrated natural science (IPA) webbed type of handout types are eligible for use in integrated science teaching. This type of research is a kind of research and development / Research and Development (R & D) with reference to the 4D development model that is (define, design, develop, and disseminate). Data analysis techniques used to process data from the results of the assessment by the validator expert, and the results of the assessment by teachers and learners while testing is limited (12 students of class VIII SMPN 10 Kendari) using quantitative descriptive data analysis techniques disclosed in the distribution of scores on the scale of five categories grading scale that has been determined. The results of due diligence material gain votes validator material in the category of "very good" and "good", of the data generated in the feasibility test presentation obtained the category of "good" and "excellent", from the data generated in the feasibility of graphic test obtained the category of "very good "and" good ", as well as of the data generated in the test the feasibility of using words and language obtained the category of" very good "and" good ", so with qualifications gained the teaching materials IPA integrated type webbed by applying discourse analysis on the theme of energy and food for Junior High School (SMP) grade VIII suitable as teaching materials. In limited testing, data generated in response to a science teacher at SMPN 10 Kendari to product instructional materials as "excellent", and from the data generated while testing is limited by the 12 students of class VIII SMPN 10 Kendari are more students who score indicates category "very good", so that the qualification obtained by the natural science (IPA) teaching material integrated type webbed by applying discourse analysis on the theme of energy and food for SMP / class VIII fit for use as teaching material.

1. Introduction

Natural science (IPA) Integrated is a natural science that is presented as an inseparable unity, meaning that students do not learn physics, biology, and chemistry separately as a stand-alone subject, but all mixed together in unity. These subjects more accurately be called the natural sciences, does not need to be given an additional "stick together" in tow, because of the former is the essence of science is true, it means the science of natural born from the union of physics, biology, and chemistry, but it was born as a science natural sciences (Depdikbud, 1996). In the natural sciences known their integrated learning approach. There are 10 varieties of learning approaches, but there are several potential approaches to be applied in learning science suggested by Vogarty, (1991), among which are connected, webbed, shared,

Content from this work may be used under the terms of the Creative Commons Attribution 3.0 licence. Any further distribution of this work must maintain attribution to the author(s) and the title of the work, journal citation and DOI.
Published under licence by IOP Publishing Ltd

and integrated. Integrated learning in science can be packaged with the theme or topic of a discourse that is discussed from various viewpoints or scientific disciplines are easily understood and recognized learners. In integrated science teaching, a concept or a theme discussed from various aspects of the field of study in the field of science studies. Thus through this integrated learning some concepts relevant to be the theme does not need to be discussed repeatedly in different fields of study, so the use of more efficient time for discussion and achievement of learning objectives are also expected to be more effective. Such an approach is called a learning approach webbed.

The success of the learning process is influenced by one factor, namely teaching materials. Integrated science teaching materials in circulation is still less effective, one of which is a book. Books that match the curriculum imposed by the government which integrated science textbooks that have been circulating in the school is not packed into the topic / theme despite being labeled specific integrated IPA. Presentation of the material on teaching materials are still separated by fields of study even though it was put together in a book. Based on observations in several junior high schools in the area of Kendari, integrated science textbooks that have been circulated in schools still vary in terms of type and quality. Science teacher at SMPN 10 Kendari states are still difficulties in determining the books used. That is because a lot of books that have not integrated IPA labeled in accordance with the concept of the integrated IPA. Therefore, if the quality of the existing books do not meet quality standards, particularly in relation to the concept and application of concepts, then there is the book would be wrong teaching materials. It is very dangerous world of education.Based on previous research, the development of teaching materials research conducted by Azmi Izati which resulted in the development of teaching materials through integrated science lesson study on the material chemicals for food. Azmi Izati developing integrated science teaching materials by applying an integrated learning approach cobwebs (webbed). This approach combines several topics subjects. Learning tied up with a theme that is known as thematic learning, because it uses a theme as the basis for learning in various disciplines of subjects (Indrawati, 2010).

Based on research that has been done before, researchers interested in conducting to develop an integrated science teaching material in different ways. Lies the difference that, in previous research to determine the theme is determined after an analysis of curriculum / basic competence (KD) by applying the approach to science learning the unified webbed or thematic, whereas in the present study determining the themes that will be used not only refers to the analysis of the curriculum but after incorporation of global macro-structure and the process of discourse analysis. Discourse analysis that discusses how language user digest what was written by the authors in textbooks, understand what penyapa delivered orally in a conversation, and the context also argued that accompanies the text (Mulyana, 2005). In discourse analysis contained first step is to do the smoothing text. Establishment of basic text or text smoothing and sharpening is done to establish the meaning of the concept does not exceed or reduce the text's meaning. Smoothing of text is done through the removal or insertion of the word. Elimination performed on the word excessive and could give meaning to confuse. While the insertion of the word done to further improve the accuracy of the material (Khamalt, 2012). After that, the establishment of global structures and macro-structure that ultimately merging global structure and macro-structure of each subject matter. After merging global and macro structure will be determined themes that formed the basis for the concepts that have been combined.

Based on the above background, the author will conduct research on the development of teaching materials IPA integrated by taking examples of linkages of the same material in an earlier study conducted by Azmi Izati, but the present study the authors apply the discourse analysis of the material book studied so as to produce materials teaching a new form of handouts.

2. Research Method

This research method is a research with development available model namely the development of teaching materials science research integrated type webbed by applying discourse analysis. The object of this study is science textbooks for SMP physics class grade VIII. Design research and development refers to the stage of development of Four-D model suggested by Sivasailam Thiagarajan, *et al.*, (1974). This model consists of four stages of development that define, design, develop, and disseminate or adapted into a 4-D models, namely the definition, design, development, and deployment.

In this development model, an analysis of the discourse on design as the initial design phase of development. The study also only to a point develop or stage of development, and at this stage of development is only at the stage of validation expert (expert appraisal) followed by a revision, because this research focuses on the preparation of learning tools and other reasons are lack of time and costs that are not tested try and deployment for this device.

From some of the data collection instruments, the process of developing these materials using questionnaire techniques with questionnaire instrument. Questionnaires are a number of written questions that are used to obtain information from respondents in the sense of personal statements or things he knew. The questionnaire used consisted of two parts: part I questionnaire assessment experts and part II trials in the form of sheets (response of students and teachers IPA). The assessment results are then analyzed using data analysis techniques were determined. Data collection instruments and questionnaire used in this study was modified from Indonesian National Standards of Education (BNSP).

In a questionnaire assessment of experts consists of four aspects, which includes the criteria of material criteria of presentation, the graphic criteria, criteria for the use of words and language, while the sheet test consists of two aspects, namely responses Science teacher integrated the teaching materials developed and feedback learners the integrated science teaching materials. Limited testing done to obtain direct input in the form of the response of teachers and students' responses. In this trial conducted in one school and have some students who were the respondents.

Analysis of the data used in this research is quantitative descriptive analysis. Quantization data is done with the sum score of every aspect and overall will be described in qualitative analysis. Scores are categorized into five criteria, with the formulation as used by Azwar (2007). Teaching materials are said to be feasible if the minimum score assessment results indicate the category of "enough".

TABLE 1, Rating Criteria.

Num.	Rate Range	Criteria
1.	Mi + 1,5 Sbi < X	Very Good
2.	Mi + 0,5 Sbi < X ≤ Mi + 1,5 Sbi	Good
3.	Mi - 0,5 Sbi < X ≤ Mi + 0,5 Sbi	Enough
4.	Mi - 1,5 Sbi < X ≤ Mi - 0,5 Sbi	Less
5.	X ≤ Mi - 1,5 Sbi	Very Less

The next step is to calculate the ideal maximum score, minimum score ideal, the ideal score and standard deviation of the ideal of every aspect. Ideal maximum score on every aspect achieved if the validator to select all the criteria with the highest score. While the ideal minimum score achieved if the validator to select all the criteria with the lowest score. Total score for each of these aspects, then substituted into the tendency of which is used as a criterion in the assessment.

3. Experimental Results and Discussion

3.1 Stage I: Define (definition)

The purpose is to establish and determine the conditions of learning that includes learning objectives and learning materials restrictions. The steps are as follows

3.1.1 Analysis of the front-end (front-end)

In the first stage is a preliminary analysis that begins with identifying problems in the form of reviews Integrated Science curriculum to learning. Based on observations at junior schools in the area of Kendari, the applicable curriculum at junior high schools in the area Kendari there are those who use the Standard based curriculum (SBC). The number of schools that still use the curriculum SBC caused by unprepared teachers and students in implementing the curriculum in 2013, and the lack of willingness of the relevant teaching materials. After identifying the problems of learning science SMP in particular curriculum, it turns out the standard of competence curriculum SBC is still fragmented between fields.

Particularly at SBC curriculum, teaching materials used by teachers in teaching integrated science mostly use textbooks. Science teaching materials used by teachers still have the presentation of the material separated. Therefore, researchers are interested in developing an integrated science teaching materials to study material that has cohesion, both in materials physics, chemistry or biology, especially the curriculum SBC. Standards of competence is taken from the standard of competence class VIII SMP curriculum SBC can be seen in Table 2

TABLE 2, Competency Standards to be Combined.

IOP Conf. Series: Journal of Physics: Conf. Series **846** (2017) 012028 doi:10.1088/1742-6596/846/1/012028

Field Of Study	Competency Standards
Physics	5. To understand the role of work, force, and energy in everyday life
Biology	1. Understanding the different systems in the human life
Chemistry	4. Understand the use of chemicals in life

3.1.2 Analysis of Students

At this stage, researchers analyzed the characteristics of the students. Characteristics of students in junior high school age are likely to experience cognitive development. According Arajoo T.V (1996) suggest that the cognitive aspects include intellectual functions such as comprehension, knowledge and thinking skills. In addition, there is an increase in intellectual functioning, memory capability in language and conceptual development.

Picture of students' characteristics, such as the level of ability or intellectual development and skills of individuals or social already owned developed to achieve the learning objectives set. The results of the study will be used as consideration for developing teaching materials.

3.1.3 Analysis of Concept

Concept analysis aims to identify the main parts in the subject matter that will be developed. In the analysis of the concept, researchers determined the basic competencies that will be used in developing teaching materials. Basic competencies selected in this study are shown in Table 3 below.

TABLE 3, Basic Competence Content Combined

Field Of Study	Basic Competence
Physics	5.3 Explaining the relations forms of energy and change, work and energy principles and their application to everyday life.
Bilology	1.4 Describe the digestive systems of humans and its relationship to health.
Chemistry	4.3 Describe natural and artificial chemicals in packaging that contained in foodstuffs.

In determining the basic competence to do with the concept of linkage analysis to the basic competence. The activities at this stage is to study the materials of effort and energy as the subject matter of physics is selected, the material that will be combined are the digestive systems of humans as the subject matter of biology and on chemicals in food as the subject matter of the chemical, which is considered the third such material independent from each other. To associate the third such material required a connecting theme. The theme chosen research as the basis for linking the material that is energy and food. Selection of the theme of energy and food due to these themes considered to be able to connect the material work & Energy, the digestive system and food additives, and researchers believe that the theme very close in the event of daily life learners. The linkage concept can be seen in Figure 1 below.

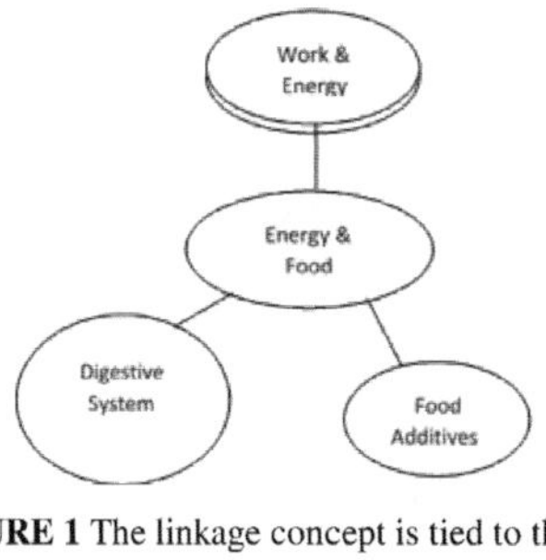

FIGURE 1 The linkage concept is tied to themes

3.1.4 Analysis of assignment

The task analysis performed to specify the content of the teaching material in the form of an outline that includes.

- Analysis of content structure,analysis of the structure of the content is aimed at detailing the assignment of part of the contents of teaching materials were developed. The results of the analysis of the structure of the content in the form of training and evaluation.
- Procedural analysis, analysis of assignment that has to load procedural activities procedure. The results of the analysis of the procedural form of a mini lab.

- Analysis of the information process, handouts should take a lot of additional information in the detailed tasks. The results of the analysis of process information such as the investigation and evaluation activities

3.1.5 Learning Objectives

Formulation of learning objectives or indicators of achievement should be based on basic competencies and indicators listed in Curriculum SBC. Here is the process of formulating learning goals based on a derivative of basic competence indicator material.

- Explaining the sense of energy
- Shows energy forms and examples in everyday life.
- Describe the types of food based on calories and other substances contained in it, as well as additives that can have an impact on health.
- Explaining the process of digestion in humans to produce energy.
- Cites the example of disorders and diseases of the digestive system that is commonly encountered in everyday life
- Explaining the link between energy and effort.
- Distinguishing the concept of kinetic energy and potential energy at a moving object.
- Applying the concept of energy and changes in daily life.
- Introduce the law of conservation of energy through the examples in everyday life

3.2 Stage 2: Design

At this stage, the draft produced teaching materials. The design phase aims to design teaching materials developed. This stage is an important stage in the study for at this stage will be developed teaching materials. Activities undertaken at this stage is the selection of teaching materials format, and preliminary design of teaching materials. The steps are as follows.

3.2.1 Selection format

Selection of the format of teaching materials intended for the design or designing learning content will be developed. The format of instructional materials selected in this study is a kind of handout teaching material has been adapted from BSNP format.

- Preliminary
  1) Cover / identity
  2) Core competencies and learning objectives
  3) Table of contents
- Contents
- Complement
  1) Example
  2) Exercise
  3) Bibliography / references

3.2.2 An initial draft

Here is a description of the preliminary design of a product that has been developed teaching materials.

3.3 Stage 3 : Development

At this stage the final form of teaching materials produced after going through revisions based on input from experts and test result data. The steps are performed at this stage is as follows.

3.3.1 Expert / Practitioner (expert appraisal) Validation

At this stage theoretically request consideration of experts and practitioners. Validator consists of expert material and linguists.

- Validator of material Once the product is produced, then the next step is to do is validate the validator material on teaching materials produced. Validation response or in the form of a questionnaire that has been made before then provided to the validator material. The first Validator material is Fahyuddin (Chemistry Education, Halu oleo university lecturer) and second validator material is Sitti Kasmiati where from Physics Education, Halu oleo university lecturer and the third validator matter is Sri Suryana Dinar from Department of Biology Education Halu Oleo University. From each of the validator at the same time provide an assessment in terms of

IOP Conf. Series: Journal of Physics: Conf. Series **846** (2017) 012028 doi:10.1088/1742-6596/846/1/012028

the presentation and graphics. Data validator votes materials to product type integrated science teaching materials webbed for SMP class VIII by applying discourse analysis on the theme of energy and food set out in Table 4.

TABLE 4,. The data Rate Validator Material

Score Group	Category	Frequency
65 < X	Very Good	2
55 < X ≤65	Good	1
45 < X ≤55	Enaugh	-
35 < X ≤45	Less	-
X ≤35	Very Less	-

- From the data generated on an expert assessment of the material to the product of teaching materials, namely from 3 validator, 2 of them who scored stated excellent category, and the first validator who scored expressed either category so that it can be concluded that the material aspect of the product of teaching science materials integrated type webbed for SMP to class VIII by applying discourse analysis on the theme of energy and food are used as teaching materials.mFeedback, biological and chemical materials should be integrated on the concepts / principles of physics, so that students learn physics as well as learn biology and chemistry.
- validator of Presentation Validator assessment data presentation of the product integrated science teaching materials webbed type for SMP class VIII by applying discourse analysis on the theme of energy and food set out in Table 5.

TABLE 5, The data Rate Validator Presentation

Score Group	Category	Frequency
29,25< X	Very Good	1
24,75 < X ≤29,25	Good	2
20,25 < X ≤24,75	Enough	-
15,75 < X ≤20,25	Less	-
X ≤15,75	Very Less	-

- From the data generated on an expert assessment of the presentation of the product of teaching materials, namely from 3 validator, 2 of them were scores expressed both categories, and the first validator who scored stated category very good, so it can be concluded on aspects of product presentation teaching science materials integrated type webbed for SMP class VIII by applying discourse analysis on the theme of energy and makaan fit for use as teaching material. Feedback, in presenting the teaching material should be added more examples of problems and exercises.
- validator of Graphics Graphics validator assessment data to the product type integrated science teaching materials webbed for SMP class VIII by applying discourse analysis on the theme of energy and food set out in Table 6.

TABLE 6, The data Rate Validator of Graphics

Score Group	Category	Frequency
65< X	Very Good	2
55 < X ≤ 65	Good	1
45 < X ≤ 55	Enough	-
35 < X ≤ 45	Less	-
X ≤ 35	Very Less	-

- From the data generated on an expert assessment kegrafisan to product instructional materials, namely from 3 validator, 2 of them who scored stated excellent category, and the first validator who scored expressed either category, so it can be concluded that the aspects kegrafisan product instructional materials science integrated type webbed to SMP class VIII by applying discourse analysis on the theme of energy and food fit for use as teaching material. Comments and suggestions, most images contained in teaching materials increased again.
- Validator Use of Words and Language Once the product is produced, then the next step is to do is validate the language validators against the teaching materials produced. Validation response or in the form of a questionnaire that has been made before then given to the language validators. Assessment questionnaire results linguists to science teaching materials unified product type webbed for SMP class VIII by applying discourse analysis on the theme of energy and food described in Table 7.

TABLE 7, Data Rate Linguist

Score Group	Category	Frequency
42,5< X	Very Good	1
35,75 < X ≤ 42,5	Good	1
29,25 < X ≤ 35,75	Enough	-
22,75 < X ≤ 29,25	Less	-
X ≤ 22,75	Very Less	-

- From the data generated on an expert assessment of the use of words and language to the product of teaching materials that score all validators declared category very well and good, so it can be concluded that the aspects of the use of words and language product instructional materials science integrated type webbed for SMP class VIII with applying discourse analysis on the theme of energy and food fit for use as teaching material. Feedback, note the use of function words between sentences

3.3.2 Limited Trial

After passing through the process of reviewing the materials experts, presentation, graph and language, then the next step is to do is to conduct limited testing of the products of teaching materials that had previously revised. The trial is given to VIII grade science teacher at SMPN 10 kendari, Tasram and Faizah and 12 respondents grade students of SMPN 10 Kendari with the aim to see the practicality of teaching materials developed can facilitate teachers bring learning materials and help students learn science subjects.

- Teacher Response

TABLE 8, The data Rate of Teacher Response

Score Group	Category	Frequency
65< X	Very Good	2
55 < X ≤ 65	Good	-
45 < X ≤ 55	Enough	-
35 < X ≤ 45	Less	-
X ≤ 35	Very Less	-

- From the data generated in response to the science teacher to product instructional materials that score all validators declared category very well, so it can be concluded that the product instructional materials science integrated type webbed for SMP class VIII by applying discourse analysis on the theme of energy and food fit for use as teaching materials.
- Student Response, the following qualitative discussion of the 25 items were given to the matter of the 12 respondents in response to integrated science teaching material type webbed for SMP class VIII by applying discourse analysis on the theme of energy and food that has been developed.

TABLE 9, The data Rate Students Response

Score Group	Category	Frequency
81,25 < X	Very Good	7
68,75 < X ≤ 81,25	Good	4
56,25 < X ≤ 68,75	Enough	1
43,75 < X ≤ 56,25	Less	-
X ≤ 43,75	Very Less	-

- From the data generated while limited trial by the 12 students of class VIII SMPN 10 Kendari are more students who score showed excellent category, so the aspect of student response teaching materials science integrated type webbed for SMP class VIII by applying discourse analysis on the theme energy and food fit for use as teaching material

4. Conclusion

Integrated science teaching materials webbed type SMPs class VIII by applying discourse analysis on the theme of energy and food are developed has gone through several stages of development with reference to the 4D development model that is (define, design, develop, and disseminate). Data analysis techniques used to process data from the results of the assessment by the validator expert (material, presentation, kegrafisan, words and the use of language), as well as assessment by teachers and learners while testing is limited (12 students of class VIII SMPN 10 Kendari) is using quantitative descriptive data analysis techniques disclosed in the distribution of scores on the scale of five categories grading scale that has been determined. Subjects considered feasible if ratings have a minimum score in the category of "enough".

From the data generated on an expert assessment of the material to the product of teaching materials, namely from 3 validator, 2 of them who scored stated excellent category, and the first validator who scored declared good category of data produced on an expert assessment of the presentation of the product of teaching materials, namely from 3 validator, 2 of them were scores expressed both categories, and the first validator who scored stated excellent category, from the data generated on an expert assessment kegrafisan to product instructional materials, namely from 3 validator, 2 of them who scored stated excellent category, and one validator the score is expressed both categories, as well as from data generated on an expert assessment of the use of words and language of the product instructional materials namely score validator stated category very well and good, so with qualifications gained the teaching materials science integrated type webbed SMP class VIII with applying discourse analysis on the theme of energy and food fit for use as teaching material. In limited testing, data generated in response to a science teacher at SMPN 10 Kendari to product instructional materials that score all validators declared excellent category, and from the data generated while testing is limited by the 12 students of class VIII SMPN 10 Kendari that more students scores showed excellent category, so that the qualifications obtained in the pilot phase it does not require a revision or an integrated science teaching materials webbed type in SMP Grade VIII by applying discourse analysis on the theme of energy and food fit for use as teaching material.

Acknowledgments
We would like to thank Directorate P2M-Ditjen DIKTI for helping in funding this research.

References

Akker, V. D. J. 1999. Principles and Methods of Development Research. Design Approaches and Tools in Education and Training (pp. 1-14). Dortrech: Kluwer Academic Publishers.

Arikunto, S. 1997. Prosedur Penelitian Suatu Pendekatan Praktek. Jakarta: Rineka Cipta.

Azwar, S. 2007. Tes Prestasi: Fungsi dan Pengembangan Pengukuran Prestasi Belajar. Yogyakarta: Pustaka Belajar.

BSNP, 2006. Panduan Penyusunan Kurikulum Tingkat satuan Pendidikan Jenjang Pendidikan Dasar dan Menengah. Jakarta : Badan Sandar Nasional Pendidikan.

Depdikbud. 1996. Kamus Besar Bahasa Indonesia. Jakarta: Balai Pustaka.

Fogarty, R. 1991. How to Integrate the Curricula. New York: The Mindful School.

Gay, L.R. 1991. Educational Evaluation and Measurement: Com-petencies for Analysis and Application, second edition. New York: Macmillan Publishing Compan.

Indrawati. 2010. Model Pembelajaran IPA Terpadu untuk SMP. Bandung: PPPPTK IPA.

Mufidah, I. 2010. Pengembangan Bahan Ajar Ipa Terpadu Berbasis Sains-Lingkungan-Teknologi-Masyarakat (Salingtemas) Untuk Smp Kelas VII Semester I. Malang: Universitas Negeri Malang.

Mulyana. 2005. Kajian Wacana:Teori, Metode, dan Aplikasi prinsip-Prinsip Analisis Wacana. Yogyakarta: Tiara Wacana.

_________. 2007. Kurikulum Tingkat Satuan Pendidikan. Bandung: Remaja Rosda Karya.

Ruseffendi. 1998. Statistika Dasar untuk Penelitian Pendidikan. Bandung: IKIP.

Salirawati, D. 2008. Metode Pembelajaran Inovatif sebagai Magnet Belajar. Makalah Lokakarya Metode Pembelajaran Inovatif dan Sistem Penilaiannya. Yogyakarta: FMIPA UNY.

Sanjaya, Wina. 2012. Perencanaan Dan Desain System Pembelajaran. Jakarta: Kencana.

Siregar, dkk. 1994. Buku Panduan Analisis dan Penulisan Buku Teks MIPA untuk Pengembangan Keterampilan Mahasiswa. Tim Penelitian dan Pengembangan. Bandung: IKIP Bandung Press.

Siregar, N. 1998. Penelitian Kelas: Teori, Metodologi, dan Analisis. Bandung: IKIP Bandung Press.

Analysis of Learning Tools in the Study of Developmental of Interactive Multimedia Based Physic Learning Charged in Problem Solving

Sondang Manurung[1,*] **, Deo Demonta Pangabean**[1]

[1]State University of Medan (UNIMED), Jln. Willem Iskandar, Pasar V Medan Estate, Medan, North Sumatra 20221, Indonesia.

[*]E-mail : sondangrina@gmail.com

Abstract. The main purpose of this study is to produce needs analysis, literature review, and learning tools in the study of developmental of interactive multimedia based physic learning charged in problem solving to improve thinking ability of physic prospective student. The first-year result of the study is: result of the draft based on a needs analysis of the facts on the ground, the conditions of existing learning and literature studies. Following the design of devices and instruments performed as well the development of media. Result of the second study is physics learning device -based interactive multimedia charged problem solving in the form of textbooks and scientific publications. Previous learning models tested in a limited sample, then in the evaluation and repair. Besides, the product of research has an economic value on the grounds: (1) a virtual laboratory to offer this research provides a solution purchases physics laboratory equipment is expensive; (2) address the shortage of teachers of physics in remote areas as a learning tool can be accessed offline and online; (3). reducing material or consumables as tutorials can be done online; Targeted research is the first year: i.e story board learning physics that have been scanned in a web form CD (compact disk) and the interactive multimedia of gas Kinetic Theory concept. This draft is based on a needs analysis of the facts on the ground, the existing learning conditions, and literature studies. Previous learning models tested in a limited sample, then in the evaluation and repair.

1. Introduction

Physics is a science that underlies the development of technology, so students need to learn in the form of General Physics (Halliday, 2011., As a physics teacher candidates are expected later they had a physics teacher professional competence and the ability to think logic high (Haryanto, 2006., Liliasari, 2005., Lawson, 1995., and Piaget, 1964). Increased competence of teachers can not be done either at the Institute of Education Workers (LPTK). Preliminary study conducted Manurung & Rustaman (2011) General Physics learning in one LPTK in Medan indicates that in general the lecturers still dominated learning. Lecturers teach the material with a lecture and question and answer. General Physics practical implementation is still verification. Other findings of the preliminary study (Manurung & Rustaman, 2011) are: (a) Methods used in the lecture General Physics has been made the students glued to listen and really boring, because the learning situation aimed at learning to know, and problems delivered tends to

Content from this work may be used under the terms of the Creative Commons Attribution 3.0 licence. Any further distribution of this work must maintain attribution to the author(s) and the title of the work, journal citation and DOI.
Published under licence by IOP Publishing Ltd

be academic (book oriented), (2) Students lack the experience to be able to solve the problems and issues are given less refers to contextual issues close to the everyday lives of the students so that learning General Physics less meaningful for students. Furthermore, several weaknesses of learning General Physics for this, namely: (a) the learning process can not bring the phenomenon, (b) the lack of discovery process, (c) lack of instructional media and tend not at all, and (d) a weak understanding of the concept. From studies conducted in the analysis of the material, it was revealed that many material physics featuring diagrams, graphs and mathematical formulas. The low quality of teaching physics, in terms of results and the process of student learning was caused by teachers teach not optimal. (Manurung, 2010).

McDermot (1990) and Slavin (2000) states that one of the important factors that affect the poor performance is the lack of good science teacher preparation process. The main factors in order to improve the quality of teaching and learning is the teacher, thus need to increase the professionalism of teachers in the field of science and technology (Reif *et al*, 1976). Departing from this reality seems to be improving the quality of teachers through teacher education candidates must be constantly carried out. One of them by providing them the knowledge and direct experience in conducting physics experiments, including experiments involving physics concepts are abstract with the use of interactive multi-media relevant because not all experiments can be conducted directly in the laboratory (.Finkelstein *et al*, 2006).

Misanchuk & Hunt (2005)., Ivers & Barron (2002).,and Santyasa. (2006). has been designing and problem solving laboratories content on Web-based Basic Physics lecture gives the result that the retention distance students as well as direct interaction. Simulation and virtual labs for a variety of basic physics concepts through project-based algebra. PhET (Physics Education Technology) undertaken and reported by Finkelsstein et al (2006), used to treat many participants basic physics courses at various universities. Furthermore, Finkelstein et al (2006) says that the computer can be used to support the implementation of good physics lab to collect data, present, and process data. Finkelstein *et al* (2006), Srinivasan & Crooks (2005) promoted a number of forms of interaction can be generated through a computer media such as the presentation of practice and training, tutorials, games, simulations, discovery, and problem solving. According to Jonassen (1997) and Jonassen & Grabinger (1990) the use of multimedia in learning to encourage students to learn the process of discovery (discovery learning process) and can solve the problem of the ill-structured problems (Cunningham, 2009., Jonassen, 2004).. Solving the problem is a complex process and is important in everyday life studying physics. Although there are many efforts to improve students' problem-solving in the education system, Gerace (2001) said that there is no standard way to evaluate a written problem solving applicable, reliable, and easy to use. Problem solving skills developed in General Physics is to present a situation where certain information is given, more often as numerical values for the variables in the situation, and the values of other variables that can be determined. According to Heller & Heller (2010), Heller & Hollabaugh (1992b), Gick (1986), and Gick & Holyoak (1980) show that there are 5 steps of problem solving strategy, namely: Focus on problem solving, Explanation Physical, Planning Implementing, Implement planning, and Evaluating of response. While the problems of physics in everyday life or physical problem that is owned physicists are ill-structured problems, problems which must be solved through innovative learning physics (Cunningham, 2009).

Lawson (1995) states the ability to think, think creatively, make decisions, and solve problems is very important in getting a job, because they have the ability to ensure its survival. The ability to think logic is a person's ability to solve problems, able to think and devise a solution with a logical sequence (sense). managing careers and work. Furthermore Liliasari (2005) suggest a high level thinking skills can be developed through the teaching of science with models of information technology-based learning in science (chemistry, physics, and biology). Based on the above, it is necessary to do the development of physics-based models of learning interactive media multi-charged problem solving to improve thinking skills and problem-solving skills that are ill-structured problems (Shin *et al*, 2003, Darmadi, 2007., Dori & Belcher, 2005). Learning conceptual understanding of physics that suppress the problem-solving activities conducted through the process so physically understand not just memorize formulas.

Originality of this research is to improve the course General Physics II based on interactive multimedia charged able to overcome the problem solving physics problems are ill-structured problems are always present in everyday life, has never been investigated (Budiman *et al*, 2008, Burke, 1998, and Gayeski, 1993)).

2. Material and Method

The research method in the first phase is the implementation of a preliminary study of the physics-based multimedia interactive learning and planning for students who attend lectures general physics I.

IOP Conf. Series: Journal of Physics: Conf. Series **846** (2017) 012029 doi:10.1088/1742-6596/846/1/012029

Location and Subjects Research in the Department of Physics, State University of Medan, involving 36 students. The subject of research that consists of these students were divided into six groups of five students in each group. The research instrument at one stage in preparing the analysis of the needs is 1) Gazette observation experience of using interactive multimedia media, At the stage of preliminary study (and Fraenkel *et al*, 2012). developed the necessary instruments for field studies as shown in **Table 1**.

Table 1. Step of Research Instrument

No	Aspect	Indicator
1	Student perspectives on the ability of facing an interactive multimedia learning in I general physics course.	Discourse move related to the problem solving
2	Attitude and viewpoint of lecturers	Attitudes and views of lecturer on the implementation of interactive multimedia based physic learning
3	Attitude and viewpoint of students	Attitudes and views of students on the implementation of interactive multimedia based physic learning
4	Problem solving ability in the implementation of the lecture	Utilization of problem solving activities in the lecture
5	Logical thinking ability	Utilization of logical thinking ability

No	Data Source	Instrument	Time
1	Lecturers	Observation sheet (conducted by education experts)	At the time of the lecture on the topic of work and Energy
2	Lecturers	Observation sheet	Before implementation of learning model
3	Students	Observation sheet	Before implementation of learning model
4	Process of lecture for lecturer and student	• observation sheet • questionnaires • Sign- posts • interview	Before implementation of learning model
5	Students	test	Before and after implementation of learning

Modification from Robert & Campbell, 2013

3. Experimental Results and Discussion

Preliminary research results indicate: (a) only 60% effective face to face lectures, because lecturers burdened with many activities, such as teaching PLP; (b) the percentage of synchronization between face-to-face lecture materials and lab resulting of interviews conducted by only 30%; (c) the percentage of student involvement in the process of building concepts in mathematical equations, and graph obtained only 32%, this is due to the role of professors who dominate the class by giving students homework without direct troubleshooting steps to train reasoning thinking. Given exercises do not follow the pattern of effective problem solving or reasoning so the thinking lines are not well-trained students. Interaction learned more in one direction i.e from the lecturer to the students; (d) The lecture very much, so that in the lecture students are burdened with many tasks; e) only 17% lab activities that can improve the understanding of the concept of student, it is because the characteristic of lab only verification using detailed instructions and a lack of benefit (e) only 15% response activities that involve students in the process of solving the problem systematically..Based on the results of preliminary studies, the characteristic of an assessment of the physical learning system is shown in **Table 2**.

Table 2. Characteristics of Assessment of Physics Learning System

No	Element	Positive Component	Negative Component
1	Effective face-to-face lectures	60%	40%
2	Synchronization of lecture material in the lab	30%	70%
3	The involvement of students in the lecture	32%	68%
4	Lecture material is quite dense	10%	90%
5	Linkage lecture materials with previously accepted concept	18%	82%
6	Practical activities to improve understanding of the concept	17%	83%
7	Response activities involving students	15%	85%

Based on data obtained from preliminary studies as in **Table 2** to be a reference and encourage to develop an interactive multimedia based learning methods performed in this study, is expected to help students, lecturers and teachers in understanding topics / concepts General Physics I or concepts more complicated in physics. Therefore, teachers or prospective teachers to master the use of interactive multimedia important to improve the professionalism, because it can enrich the user experience in speaking, thinking, and even in the formation of academic culture. In addition, teachers can also collect, organize, and analyze empirical evidence and theoretical basis for a more comprehensive answer tough questions about the understanding of the concept.

Problem solving initial response sheets is given to the three experts who are competent in the field of assessment and general physics of matter, and obtained some important notes about the suggestions and responses to the development of problem solving model as shown in **Table 3**, which summarized the results of interviews and feedback sheets

Table 3. Responses and expert advice on scientific arguments based learning model

No	Substantion	Expert 1
1	Syntax of problem solving	More precisely if compiled syntax of problem solving that corresponded to the teaching theory and lecture
2	Latticework	Be required of indicator problem solving skills
3	Guide to do problem solving in general physics I course	Necessary introduction and direction signs in the problem solving
4	Other responses	Implementation in the poblem solving lecture is conducted based lectures highly anticipated as a result of standardized
No	**Expert 2**	**Expert 3**
1	Try using a more efficient sentence	Integrate multimedia avtivities based learning in the RPP. Adjust the allocation of time lecturing
2	Latticework in terms of problem solving adjusted to basic competence in I general physics courses	Latticework in terms of problem solving corresponding with learning syntax
3	Should be implanted and cultivated in every lecture commonly	As an initial stage is expected to be a model for all science-based lecture
4	Preliminary studies are expected to be sharper explored further is done in-depth analysis through discussion with several experts and other specialists. There should also be tested against several subjects to gain	This is still an preliminary study. Required a long way to earn academic recognition

	sharpness models.	

Generally, validator is considered that the problem solving based learning in the classroom is very inquiry is expected to change the paradigm of thinking about college physics student general I so the concepts contained in the course will be more easily understood. Of the five substances proposed to be taken validator researchers showed that the problem solving is very need to be developed in the implementation of public lectures I general physics spesifically. Validator also expect there is a specific guidance on problem solving on Igeneral physics course I. In this case, the problem solving will be developed in order to be adapted to the basic competencies in I general physics. Required material also has to use indicators of problem solving so the implementation can achieve the goal of learning about conceptual understunding understanding in general physics I course. To be able to measure an activity can be expressed in accordance with the development probem solving is certainly necessary observation sheet. This observation sheet dedicated to the observation group was formed. The observation sheet must be prepared in accordance to the conditions of the class, and can facilitate observer groups provide an assessment of the observed activity. Generally, validator is considered that the scientific arguments based learning in the classroom is very inquiry is expected to change the paradigm of thinking about college physics student general I so the concepts contained in the course will be more easily understood. Of the five substances proposed to be taken validator researchers showed that the scientific argument is very need to be developed in the implementation of public lectures particularly physics I. Validator also expect there is a specific guidance on scientific arguments on general physics I course. In this case, the argument science will be developed in order to be adapted to the basic competencies in general physics I. Required material also indicators argue for scientific skills so that the implementation can achieve the goal of learning argues that understanding the concept of general physics I. To be able to measure an activity can be expressed in accordance with the development scientific argument is certainly necessary observation sheet. This observation sheet dedicated to the observation group was formed. The observation sheet must be prepared in accordance to the conditions of the class, and can facilitate observer groups provide an assessment of the observed activity.

According to analysis of the textbook, characteristic of the multimedia activities based physics learning charged in problem solving, in particular, the management and the learning environment "thinkers" noticed aspects are as follows: 1) handle situations students multi-task; 2) adjust to the speed of completing different tasks; 3) monitor and handle the student work; 4) management of the equipment and materials; 5) regulate the movement of students (in the computer lab and science lab) and behavior happen outside the room (for the solution of which requires vasalah outdoor activities). Lecturer role as facilitators directly involved in the process of group (assist students in formulating a plan, act, and organize group), some of the needs in a research (knowledge of vetode used), and also function as an academic counselor (.Yahya, 2008). In this study, students were assigned to solve the physical problems that result 1 is shown in table 4.

Table 4. Data description scores in the domains of problem solving aspects

	Focusing on problem solving	**Explanation Physical**
Mean	1.01	1.00
Std. Deviation	.58	.53
Minimum	.00	.00

	Planning Implementation	**Evaluating of response**
Mean	.83	.83
Std. Deviation	.61	.61
Minimum	.00	.00

In learning that utilizes the MMI, the problems related to the limitations of working memory is how to build a complex understanding that integrate information from visual sources (pictures, diagrams, graphs, films) and verbal (text, lecture). Woolfolk (2009b) provides guidance: make sure the information-information is available at the same time or focused on small pieces; give students many ways to understand (images and description), but do not overwhelm the working memory by means of visual and verbal information pack together with pieces 26 pieces the size of a "one bite" (or the size of the memory). This statement would be used as the basis of MMI development for the lecture Physics for prospective teachers of Physic Education LPTK. In this study, the steps to follow the development of MMI project-based learning development measures proposed MMI Ivers and Baron (2002) with the adjustment of the development of project-based learning MMI to MMI development. Evaluate this phase occurs in each process. Identification Finkelstein et al. (2006) on the characteristics of the MMI that contains simulated physics that supports student learning need to be considered in designing the MMI for physics. These characteristics include: (1) engaging and interactive approach; (2) the existence of a dynamic feedback; (3) using a constructivist approach; (4) the existence of space to play and do something; (5) the existence of a visual model or access to visual models; (6) the existence of a barrier for the purposes of improving the productivity of the student. Based on preliminary studies, planning, observation (observation) and a questionnaire to all students who attend public physics lectures I obtained the results illustrate that the implementation of classroom based learning arguments for inquiry is still new and needs to be explored more deeply and continuously to obtain alternative methods to improve understanding the concept of students, especially students of the first semester. Follow up on this situation (Gerace, 2001). Based on preliminary studies from various sources shows that interactive multimedia based learning physics require their initial statement as a , in accordance with that shown by Clark & Sampsonn (2006), Driver et al (2000) stated that the process then performed through the interactive muti media could improve thinking ability of students.

4. Conclusion

Produce needs analysis, literature review, and learning tools in this study could improve thinking ability of physic prospective student. Product of research has an economic value on the grounds: (1) a virtual laboratory to offer this research provides a solution purchases physics laboratory equipment is expensive; (2) address the shortage of teachers of physics in remote areas as a learning tool can be accessed offline and online; (3). reducing material or consumables as tutorials can be done online. Application of physics learning based on interactive multimedia to increase effectiveness learning, both in terms of process and of learning outcomes.

Acknowledgments

The research is funded by the Competitive Grants DP2M Director General of Higher Education Ministry of Education and Culture. Therefore, researchers who receive grants DP2M to thank the Director General of Higher Education which has provided funds, and the opportunity for researchers to conduct research in Physics Education Program of State University of Medan. On this occasion, the authors would like to thank the Rector and Chairman of the Research Institute of the State University of Medan which has given opportunity to the team of researchers to conduct research. This has helped the research.

References

[1] Halliday,D.,Resnick, R. & Walker,J. (2011*). Fundamental of Physics.*10th ed. Singapore:John Wiley & Sons, Ptc Ltd
[2] Z. Haryanto (2006) "Tahap perkembangan intelektual siswa smp dan sma dengan kaitanya dalam pembelajaran fisika dan kemampuan pemecahan masalah," Magister dissertation, Bandung, Indonesia University of Education, unpublished.
[3] Liliasari (2005), Membangun keterampilan berpikir manusia indonesia melalui pendidikan sains.pada fakultas FMIPA, B.Sc. thesis, Indonesia University of Education,
[4] A.E. Lawson (1995), *Science Teaching and the Development of thinking,* California, Wadsworth Publishing Company, Belmont.
[5] J.Piaget (1964), The Development of Thought: The Equilibrium of cognitive Structures, New York: Viking.
[6] S.R. Manurung, & N.Y., Rustaman (2011), Field Study Report. Bandung, UPI, unpublished.
[7] S.R. Manurung (2010) , "Pengembangan pembelajaran fisika dasar untuk meluruskan kesalahan konsep mahasiswa jurusan fisika Universitas Negeri Medan," in *Research Report funded by P3M Dikti Depdiknas*, unpublished
[8] L. C. McDermott (1990), "A perspective on teacher preparation in physics and other science: the need for special science course for teachers," *American Journal of Physics.***58**(8), 734-742.
[9] F. Reif, H. Larkin, and C. Brackett (1976) "Teaching general learning and problem-solving skills," *American Journal of Physics* **44**, 212-217.

[10] N.D.Finkelstein, W. Adam, C. Keller, K. Perkins, and C. Wieman (2006), "High tech tools for teaching physics: the physics education technology project," *Merlot Journal of Online Learning and Teaching* **2**(3).

[11] M. Misanchuk & J.L Hunt (2005), "Designing problem-solving and laboratory content for a web-based distance education course in introductory general physics," Available on: http://www.physics.uoguelph.ca/phyjlh/morph/paper2_3MelJim_1pdf [Oct 29th, 2007]

[12] K. S. Ivers, & A. Barron (2002) , Multimedia Projects in Education: Designing, Producing, and Assessing (Teacher Ideas Press, Wesport.

[13] I. W. Santyasa (2006), "Pengembangan Pemahaman Konsep dan Kemampuan Pemecahan Masalah Fisika bagi Siswa SMA dengan Pemberdayaan Model Perubahan Konseptual Berseting Investigasi Kelompok. *Laporan Penelitian RUKK Menristek tahun Kedua*. Lembaga Penelitian Universitas Pendidikan Ganesha.

[14] S. Srinivasan. & S, Crooks (2005), "Multimedia in a science learning environment," *Journal of Educational Multimedia and Hypermedia* **14**(2), 151-167.

[15] D. H. Jonassen (1997), "Instructional design models for well-structured and ill-structured problem solving learning outcomes," *Educational Technology Research and Development,* **45**(1) 65-94.

[16] Jonassen, D.H., Learning to solve problems: An instructional design guide (Jossey-Bass , San Francisco, 2004)

[17] D. Cunningham (2009), Using ill-structured problems to develop metacognitive strategies," in *Annual Conference of the International Society for Exploring Teaching and Learning* (Philadelphia, Pennsylvania, 8-10.

[18] W. J. Gerace (2001), "Problem Solving and Conceptual Understanding", in *Proceedings of the 2001 Physics Education Research Conference,* edited by S. Franklin, J. Marx & K. Cummings (Eds), (PERC Publishing, New York,) 33 -45.

[19] K. Heller, & P. Heller (2010), "Cooperative Problem Solving in Physics A User's Manual," (Available on: http://groups.physics.umn.edu/physed [Jan 10th, 2011]

[20] P. Heller, & M. Hollabaugh (1992), "Teaching problem solving through cooperative grouping: designing problems and structuring groups". *American Journal of Physics* **60**(7), 637-644.

[21] M.L.Gick (1986), "Problem-solving strategies", *Educational Psychologist* **21**, 99-120.

[22] M.L. Gick, & K. J. Holyoak (1980), "Analogical problem solving," *Cognitive Psychology* **12**, 306-355

[23] I.W .Darmadi (2007), "Pembelajaran berbasis teknologi informasi untuk meningkatkan penguasaan konsep fisika mahasiswa calon pengajar," *Jurnal Penelitian Pendidikan IPA* **1**(1).

[24] Y.J. Dori, & J., Belcher (2005), "How does technology-enabled active learning affect undergraduate students' understanding of electromagnetism concepts?," *The Journal of Learning Science*, **14**(2), pp. 243-279.

[25] J. R. Fraenkel & N.E Wallen (2012), *How to Design and Evaluate Research in Education, Book 2,"* Boston, McGraw Hill, Boston.

[26] D. M. Gayeski (1993), *Multimedia for Learning*, NJ, Educational Technology Publications, Englewood Cliffs.

[27] . N. Shin, D.H. Jonassen, & S. McGee (2003),"Predictors of well-structured and ill-structured problem solving in an astronomy simulation," *Journal of Research in Science Teaching* **40**(1), 6-33.

[28] Budiman, A.Suhandi, and A. Setiawan (2008), "Model pembelajaran multimedia interaktif dualism gelombang partikel untuk meningkatkan pemahaman konsep dan keterampilan berpikir kritis pebelajar, *Jurnal Penelitian Pendidikan IPA* **2**(1).

[29] K. A. Burke (1998), Developing and using conceptual computer animation for chemistry instruction, *Journal of Chemical Education* **75** (Iowa State University, 1998).

[30] R. Robert, and T. Campbell, T (2013), "Constructing arguments: Investigating pre-service science teachers' argumentation skills in a socio-scientific context," *Science Education International* **24**(2), 195-211.

[31] R.E. Slavin (2000), R. E., *Educational physichology theory and practice.* Sixth Edition, Boston, Allyn & Bacon Publisher.

[32] S. Yahya, A. Setiawan, and A. Suhandi (2008), "Model pembelajaran multimedia interaktif optik fisis untuk meningkatkan penguasaan konsep, keterampilan generic sains, dan keterampilan berpikir kritis pengajar fisika," *Jurnal Penelitian Pendidikan IPA* **2**(1).

Effect of Tutorial Giving on The Topic of Special Theory of Relativity in Modern Physics Course Towards Students' Problem-Solving Ability

Hartatiek[1,*], Yudyanto[1], Dwi Haryoto[2]

[1]Department of Physics, Faculty of Mathematics and Natural Sciences, Malang University, Malang, 65145, Indonesia
[2]Department of Physics Education, Faculty of Mathematics and Natural Sciences, Malang University, Malang, 65145, Indonesia

* Email: hartatiek.fmipa@um.ac.id

Abstract. A Special Theory of Relativity handbook has been successfully arranged to guide students tutorial activity in the Modern Physics course. The low of students' problem-solving ability was overcome by giving the tutorial in addition to the lecture class. It was done due to the limited time in the class during the course to have students do some exercises for their problem-solving ability. The explicit problem-solving based tutorial handbook was written by emphasizing to this 5 problem-solving strategies: (1) focus on the problem, (2) picture the physical facts, (3) plan the solution, (4) solve the problem, and (5) check the result. This research and development (R&D) consisted of 3 main steps: (1) preliminary study, (2) draft 1 product development, and (3) product validation. The developed draft product was validated by experts to measure the feasibility of the material and predict the effect of the tutorial giving by means of questionnaires with scale 1 to 4. The students problem-solving ability in Special Theory of Relativity showed very good qualification. It implied that the tutorial giving with the help of tutorial handbook increased students problem-solving ability. The empirical test revealed that the developed handbook was significantly affected in improving students' mastery concept and problem-solving ability. Both students' mastery concept and problem-solving ability were in middle category with gain of, 0.31 and 0.41, respectively.

1. Introduction

Special theory of relativity consisted of the following topics: Newtonian relativity, Einstein's postulates, Lorentz coordinate transformation, simultaneity and the relativity of time, length contraction, time dilation, twins paradox, relativistic doppler shift, Lorentz velocity transformation, relativistic momentum and relativistic energy [1,2,3].As the matter of fact, many students face difficulty in solving problems on the topic of the special theory of relativity. When the students were given problems that were slightly different from the given examples, they are generally unable to solve those problems. Therefore, it is important to introduce explicit solving strategy-based tutorial to help students' difficulty in solving problems.The tutorial underlines the systematic strategies to solve the problems. It does not only stress on the "given" and "questioned", but also lay emphasis on (1) focus on the problem, (2) picture the physical facts, (3) plan the solution, (4) solve the problem, and (5) check the result [4]. An excellent approach to increase the students problem-solving ability is tutorial giving [5,6]. A tutorial is learning and instruction program to help students during taking lectures by means of guided self-learning.

Some previous studies showed that to simultaneously (a) make students become effective problem solvers and (b) master the concepts are a two-goal that is hard to reach. Even by instructional model,

Content from this work may be used under the terms of the Creative Commons Attribution 3.0 licence. Any further distribution of this work must maintain attribution to the author(s) and the title of the work, journal citation and DOI.
Published under licence by IOP Publishing Ltd

there are still many students find difficulties in solving problems [7], and their mastery concepts are relatively low [8]. Combining the active learning and problem-solving strategies to the modern physics students can enhance the students mastery concept and their problem-solving ability [9,10]

This research and development are mainly aimed to (1) develop tutorial handbook on the topic of Special Theory of Relativity, (2) investigate the effect of the tutorial giving on the students mastery concept, (3) investigate the effect of the tutorial giving on the student's problem solving ability, and (4) measure the student's problem solving ability at Special Theory of Relativity.

2. Material and Method

2.1 Development Tutorial Handbook

To develop tutorial handbook product, a research design by means of research and development (R&D) was applied. Basically, there are 3 main steps for R&D approach [11]: (1) preliminary study which consists of literature study and survey, (2) product development in terms of draft product arrangement, judgment (expert validation), and trial field testing, (3) product validation. In this present study, we only restricted to the step (1) and (2).

The experimental procedures are described as follows: (1) preliminary study; in this step, an investigation was focused on the learning problems that were faced by students during taking the modern physics course. From the investigation, it was known that 80% of the students found difficulties in solving problems of modern physics. According to this preliminary study, a handbook to guide students in the tutorial was then chosen to implement. (2) draft product development; in this step, a tutorial handbook was created with the following 10 component specifications: (a) introduction, (b) learning objectives, (c) explanation, (d) problem examples and the solutions based on 5 explicit solving strategies, (e) summary, (f) problems, (g) solutions to the problems, (h) scoring method for problem solving, (i) physical constants and important, and (j) references. The tutorial handbook was also completed with "for readers" section in order to help students be easy to use and learn. (3) judgment (validation by experts); this is an assessment activity to evaluate the developed tutorial handbook. This was done to increase the theoretical validity of the products. The judgment was done by 2 physics experts by means of 1-4 scoring scale questionnaires. 4 means very feasible, 3 means feasible, 2 means less feasible, and 1 means unfeasible. The judgment was performed before the trial field testing. (4) trial field testing; in this step, the tutorial handbook was tried to limit the number of students to know the students responses including clarity of language usage, a sequence of materials, consistency of notation usage, availability of images to make the concept clearer. The students responses to the tutorial handbook content were gathered by questionnaires with scale 1 to 4; 4 = excellent, 3 = good, 2 = fair, and 1 = poor. The vision for the final product was based on the students' responses.

The collected data in this present study are material feasibility, students' response, and problem-solving ability. The value of the data was analyzed by the following mathematical expression.

$$N = \frac{SD}{SM} \times 100 \quad (1)$$

where N = grade, SD = score, dan SM = maximum score.

The grade criteria for those data are given bellow.

85-100 = Excellent
75- 84,9 = Good
65 – 74.9 = Satisfactory
55 – 64.9 = Poor
< 55 = Worst

2.2 Trial Field Testing of Tutorial Handbook

The effect of tutorial giving on the students' mastery concept and problem-solving ability was evaluated by means of Pretest-Posttest Control-Group Design [11] and designed as depicted in Table 1.

TABLE 1. Quasi-experimental Design of Pretest-Postes Control-Group

Group	Pretest	Treatment	Posttest
E	X_1	T	Y_1
C	X_2		Y_2

E = Experimental group
C = Control group

IOP Conf. Series: Journal of Physics: Conf. Series **846** (2017) 012030 doi:10.1088/1742-6596/846/1/012030

X_1 = Pretest score for experimental group
X_2 = Pretest score for control group
Y_1 = Posttest score for experimental group
Y_2 = Posttest score for control group
T = Treatment by tutorial giving

Acceptance/rejection of null hypothesis is based on the criteria as given in Table 2.

TABLE 2. Acceptance/rejection of the null hypothesis

Group	t-test result	α=0,05 db = N_1+N_2-2	Note
Pretest Exps.-control	t_0	$-1.998 < t_0 < 1.998$	H_0 is accepted (H_a is rejected)
Posttest Exps.-control	t_0	$-1.998 < t_0 < 1.998$	H_0 is accepted (H_a is rejected)

The amount of increment of the students' problem-solving ability was analyzed by average normalized gain, i.e. the ratio of average actual gain and maximum of average gain given by the following equation[12]:

$$< g > = \frac{\%<gain>}{\%<gain>_{max}} = \frac{\%<posttest> - \%<pretest>}{100 - \%<pretest>} \quad (2)$$

With category:
$< g >$ is high for $< g >$ larger than 0.7
$< g >$ medium for $< g >$ of 0.3 to 0,7
$< g >$ low for $< g >$ smaller than 0.3

3. Experimental Results and Discussion

3.1 Expert Validation Result

The draft of the special theory of relativity handbook was validated by 2 experts to examine its feasibility. The validation result is depicted in Table 3.

TABLE 3. Validation Result for Material Feasibility & Its Prediction

Sub	Component	Score	Maximum Score	Grade	Qualification
1	**MATERIAL FEASIBILITY**				
A	Introduction	7	8	87.5	Excellent
B	Learning Objective	16	16	100	Excellent
C	Content				
	1.1 Newtonian Relativity	44	56	91.7	Excellent
	1.2 Einstein's Postulates	24	24	100	Excellent
	1.3 Lorentz Transformation	45	48	93.8	Excellent
	1.4 Synchronized Clock and Simultaneity	38	40	95	Excellent
	1.5 Length Contraction	45	48	93.8	Excellent
	1.6 Time Dilation	38	40	95	Excellent
	1.7 Twins Paradox	37	40	92.5	Excellent
	1.8 Relativistic Doppler Effect	38	40	95	Excellent
	1.9 Lorentz Velocity Transformation	39	40	97.5	Excellent
	1.10 Relativistic Momentum	46	48	95.8	Excellent
	1.11 Relativistic Energy	38	40	95	Excellent
Sub	Component	Score	Maximum Score	Grade	Qualification
D	Examples and Explicit Solutions	54	56	96.4	Excellent
E	Summary	16	16	100	Excellent
F	Problems	29	32	90.6	Excellent
G	Solutions to Problems	24	24	100	Excellent
H	Problem Solving Scoring	24	24	100	Excellent
	Average			95.5	Excellent
2	**Effect Prediction**				
A	The whole content of the tutorial handbook may make students' opportunity exercise to solve problem systematically and structurally following the given 5 steps of problem-solving	8	8	100	Excellent
B	The whole content of the tutorial handbook may	8	8	100	Excellent

	give raise students' problem-solving ability				

According to the validation result, it can be inferred that the average material feasibility was 95.5 with excellent qualification. The effect prediction of the tutorial giving on providing an opportunity to the students to exercise to solve problem systematically and structurally following the given 5 steps of problem-solving was 100 with excellent qualification. The effect prediction of the tutorial giving on the students' problem skill ability was perfectly 100 with excellent qualification.

3.2 Students' Response Result

Students' response on the handbook content is given in Table 4.

TABLE 4. Students' Response on The Handbook Content and Its Effect Prediction

Sub	Component	Score	Maximum Score	Grade	Qualification
1	**HANDBOOK CONTENT**				
A	Introduction	32	40	80	Good
B	Learning Objective	35	40	87.5	Excellent
C	Content				
1.1	Newtonian Relativity	135	160	84.4	Good
1.2	Einstein's Postulates	65	80	81.3	Good
1.3	Lorentz Transformation	141	160	88.1	Excellent
1.4	Synchronized Clock and Simultaneity	106	120	88.3	Excellent
1.5	Length Contraction	135	160	84.4	Good
1.6	Time Dilation	107	120	89.2	Excellent
1.7	Twins Paradox	103	120	85.8	Excellent
1.8	Relativistic Doppler Effect	103	120	85.8	Excellent
1.9	Lorentz Velocity Transformation	101	120	84.2	Good
1.10	Relativistic Momentum	140	160	87.5	Excellent
1.11	Relativistic Energy	103	120	85.8	Excellent
D	Examples and Explicit Solutions	186	200	93	Excellent
E	Summary	69	80	86.3	Excellent
F	Problems	140	160	87.5	Excellent
G	Solutions to Problems	80	80	100	Excellent
H	Problem Solving Scoring	115	120	95.8	Excellent
Average				87.5	Excellent
2	**EFFECT PREDICTION**				
A	The whole content of the tutorial handbook may make students' opportunity exercise to solve problem systematically and structurally following the given 5 steps of problem-solving	36	40	90	Excellent
B	The whole content of the tutorial handbook may give raise students' problem-solving ability	34	40	85	Excellent
C	The whole content of the tutorial handbook may make students' opportunity exercise to solve problem systematically and structurally following the given 5 steps of problem-solving	35	40	87.5	Excellent

Based on Table 4, it is clearly seen that the students' response on the special theory of relativity handbook content was 87.5 with excellent qualification. The effect prediction of the tutorial handbook on giving the students opportunity to exercise their problem-solving ability was 90 with excellent qualification. The effect prediction of the tutorial handbook on the students' motivation was 85 with excellent qualification. The effect prediction of the tutorial handbook on the students' self-learning was 87.5 with excellent qualification.

3.3 Students'Problem Solving Ability

Students' problem-solving ability in the special theory of relativity is tabulated in Table 5.

TABLE 5. Students' Probem-Solving Ability in Special Theory of Relativity

No.	Problem-Solving Ability	Average	Qualification
1	Problem focuses	84.5	Good
2	Physical fact picturing	66.3	Fair
3	Solution plans	100	Excellent
4	Problem-solving	93.3	Excellent
5	Result checking	95.8	Excellent
	Average	88	Excellent

From Table 5, the students' problem-solving ability in the special theory of relativity was on average of 88 sith excellent qualification. The excellent handbook feasibility indicated that the handbook has good reliability. Both experts and students were in good agreement that the handbook tutorial on improving students' problem ability had a positive effect, which were indicated by excellent qualification. Therefore, the tutorial giving can increase the students' problem-solving ability as previously proposed by Pride, et al.

3.4 Students'Mastery Concept

The effect of the tutorial giving on the student's mastery concept was evaluated by comparing their grades for the experimental and control classes as depicted in Table 6.

TABLE 6. Students' Grade for Conceptual Understanding in Special Theory of Relativity

No.	Aspect	Experimental Group	Control Group
1	Average Pretest	66.9	66.4
2	Average Posttest	77.3	69.2
3	Minimum Grade for Pretest	40.0	35.0
4	Maximum Grade for Pretest	85.0	85.0
5	Minimum Grade for Postest	65.0	40.0
6	Maximum Grade for Posttest	92.5	85.0
7	Increasing	10.4	2.8
8	Gain	0.31	0.08

From Table 6, it can be revealed that the experimental and control classes had the similar initial ability. The experimental and control groups had increased of conceptual understanding of 10.4% and 2.8%, respectively. The experimental group had a gain of 0.3 with medium classification, meanwhile the control group had a low classification. It is because the students in the experimental group had extra time to study during tutorial giving.The effect of the tutorial handbook usage on the increasing of the students' conceptual understanding was evaluated by t-test analysis, as summarized in Table 7.

TABLE 7. t-test Result for the Students' Conceptual Understanding in Special Theory of Relativity

Group	Average Posttest	$t_{calculation}$	t_t (α=0.05, db= 65)	Note
Experiment	77.3	4.14	1.998	H_0 is rejected (H_a is accepted)
Control	69.2			

Table 7 showed a significant different between the experimental and control groups in terms of the students' conceptual understanding. It revealed that the tutorial handbook usage in the modern physics course may give a positive effect on the students' conceptual understanding, even though in medium classification. It can be increased by means of providing students with the summary task before starting the class.

The effect of the tutorial giving on the student's problem-solving ability was evaluated by comparing their grades for the experimental and control classes as depicted in Table 8.

Table 8 Students' Grade for Problem Solving Ability in Special Theory of Relativity

No.	Aspect	Experimental Group	Control Group
1	Average Pretest	60.4	67.8
2	Average Posttest	76.5	65.2
3	Minimum Grade for Pretest	51.0	56.0
4	Maximum Grade for Pretest	74.0	74.0
5	Minimum Grade for Postest	64.5	39.8
6	Maximum Grade for Posttest	95.5	96.3
7	Increasing	16.1	-2.6
8	Gain	0.41	-0.08

According to the information in Table 8, the increasing of students' problem-solving ability in experimental group is 16.1% and control groups decrease 2.6%, respectively. It is because the more time available for students of the experimental group to do some exercises than that of the control group. The gain for the experimental group was 0.41, meanwhile, a negative value of gain was detected for the control group.

The effect of the tutorial handbook usage on the topic of the special theory of relativity on the students' increasing of problem-solving ability was evaluated by t-test as depicted in Table 9.

Table 9 t-test Result for the Students' Problem-Solving Ability in Special Theory of Relativity

Group	Average Posttest	$t_{calculation}$	t_t (α=0.05, db= 65)	Note
Experiment	76.5	3.95	1.998	H_0 is rejected (H_a is accepted)
Control	65.2			

The t-test result on the students' problem-solving ability in the special theory of relativity showed significance different between experimental and control groups. It revealed that the tutorial giving can give positive effect in increasing the students' problem-solving ability with medium qualification. More exercises can be given to the students to provide them better problem-solving ability.

4. Conclusion

A tutorial handbook on the topic of the special theory of relativity with excellent feasibility and response from students has been successfully developed. The tutorial handbook was arranged to increase students' problem-solving ability and mastery concept with excellent qualification. In addition, the tutorial handbook could motivate students to have self-learning awareness. The empirical t-test showed that the tutorial of the special theory of relativity assigned a positive effect to the students mastery concept and problem-solving ability. The students' mastery concept and problem-solving ability obtained a gain of 0.31 and 0.41 with medium classification, respectively.

Acknowledgments

The work was financially supported by the Ministry of National Education, Republic of Indonesia, Competition Research Grant 2015 (Hibah Bersaing), Universitas Negeri Malang which is gratefully acknowledged.

References

1. Beiser, A, "Fisika Modern "(Terjemahan Liong TH), edisi ke 4, Erlangga, Jakarta, 2004.
2. Krane, K, "Fisika Modern "(Terjemahan Wospakrik, HJ), Erlangga, Jakarta, 1992.
3. Serway, Moses.Moyer, "Modern Physics", Thomson Learning, Australia, 2005.
4. Huffman, D. "Effect of Explicit Problem Solving Instruction on High School Student'Problem Solving Performance and Conceptual Understanding of Physics". Journal of Research in Science Teaching, 34(6) , 1997: 551-570.
5. Pride, T.O., Vokos, S., and McDermott L.C. "The Challenge of Matching Learning Assessments to Teaching Goals: An example from the work-energy and impulse-momentum theorems". American Journal of Physics, 66 (2), 1998: 147-157.
6. Gamze, S.S, Serap, C and Mustafa, E. "The Effects of Problem Solving Instruction on Physics achievement, problem-solving performance and strategy Use", *Am. J. Phys. Educ*. 2, 2008: 3
7. Maloney, D.P., " Research on problem-solving physics ". Handbook. Washington D.C. 1994.
8. Wandersee, J., Mintzes, J.and Novak, J. "Research on Alternative Conception Science". Handbook. Washington D.C.1994.

IOP Conf. Series: Journal of Physics: Conf. Series **846** (2017) 012030 doi:10.1088/1742-6596/846/1/012030

9. Hartatiek, "Pengaruh Paduan Pembelajaran Aktif dan *Problem Solving* terhadap Peningkatan Pemahaman Konsep dan Kemampuan Penyelesaian Masalah Mahasiswa ". Proseding Seminar Nasional MIPA dan Pembelajarannya.FMIPA UM. 2012.
10. Hartatiek, Handayanto, S.K., Yudyanto, " Pengembangan Paket Tutorial Teori Realtivitas Khusus Berbasis Penyelesaian Eksplisit untuk Meningkatkan Kemampuan Problem Solving Mahasiswa". Proceding Simposium Riset 1, LP2M Universitas Negeri Malang, 2014.
11. Sukmadinata, N. Y., "Metode Penelitian Pendidikan". Program. Pascasarjana Universitas Pendidikan Indonesia.2007.
12. Hake, R. "Interactive-engagement vs traditional methods: a six-thousand-student survey of mechanics test data for introductory physics courses". Am. J, Phys. 1988:64-74.